W9-AEJ-425

Writing for Professional and Technical Journals

WILEY SERIES ON HUMAN COMMUNICATION

W. A. Mambert
PRESENTING TECHNICAL IDEAS: A Guide to Audience Communication

William J. Bowman
GRAPHIC COMMUNICATION

Herman M. Weisman
TECHNICAL CORRESPONDENCE: A Handbook and Reference Source for the Technical Professional

John H. Mitchell
WRITING FOR TECHNICAL AND PROFESSIONAL JOURNALS

Writing for Professional and Technical Journals

JOHN H. MITCHELL
University of Massachusetts

JOHN WILEY & SONS, INC.
NEW YORK / LONDON / SYDNEY

Library of Congress Catalog Card Number: 67-31374
GB 471 61170X
Printed in the United States of America

PREFACE

This book does two things. First, it defines and illustrates the recognizable characteristics of professional and technical articles. Second, it anthologizes conflicting sections of the style guides issued for representative disciplines and illustrates those style guides by reprinting and reviewing representative articles. No attempt has been made to define or to illustrate the techniques of popular writing employed in mass media.

The arrangement of the first four chapters is chronological; that is, the standard sequence for writing a professional article is shown. Chapter 1 discusses the planning that must precede the writing phase. Chapters 2 and 3 discuss the processes of data collection, correlation, selection, and arrangement that must be followed to ensure the depth of content appropriate to and characteristic of professional writing. Chapter 4 reviews, in their proper sequence, the standard sections of a journal article. This arrangement of chapters permits a beginning writer to use the book as a text; it also permits an experienced writer to use it as a reference book, to proceed at his own pace, and to omit sections of the book that are not pertinent either to the topic he is developing or to the field or journal whose style guide he is following.

Chapter 5 is an anthology of representative style guides and articles arranged by discipline or subject field. Mutually exclusive aspects of the style guides are compared. Articles taken from representative journals and illustrative of the guides are reprinted as examples. Comments on these articles show the characteristics of writing that made the articles appropriate to the journal that printed them.

JOHN H. MITCHELL

Amherst, Massachusetts
March, 1968

For Larayne

This—and all my books

CONTENTS

INTRODUCTION

Professional and technical articles are essential to all fields of science, to the humanities, and to an industrial society. Their evolution—as well as their complexity—is linked directly to formalized scientific methodologies. In the usual but certainly not the only version of scientific method the first step is to define the problem; the second step is to search the literature bearing on the problem; the third step is to do some independent work on the problem; and the fourth step is to report the success or failure of the independent work on the problem. Three of the four steps involve communication skills; all of the steps are essential to professional articles. Without professional communication, one can fairly state, there would be no scientific, technological, or humanistic advance.

There is neither a single format nor a single formula for professional articles. Articles may be review, tutorial, documentary, theoretical, or descriptive in nature. Because of this variety, it is impossible to hypothesize a formula, as, for example, the formulas for writing engineering reports, which will govern format and style for all articles. For that reason, Chapters 2, 3, and 4 of this book discuss general techniques common to all article writing.

Chapter 5, however, discusses specific disciplines and journals. In some disciplines the appropriate national professional society has fixed upon a pattern for articles. In many of the scholarly journals the editor has established a formula for articles acceptable for publication in his journal. Because of this general/specific condition in professional publishing, writers should be aware of the variations illustrated in Chapter 5. Neither societies nor editors are to be blamed. The former know the needs of the professions; the latter realize that, according to basic information theory, a communication in an expected or accustomed form will transfer information faster and more completely than one in a random pattern.

In the several cases in which professional societies or journal editors have established a formula that formula has usually been determined by either the "language" of a discipline or by the subject matter of a journal. An equation for atomic fission is in mathematical language. An article that derives and defends the equation can best communicate in that language. For this reason the American Institute of Physics requires articles published in its journals to conform to elaborate glossaries defining mathematical terms and symbols. These symbols are not always compatible with those of the American Chemical Society, the American Mathematical Society, or the National Aeronautics and Space Administration.

Similarly, subject matter and custom can fix the format of communication. A telephone book is an obvious example: it contains an amazing number of discrete bits of information, but, because that information is in an expected form, a child can find the bit he wants quite easily. There are other ways of arranging a telephone book, but none is more appropriate to both subject matter and custom.

There are a tone and perhaps a style common to all professional writing, and for these a conditioned audience exists. Just as readers of formula-written journals will resist a nonformula article in those journals no matter how literate and well organized the article may be, so will readers of all professional and technical journals resist articles whose tone is not convincing and whose style does not reflect an inexorable logic. This is not to say that professional writing must be ponderous and dull. It is simply to indicate that most readers of professional journals are not seeking pleasure and amusement but information and understanding. As Einstein once put it, "If you are out to describe the truth, leave elegance to the tailor."

Two groups produce almost all professional articles. The first group is composed of research personnel and educators who achieve identity in a field or promotion in academic rank in almost direct proportion to the caliber and frequency of their publications. The second group is composed of a growing number of professional science writers and journalists who write with knowledge of a specific field and communicate this knowledge at various levels of understanding. Both groups employ the techniques discussed in this book, and both groups write within the framework, discussed in Chapter 5, appropriate to the discipline or editor that they serve.

Rarely, however, can a professional communicate within the same framework every time he writes an article. Much of the research conducted today is interdisciplinary. This means that it must be reported in the literature of more than one field. It is probable that a writer at some time in his professional career will be required to present information via almost every type of communication illustrated in this book.

As opposed to the writing of research personnel and educators, the work of professional science writers and journalists is essentially a service function. These writers do not usually conduct research themselves; they report the work of scientists and scholars to others. They are essentially specialists skilled in translating technical information into a language and form that are appropriate to readers with specialized but sometimes different backgrounds. The value of their service is dual: first, they release highly trained—and highly paid—research personnel from some reporting obligations; second, they are usually more capable than research specialists in presenting complex information in an understandable form.

1

DESIGN AND APPROACH

"If a man can group his ideas, he is a good writer."
Robert Louis Stevenson

The organization of an article may vary with the logic of the writer's argument, with the requirements of the subject or discipline, or with the regulations of an editor. Good organization is simply arranging data in the way that will best communicate a given subject to a specific audience.

There are two types of logical reasoning common to professional articles: inductive and deductive. Inductive reasoning is used to arrive at new principles from known data; the writer goes from experience to general rules. Deductive reasoning, on the other hand, is not creative; it is the application of accepted rules to specific cases. Here the writer finds the right rule for the instance and then applies it correctly. The various processes of inductive and deductive reasoning are discussed in this chapter. Either type of reasoning is acceptable in writing, but *the types must be kept separate*. A reader will be confused by logic that proceeds from the general to the particular and then reverses itself.

Logic within an article must be unidirectional, and logical organization exists when the reader is led by an inexorable progression of facts to an inevitable conclusion. Organization based upon such a progression does not permit too frequent a use of analogies and statements by authorities. Good articles are never a patchwork of comparisons and quotations; data alone are important.

The requirements of a subject or discipline affect organization for a different reason. *Most scientific and human situations are predictable.* For this reason, standard practices in communication have evolved that give similar situations similar treatment. Indeed, it is safe to say there could be no communication in science without agreement upon terminology and categorization. Organization in scientific writing is frequently a reflection of a specific scientific method.

The predictability of the reader also affects organization. Because almost all readers have colossal self-esteem, they approach professional articles with an attitude of "What's in it for me?" They feel that their time is valuable. They want to be informed; they do not read articles for amusement. This situation forces writers to present a meaningful title and summary at the outset. The reader is attracted only when he finds something pertinent to him in the title and summary. He remains attracted unless the tone of the writer offends him in one of two ways; that is, unless the writer offends him either by a lack of seriousness or by "talking down" to him. Facetious writing marked by witticisms and whimsical asides will repel a reader who wants only to be informed. Patronizing writing which labors the obvious and proceeds with a "John has two apples; he gives Mary one" sort of logic will so antagonize a reader that he will unconsciously set up an emotional block to any further communication. Few readers completed a recent article which actually began: "There are five living economists of importance, and we think . . . ."

Style and mechanics are important to organization because they are the invisible elements of organization. Style constitutes the theory, and mechanics are the application of the theory. Style involves tone, taste, and word handling; mechanics involve conventions. The conventions of language may be found in scores of reference books and style guides. Style, however, is subjective. It refers to the total effort that goes into the research and production of a given piece of writing. The characteristics of style are contained in the following passage by Alfred North Whitehead.[1]

"Finally there should grow the most austere of all mental qualities; I mean the sense for style. It is aesthetic sense, based on admiration for the direct attainment of a foreseen end, simply and without waste. Style in art, style in literature, style in science, style in logic, style in practical execution have fundamentally the same aesthetic qualities, namely, attainment and restraint.

". . . Style, in its finest sense, is the last acquirement of the educated mind; it is also the most useful. It pervades the whole being. The administrator with a sense for style hates waste; the engineer with a sense for style economizes his material; the artisan with a sense for style prefers good work. Style is the ultimate morality of mind.

". . . With style the end is attained without side issues, without raising undesirable inflammations. With style you attain your end and nothing but your end. With style the effect of your activity is calculable, and foresight is the last gift of gods to men. With style your power is increased, for your mind is not distracted with irrelevances, and you are more likely to attain

[1] A. N. Whitehead, *The Aims of Education and Other Essays* (New York: Macmillan, 1929), pp. 19–20. Copyright 1929 by The Macmillan Company, renewed 1957 by Evelyn Whitehead. Used with permission of The Macmillan Company and Ernest Benn Limited.

your object. Now style is the exclusive privilege of the expert. Whoever heard of the style of an amateur painter, of the style of an amateur poet? Style is always the product of specialist study, the peculiar contribution of specialism to culture."

The attainment and restraint which Whitehead said characterize style are inherent in professional writing. Patterns of logic and reasoning are preoccupied with exactness and brevity, and exactness requires the use of professional terminology or "jargon." The objections to jargon rise essentially from readers untrained in a discipline. These objections are valid when they are levelled at writing in mass media that attempts to popularize an esoteric area. They are invalid, however, when levelled at professional articles written for trained and professional readers. Exactness in writing is difficult to achieve unless the accepted terms of a discipline are applied. The writer of professional articles can assume an intelligent audience and can use the exact or jargon terms; only when he coins new terms does he need to define them.

This is not to say that a writer may use all the catchwords, trite words, stock responses, unthinking approaches, and automatic assumptions that exist in every discipline. "Good" jargon words read like the cards of a poker hand; only one interpretation is possible. Poker is a game with nice mathematical rules and fairly predictable probability only when a nine is a nine and a three is a three. Making those nines and threes wild changes the predictability. It is the confusion resulting from the change that sometimes makes players—and readers—drop out.

Techniques exist for determining readability, and almost all are preoccupied with brevity and exactness. The Flesch Check System[2] (see p. 152), the Dale-Chall Formula,[3] and the Gunning Fog Index[4] are most significant. All have limitations; the first two can only be applied to writing *post facto*, and the Gunning Index involves only principles. Gunning's "Ten Principles of Clear Writing," however, are widely accepted and have become the decalogue governing the immense amount of writing done within the Department of Defense.

1. Write to express, not impress.
2. Make full use of variety.
3. Keep sentences short.
4. Use the familiar word.

[2]Rudolph Flesch, *How to Write, Speak, and Think More Effectively* (New York: Harper & Brothers, 1960).

[3]Edgar Dale and Jeanne Chall, *A Formula For Predicting Readibility* (Columbus: The Ohio State University, Bureau of Educational Research, 1948).

[4]Robert Gunning, *The Technique of Clear Writing* (New York: McGraw-Hill, 1952).

5. Prefer the simple to the complex.
6. Avoid unnecessary words.
7. Put action in your verbs.
8. Write the way you talk.
9. Use terms your reader can picture.
10. Tie in with your reader's experience

It should be noted that Gunning's Principles aim at clarity in writing. They would be adequate if communication were a science, if words were exact, and if writers and readers were entirely predictable. None of these situations pertains.

A final aspect of style involves the extent to which *content* determines organization. A writer should always present his ideas in a written form that will reflect their logical relationship. For example:

- An unqualified observation requires a *simple sentence.*

  Propagation of rust-resistant white pine by grafting is feasible.

  The propeller failed at 5200 rpm.

  The rate of flow varies inversely with temperature.
- Coordinate ideas expressed in comparison, contrast, or balance require a *compound sentence.*

  Propagation of rust-resistant white pine by grafting is feasible, but processing costs are prohibitive.

  The propeller cavitated at 4200 rpm, and it failed at 5200 rpm.

  The rate of flow varies inversely with the temperature; it varies directly with the pressure.
- Subordinate ideas require a *complex sentence.*

  When cost is not important, propagation of rust resistant white pine by grafting is feasible.

  Although the plane was put through a suitable warm-up period, the propeller failed at 5200 rpm.

  When J-4 fuel is used, the rate of flow varies directly with the temperature.
- Coordinate ideas, either or both of which are qualified, require a *compound-complex* sentence.

  When cost is not important, propagation of rust-resistant white pine is feasible, but processing costs are prohibitive except for certain ornamental plantings.

  Even after a suitable warm-up period, the propeller failed at 5200 rpm; however, when ground effect was absent, it cavitated at 4200 rpm.

  When J-4 fuel is used, the rate of flow varies directly with temperature; but when a solid propellent is used, temperature has no effect.

By extension, written elements other than the sentence should reflect logical organization. Paragraphs, subsections, sections, and chapters should all be constructed to show an easy and obvious logic.

A cliché about writing says, "Grammar involves right and wrong, and rhetoric involves better or worse." The generality is reasonably accurate; certainly the principles of grammar and mechanics distinguish right from wrong. Because rhetoric involves better or worse, it is essentially a matter of style. The preceding paragraphs discuss the theories on which a good writing style is based, but there remains something to be said, however arbitrary it may seem, about the application of those theories within the framework of accepted principles of grammar.

Simplicity, to Whitehead, is the fundamental characteristic of style. Variety should be used only to relieve a particular passage. Whitehead would have disapproved of the preceding paragraph. That paragraph contains a simple, a compound, a complex, and a compound-complex sentence. It uses too much variety. The present paragraph uses only simple sentences. It needs variety for relief.

Clarity is involved when a writer follows the advice of Gunning and Whitehead and prunes a sentence to its simplest form. The following condensations, which delete the vague phrases *there may be, it is possible to state, in the case of,* and *it has been decided that,* increase the strength of the sentences and achieve some dignity in simplicity. (In most of these condensations an *active* verb replaces the *passive* verb of the original version. A change from passive to active construction usually produces greater clarity and simplicity in a sentence.)

There may be overspeeding of the propeller if there is a rapid increase of power.
The propeller may overspeed with a rapid increase of power.

It is possible to state that the completion of the reaction will be achieved in the presence of platinum.
Platinum acts as a catalyst, and the reaction goes to completion.

In the case of corrosive synthetic fuels, injection carburetors with diaphragms cannot be used.
Synthetic fuels corrode the diaphragms of injection carburetors.

It has been decided that National Science Foundation support will be cut back.
The NSF support will be cut back.

Condensing and pruning, however, do not always improve clarity.

The following examples are longer, but they are perhaps more clear than the condensations above. Whenever it is possible to be more clear, a writer of professional articles should always abandon a fine rhetorical effect and rephrase for the sake of clarity.

> A rapid increase of power will unseat the governor. When the governor in unseated, the propeller will overspeed.
>
> Only the catalytic effect of platinum in the presence of nitric acid permits the dissolution of cesium sulfate.
>
> Synthetic fuels cannot be used in injection carburetors; aromatics in the fuel corrode the carburetor diaphragm.
>
> The NSF announced that support for the study of gull sterilization by sulfur black will be cut by 50 percent in 1968.

Brevity and restraint are more than desirable stylistic aspects; they are also economic requirements. Biological publication, for example, today costs six cents a word. It is unlikely that an editor confined to a budget will accept an article that is circular or verbose.

Most research and writing projects begin with a problem and end with a solution. The solution may be tentative or it may be conclusive, but whether the project achieves a QED or not, the work begins with facts that generate a question. The only exceptions are articles that are descriptive or enumerative.

Facts therefore exist before the question or hypothesis is formulated. Creative research organizes them into relationships which may or may not be true, but which are plausible enough to serve as a basis for investigation. When they are organized into a specific problem or hypothesis, the solution of the problem of the substantiation of the hypothesis generally determines the organization of the article. Defining the problem or stating the hypothesis, however, is the essential first step. Only when the problem has been defined can relevant data be collected.

Problem definition always precedes both research and writing, and as the definition keeps the work of a scientist at a bench relevant, so does it keep the work of a writer at a desk relevant. This is not to say that there are no blind stabs and lucky accidents made in laboratories and libraries. The work of the Curies with radium and of Fleming with penicillin are delightful examples of chance. However, the data they discovered by accident were relevant to trained scientists of their caliber, and these data became discrete bits in the solution of the problem as defined. The source of true data is unimportant; the relevance of the data is all-important.

It is apparent that the definition of a problem serves as a guideline to both research and writing, and that data are collected, selected, and arranged to solve the problem as defined. It is precisely for this reason that the definition must be made with both knowledge and care. If a problem seems easy to label and describe, it is probably not a true problem or a matter worthy of study and research. Indeed, work at the Systems Coordination Division of the Naval Research Laboratory showed that in two cases out of every five the application of symbolic logic to a definition showed a tautology or—worse still—a problem definition that contained its own solution.

All problem definitions should be formulated; that is, they should be reduced to or expressed in a formula or set forth in a definite and systematic statement. Because "a definite and systematic statement" is a description of an article or report in its entirety, it is apparent that a good definition anticipates or predicts the organization of the article or report. Further, the statement anticipates the problems of searching a library indexed by links, roles, keywords, or by any meaningful or logical system. Writing a statement that will resolve these problems in advance requires an awareness of some basic difficulties in language.

There are four areas of difficulty in all language systems—semantics, generics, syntactics, and viewpoint. Semantics involves *word meanings.* Words meaning the same thing, or seeming to mean the same thing, are often used by library indexers to file or store information. These synonyms, near-synonyms, and homographs lead to the loss or irretrievability of a significant amount of data stored in a library; for example, one index librarian will store data under the term "imperfections." Another will use the term "defects." A writer searching that library must use both terms if he is to recover all data so indexed. Only if his search statement or problem definition is adequate can he hope to do so.

Generics involves *hierarchical word families;* for example, "nickel cadmium batteries" and "nickel zinc batteries" are members of the class "nickel batteries." "Nickel batteries" are members of the class "batteries." This generic tree, if allowed to branch out indefinitely, becomes a jungle of verbiage that effectively hides data.

Syntax or syntactics involves *word order;* for example, putting the word "only" anywhere in the sentence "She said that she loved me" changes that sentence. At least seven meanings are possible; all depend on the syntax of the sentence.

Viewpoint involves *individual orientation.* For example, in the nickel cadmium battery mentioned above, is the nickel a base element, an alloying agent, or a catalyst? The viewpoint of the indexer—which is not always that of the writer—can result in individual interpretations that may seem random and perverse.

Writing a definite and systematic statement which avoids the pitfalls of language and serves as both a problem definition and a search statement compatible with a library's indexing system requires a high degree of exactness and specificity. Consider the following examples.

To determine a good insecticide for houseflies.

This is obviously too broad. What kind of insecticide? What does "good" mean? Is *Musca domestica* what is meant by "housefly"? The statement might make an adequate title for a *popular* article, but it is meaningless to a professional.

To determine the effect of dichlorodiaphenyltrichloroethane on humans and houseflies.

This is still too broad. What concentration of DDT is involved? How is it carried? What kind of effect? All houseflies? And so forth.

To determine the physiological effect of skin penetration, inhalation, and ingestion of DDT in a 5 percent wettable powder upon humans and domestic animals.

This is better. There are still generic problems with the words "physiological effect" and semantic problems with the words "domestic animals." One hopes that "skin penetration, inhalation, and ingestion" will resolve the former. Viewpoint and intelligence may resolve the latter. Pet parrakeets, fish, and snakes are not "animals," and anything more exotic is not usually "domestic."

The last statement is adequate to guide both the writer and the searcher of literature. The writer would not digress into a discussion of the preparation and production of DDT because those areas are not relevant to his statement. The searcher of literature would be guided by the keywords in the statement. These would keep him from being sidetracked into irrelevant areas. A discussion of keywords, links, roles, and descriptors is given in the following chapter.

2

DATA COLLECTION AND CORRELATION

Data are to be found in three ways—by doing original research, by searching the literature, and by interviewing individuals with unpublished opinions and ideas. Original research, except as it grows from problem definition, is not the concern of this book. A search of the literature, however, is central to the writing of articles. Because the possibility of chance discovery of data in the literature is slim, elaborate techniques for data collection must be used. However, even sophisticated techniques are inefficient, and it has been estimated that only 60 percent of the information in standard libraries can be recovered.

The major reason for problems in information retrieval is that there is simply too much information. It is common to hear statements such as "The number of scientific periodicals has been doubling every 15 years since 1750," or "Ninety percent of the scientists who have ever lived are living and writing now."

Mechanical and electronic techniques for collecting and collating information have not been developed as rapidly as the information they are designed to process. The result of this cultural lag has been the duplication of expensive research and the overlooking or loss of important information. "Under present conditions in the research and development field, it has been estimated that over one billion dollars is wasted annually by duplication of previous but unknown accomplishments. A much larger amount (over half of the research and development effort) has been estimated as the indirect waste of work that would not have been undertaken, or would have been approached differently, if available information had been properly researched and evaluated."[1] It is only now known that the original patent for the transistor was issued in the 1930's and had expired

[1]T. C. Pritchard, "Future Trends in Technical Communication," *Graphic Science*, 7, No. 3 (March 1965), 18.

before the post—World War II applications for transistor patents were filed. Perhaps the most startling lack of correlation of data occurred within a major electronics company. The tube division of that company had been working for six months to develop a tube with specific characteristics when a competitor wrote requesting licensing rights to precisely that tube, which had been patented already *by another division of the company*.

Computer techniques have long been applied to the retrieval of technical information but, as of this writing, they are regarded by the American Library Association as only 60 percent effective—the same effectiveness as that enjoyed by the classic library techniques outlined below. Because computer systems are expensive and because the coding of information for use by present computers is time-consuming, there has been no wholesale adoption of computer techniques by science and industry. There is, however, considerable work presently being done by computer manufacturers to improve the effectiveness of systems for the machine retrieval of information. When an efficient system, compatible with existing library techniques, is perfected, there can be little doubt that the machines will replace whole armies of librarians presently engaged in identifying, retrieving, and correlating information. There can be little doubt about the expense: converting the 40 million cards in the Library of Congress catalog for computer use will cost $40 million. However, in prevention of duplication of research alone, conversion would pay for itself in six months; yet, problems of machine compatibility must first be solved.

Money is not a limiting factor; present funds permit the expansion of data sources at the rate of 1000 to 1500 new journals, 6000 scientific books, and over 100,000 technical reports annually. Between $2 and $8 billion of defense funds are annually expended for engineering data alone. By 1970 the projected amount to be spent on acquiring new engineering data will be $17 billion a year. The Department of Defense willingly allocates a significant portion of these funds to computer research, but the research goal is not the organization of data as we presently know it but the utilization of a computer as a design aid with the communication falling out as a by-product.

An excellent discussion of research goals in the application of computers to information retrieval is available from the Clearinghouse for Federal Scientific and Technical Information, U. S. Department of Commerce, Springfield, Virginia, 22151, in document Order AD 615 718N, "The Conceptual Foundation of Information Systems," ($2.00).

Existing retrieval systems vary with the type of computer involved. Punched cards, magnetic memory cores, and a variety of tapes scanned by a variety of techniques are all in present use. When that use is specialized, an existing computer technique can be modified to a form entirely adequate for a specialized area.

It is more difficult, however, to design a retrieval system that is compatible with the techniques ideal for other specialized areas. For example, the American Chemical Society maintains in its Washington office an electromechanical system which uses punched IBM cards. These cards contain information on the physiological effects of some 2 million known elements, compounds, and mixtures. The system is capable of furnishing its specialized information rapidly, but it is incapable, without expensive modification and laborious punching of other IBM-card "libraries," of showing the existence, occurrence, and sources of those compounds or the methods for their manipulation, preparation, and production. All this information exists, but it does not exist *in the same place* or *in the same form*. This is not to fault the system which *for its purposes* is outstanding. The system simply was not designed to correlate other types of information. The same is true of the *Index Medicus* collated by two Honeywell computers from 150,000 citations annually. For its purposes it is outstanding. It enjoys an international reputation. Direct links with other depositories, however, are lacking as yet.

When a writer knows what data he wants and where those data may be found, machine retrieval is hugely effective. It can only be effective, however, if data exist in a form that may be searched by machine techniques. Thus it is necessary to prepare "libraries" of information with which the machine may work. Whether these libraries are punched cards or magnetic cores is unimportant; what is important is that they are different from or are modifications of the Library of Congress or Dewey Decimal systems discussed below. In other words, a specialized library or storage area must be prepared. Because a new storage area duplicates one that already exists in another form, expensive new storage systems are created only when there is a need for high-speed or repetitive searches.

It is difficult to decide on a technique for indexing new material; for example, the U. S. Air Force stores the world's finest collection of Chinese physics periodicals at Hanscom Field. No one really knows what is in the collection, because no adequate cataloguing system exists that is compatible with the one in the Defense Documentation Center, the one in the Central Intelligence Agency—and the Chinese language.

Until machine-retrieval techniques are perfected, the professional societies will not abdicate their responsibilities for data *in their own fields*. So far their techniques have been adequate. The indexing and abstracting services of the American Institute of Biological Sciences (AIBS) and the

American Chemical Society (ACS) deserve their international reputation in the areas which concern them. Other disciplines are following their examples, and the procedures recently adopted by the Engineers Joint Council (EJC) are a remarkably effective solution to problems of information control *in the applied sciences*.

Because the EJC procedure is both flexible and compatible with most machine systems, its adoption will extend beyond the applied sciences and into the basic sciences, where it is being adopted in whole or in part. For this reason the coordinate indexing and abstracting systems of the EJC are discussed in detail. The discussion is based on reprints of documents released by the Battelle Memorial Institute, which, under contract with the EJC, offers on a nationwide basis a course in the techniques. The remainder of this discussion was written by Battelle staff scientists Carol A. (Penn) Tippett and Margery J. McCauley. It is taken from their article entitled "How to Use Abstracts and Keywords," *Journal of Metals*, January 1965.

"The EJC proceeded on five premises.

1. That no scientist could monitor all current information in his field without abstracts and abstracting/announcement services.

2. That a natural language—English as opposed to artificial or synthetic code schemes—would remain the standard method of communication among scientists, and that interposition of a code between a scientist and the information he needs would aid only a machine technique.

3. That a thesaurus could be written that would in no way restrict the choice of words by authors and would also ensure that others would find the author's publications even though different words were involved.

4. That an acceptable system would have to be compatible with both a card file on a scientist's desk and a large, computer-based storage and retrieval system such as that at the Defense Documentation Center.

5. That the cost of abstracting and indexing information *at the time of publication* is but a fraction of the cost of abstracting and indexing by everyone storing information.

"The applications of the EJC's premises have been made. Because abstracting presupposes reading and understanding articles, the editors of professional journals who read and edit all articles they accept now either require authors to submit an abstract of each article or the editors write the abstract themselves. Either descriptive or informative abstracts are acceptable.

"A *Thesaurus of Engineering Terminology* was the first item 'written' for use in the EJC system. It was compiled by merging eighteen vocabularies submitted by member professional societies, DDC, and NASA.

Because the *Thesaurus* is fundamental to EJC procedure, its application and capabilities should be understood.

"Essentially, the EJC *Thesaurus* lists keywords or uniterms selected, first, to permit indexers of documents to 'describe' at *different levels of generality* and from *different technical points of view* the information contained in a document and, second, to permit searchers for information to phrase an inquiry appropriate *to the scope and degree* of their immediate interests. In practice, the indexer assigns to a given document not only index terms brought to mind by the words of the author but also terms listed in the *Thesaurus* as words used by technologists with different interests and viewpoints. Similarly, the searcher phrases his enquiry using terms from the *Thesaurus* that *might* have been used to index the document from other points of view.

"Flexibility is inherent in this system. A searcher looking for 'a few good references' can choose from the *Thesaurus* only the terms that most precisely describe his area of interest. A searcher wishing 'complete information' can choose all terms and all points of view pertinent to the larger area.

"Links and roles are used in the EJC system to minimize false retrieval of nonpertinent articles. The author or editor adds them to the abstract as illustrated in Figure 1. The searcher uses them in retrieving only the information pertinent to his interest and point of view.

"For example, the abstract shown in Figure 1 (printed with other abstracts both in the issue containing the original article and in publications by various abstracting services) shows selected keywords printed after capital letters. These letters represent *links*. Each link contains a list of keywords for one subject or intellectual relationship discussed in the article. Because the article discusses five subjects, it is indexed with five links. Link title (or chapter titles) for this article might be

Link A: Production of titanium slag.
Link B: Chlorination of titanium slag.
Link C: Purification of titanium tetrachloride.
Link D: Reduction of titanium tetrachloride.
Link E: Electrolysis of magnesium chloride.

"Roles are used to show the functions or contexts in which the keywords are used in describing the content of the article. Just as links are represented by capital letters, *roles are indicated by* arabic numbers. In a sense, role numbers are substituted for the grammar used in the abstract. As numbers, however, they lend themselves readily to both human and machine retrieval systems.

Titanium From Slag in Japan, Toshio Noda, JOURNAL OF METALS, 1965, vol. 17, no. 1, pp. 25

Abstract: The author describes the production of titanium sponge by the Osaka Titanium Co. Ltd. using the Kroll process. Titanium slag is produced from ilmenite by arc melting in an electric furnace; the lining of the furnace and means for tapping oxygen by blowing are described. The production of $TiCl_4$ from the titanium slag is accomplished by mixing with petroleum coke and pitch, and sintering and chlorinating the briquettes; purification of the product by continuous distillation is described. Also discussed is the reduction of $TiCl_4$ by magnesium to produce the sponge and the vacuum distillation of the product. The $MgCl_2$ produced as a by-product is subjected to electrolysis in order to recover magnesium and chlorine. The yield of each reaction and the quality of the titanium sponge are discussed.

Keywords: A. Production 8; Reaction (Chemical) 8; Yield 8; Reduction 8,10; Titanium Ores 1; Ilmenite 1; Slags 2,9; Titanium Dioxide 2,9; Osaka Titanium Co., Ltd. 9; Electric Arc Furnaces 10; Tapping 10; Gas Injection 10; Oxygen 10.

Keywords: B. Production 8; Chlorination 8,10; Yield 8; Coke 1; Slags 1; Titanium 1; Titanium Dioxide 1; Pitch 1; Briquettes 1; Osaka Titanium Co., Ltd. 9; Titanium Tetrachloride 2,9; Mixing 10; Sintering 10.

Keywords: C. Purification 8,10; Titanium Tetrachloride 1,2; Osaka Titanium Co., Ltd. 9; Distillation 10.

Keywords: D. Production 8; Reduction 8,10; Quality 8; Yield 8; Titanium Tetrachloride 1; Magnesium 1; Sponge Metal 2,9; Titanium 2,9; Magnesium Chloride 2,3; Osaka Titanium Co., Ltd. 9; Vacuum Distillation 10.

Keywords: E. Recovery 8; Electrolysis 8,10; Magnesium Chloride 1,3; Magnesium 2; Chlorine 2; Osaka Titanium Co., Ltd. 9.

Figure 1

"There are eleven roles in the EJC system, and they are designated by numbers from 0 to 10. Table 1 shows the standard meanings for each role. The following examples show how the roles are applied to keywords *during the indexing operation*.

Role 8: If an article discusses the production of titanium sponge, the term 'production' is indexed in role 8 (PRODUCTION – 8) because it is the principal subject under discussion.

Role 1: If an article discusses the reduction of molybdic oxide to molybdenum, molybdic oxides is indexed in role 1 (MOLBYDIC OXIDES – 1) because it is the input to the reaction or the reactant.

Role 2: For the same article, molybdenum is indexed in role 2 (MOLYBDENUM – 2) since it is the output or product.

Role 3: If an article discusses the mechanical properties of cadmium in which there are trace amounts of antimony present as impurities, antimony is indexed in role 3 (ANTIMONY – 3) because it has an undesirable context.

Role 4: If the production of steel for use in making pumps is discussed, pumps is indexed in role 4 (PUMPS – 4) because it is the intended application.

Role 5: If the process of arc welding is carried out in an atmosphere of helium, helium is indexed in role 5 (HELIUM—5) because it is the atmosphere in which the process is carried out.

Roles 6 and 7: When a cause-effect relationship is discussed in an article, such as the effect that temperature has on the rate of production of an alloy, temperature is indexed in role 6 (TEMPERATURE—6) and the rate is indexed in role 7 (RATE—7). Roles 6 and 7 are always used together; neither role 6 nor role 7 may be used alone in any given link. Role 6 is used for the independent variables; role 7 is used for the dependent variables.

Role 9: Role has three distinct meanings or contexts.

First, if a property of a material is being discussed, such as ductility of aluminum, the keyword 'aluminum' is indexed in role 9 (ALUMINUM—9) as the object of the preposition *of* because it possesses the property being discussed. In this case, aluminum possesses ductility.

Second, role 9 is also used on terms which identify locations.

Table I. The EJC System of Roles

Role No.	Definition
8	The primary topic of consideration is; the principal subject of discussion is; the subject reported is; the major topic under discussion is; there is a description of
1	Input; raw material; material of construction; reactant; base metal (for alloys); components to be combined; constituents to be combined; ingredients to be combined; material to be shaped; material to be formed; ore to be refined; sub-assemblies to be assembled; energy input (only in an energy conversion); data and types of data (only when inputs to mathematical processings); a material being corroded
2	Output; product, byproduct, coproduct; outcome, resultant; intermediate product; alloy produced; resulting material; resulting mixture or formulation; material manufactured; mixture manufactured; device shaped or formed; metal or substance refined; device, equipment, or apparatus made, assembled, built, fabricated, constructed, created; energy output (only in an energy conversion); data and types of data (only as mathematical processing outputs)
3	Undesirable component; waste; scrap; rejects (manufactured devices); contaminant; impurity, pollutant, adulterant, or poison in inputs, environments, and materials passively receiving actions; undesirable material present; unnecessary material present; undesirable product, byproduct, co-product
4	Indicated, possible, intended present or later uses or applications. The use or application to which the term has been, is now, or will later be put. To be used as, in, on, for, or with; for use as, in, on, for or with
5	Environment; medium; atmosphere; solvent; carrier (material); support (in a process or operation); vehicle (material); host; absorbent, adsorbent
6	Cause; independent or controlled variable; influencing factor; *X* as a factor affecting or influencing *Y*; the *X* in *Y* is a function of *X*
7	Effect; dependent variable; influenced factor; *Y* as a factor affected or influenced by *X*; the *Y* in *Y* is a function of *X*
9	Passively receiving an operation or process with no change in identity, composition, configuration, molecular structure, physical state, or physical form; possession such as when preceded by the preposition *of*, *in*, or *on* meaning possession; location such as when preceded by the prepositions *in*, *on*, *at*, *to*, or *from* meaning location; used with months and years when they locate information (not bibliographic data) on a time continuum
10	Means to accomplish the primary topic of consideration or other objective
0	Bibliographic data, personal names of authors, corporate authors and sources, types of documents, dates of publication, names of jounals and other publications, other source-identifying data, and adjectives

A discussion of steelmaking in the United States is indexed with United States in role 9 (UNITED STATES—9).

Third, if a material is the object of an operation which does not involve an input that will undergo change, such as transporting pipe, the material is indexed in role 9 (PIPE—9).

Role 10: If an article discusses the means for accomplishing an operation (such as refining by means of electrolysis), the means (in this case electrolysis) is indexed in role 10 (ELECTROLYSIS—10).

Role 0: Descriptive adjectives such as 'pilot-plant' or 'batch' which modify a process are indexed in role 0 (PILOT-PLANT—0, BATCH—0).

"Of the various component steps involved in preparing articles for input to a retrieval system, preparing the index and searching the index, the abstracting and indexing are the most costly and require the greatest investment of time. The abstracts and keywords for *Journal of Metals* articles are prepared by technically trained information specialists at Battelle who are experienced in the EJC system of indexing and abstracting. After proper training, this function will be assumed by members of the staff of *Journal of Metals*.

"With little investment of time, a reader can use the printed abstracts and keywords provided in the *Journal of Metals* to establish his own index which will provide him with rapid retrieval of references pertaining to the areas of interest he may select. When he identifies the articles which he would like to enter in his index, the reader clips the abstract and keyword form found in the journal. For ease of handling, these abstract forms can be pasted on 3″ x 5″ or larger 4 x 6 in. cards. Each abstract card must be assigned a number which is written in the block found in the upper right-hand corner of the abstract form. This number becomes the article number or *accession number*. Accession numbers are assigned in simple ascending order. The abstract cards are then filed in ascending numerical order, thus creating a file of abstracts of articles.

"The index is prepared from the list of keywords which follows the abstract. A 5 x 8 in. card is created for each term-role (a keyword plus its role number) in the abstract. The term-role is written at the top and the card is divided into ten columns. The accession number of each abstract indexed by that term-role is entered in the column representing the terminal digit of the accession number. The accession number is followed by the letter representing the link in which the term-role was found on the abstract card. Figure 2 shows how this would be done for the term-roles in links C and E of the abstract shown in Figure 1. On the card for Osaka

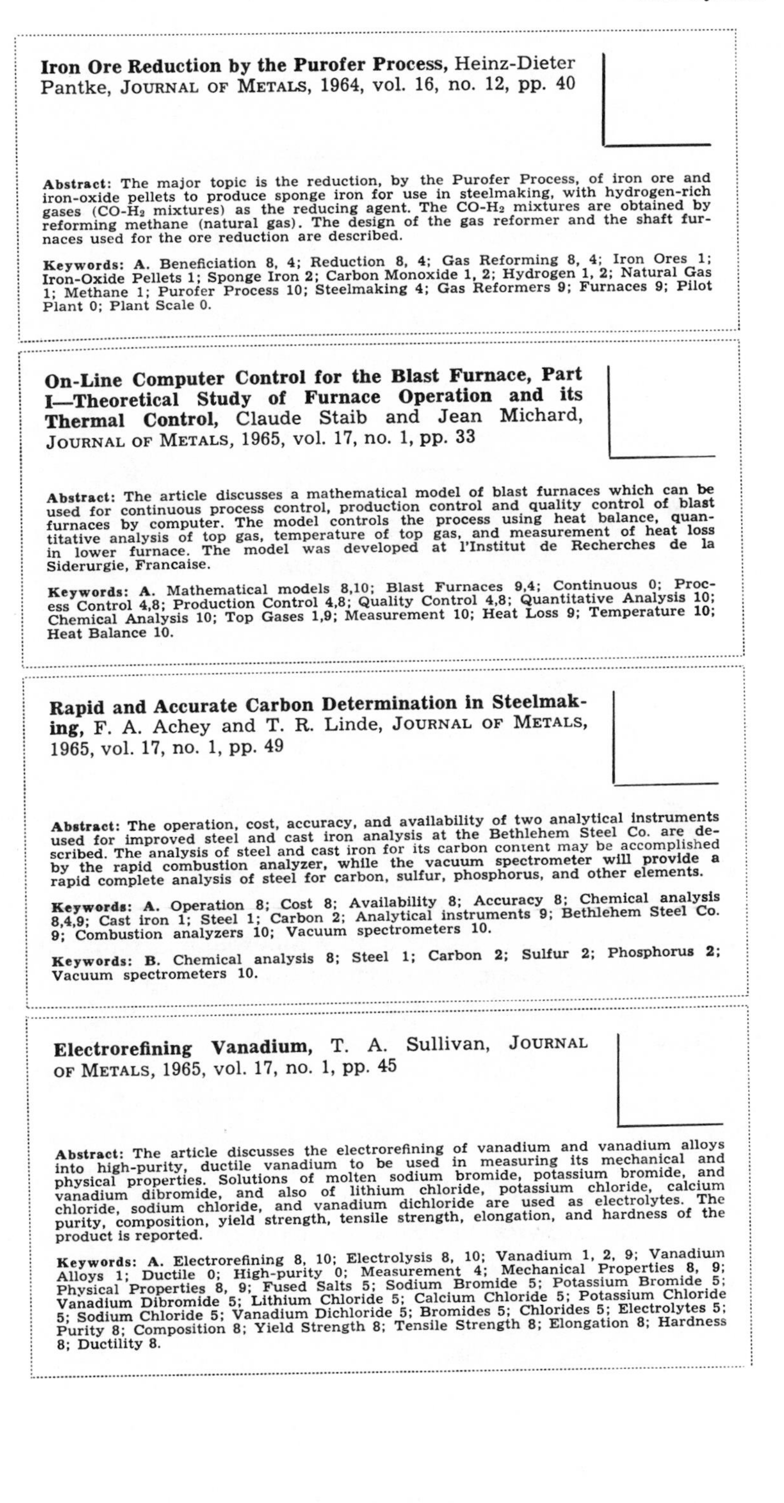

Iron Ore Reduction by the Purofer Process, Heinz-Dieter Pantke, JOURNAL OF METALS, 1964, vol. 16, no. 12, pp. 40

Abstract: The major topic is the reduction, by the Purofer Process, of iron ore and iron-oxide pellets to produce sponge iron for use in steelmaking, with hydrogen-rich gases (CO-H_2 mixtures) as the reducing agent. The CO-H_2 mixtures are obtained by reforming methane (natural gas). The design of the gas reformer and the shaft furnaces used for the ore reduction are described.

Keywords: A. Beneficiation 8, 4; Reduction 8, 4; Gas Reforming 8, 4; Iron Ores 1; Iron-Oxide Pellets 1; Sponge Iron 2; Carbon Monoxide 1, 2; Hydrogen 1, 2; Natural Gas 1; Methane 1; Purofer Process 10; Steelmaking 4; Gas Reformers 9; Furnaces 9; Pilot Plant 0; Plant Scale 0.

On-Line Computer Control for the Blast Furnace, Part I—Theoretical Study of Furnace Operation and its Thermal Control, Claude Staib and Jean Michard, JOURNAL OF METALS, 1965, vol. 17, no. 1, pp. 33

Abstract: The article discusses a mathematical model of blast furnaces which can be used for continuous process control, production control and quality control of blast furnaces by computer. The model controls the process using heat balance, quantitative analysis of top gas, temperature of top gas, and measurement of heat loss in lower furnace. The model was developed at l'Institut de Recherches de la Siderurgie, Francaise.

Keywords: A. Mathematical models 8,10; Blast Furnaces 9,4; Continuous 0; Process Control 4,8; Production Control 4,8; Quality Control 4,8; Quantitative Analysis 10; Chemical Analysis 10; Top Gases 1,9; Measurement 10; Heat Loss 9; Temperature 10; Heat Balance 10.

Rapid and Accurate Carbon Determination in Steelmaking, F. A. Achey and T. R. Linde, JOURNAL OF METALS, 1965, vol. 17, no. 1, pp. 49

Abstract: The operation, cost, accuracy, and availability of two analytical instruments used for improved steel and cast iron analysis at the Bethlehem Steel Co. are described. The analysis of steel and cast iron for its carbon content may be accomplished by the rapid combustion analyzer, while the vacuum spectrometer will provide a rapid complete analysis of steel for carbon, sulfur, phosphorus, and other elements.

Keywords: A. Operation 8; Cost 8; Availability 8; Accuracy 8; Chemical analysis 8,4,9; Cast iron 1; Steel 1; Carbon 2; Analytical instruments 9; Bethlehem Steel Co. 9; Combustion analyzers 10; Vacuum spectrometers 10.

Keywords: B. Chemical analysis 8; Steel 1; Carbon 2; Sulfur 2; Phosphorus 2; Vacuum spectrometers 10.

Electrorefining Vanadium, T. A. Sullivan, JOURNAL OF METALS, 1965, vol. 17, no. 1, pp. 45

Abstract: The article discusses the electrorefining of vanadium and vanadium alloys into high-purity, ductile vanadium to be used in measuring its mechanical and physical properties. Solutions of molten sodium bromide, potassium bromide, and vanadium dibromide, and also of lithium chloride, potassium chloride, calcium chloride, sodium chloride, and vanadium dichloride are used as electrolytes. The purity, composition, yield strength, tensile strength, elongation, and hardness of the product is reported.

Keywords: A. Electrorefining 8, 10; Electrolysis 8, 10; Vanadium 1, 2, 9; Vanadium Alloys 1; Ductile 0; High-purity 0; Measurement 4; Mechanical Properties 8, 9; Physical Properties 8, 9; Fused Salts 5; Sodium Bromide 5; Potassium Bromide 5; Vanadium Dibromide 5; Lithium Chloride 5; Calcium Chloride 5; Potassium Chloride 5; Sodium Chloride 5; Vanadium Dichloride 5; Bromides 5; Chlorides 5; Electrolytes 5; Purity 8; Composition 8; Yield Strength 8; Tensile Strength 8; Elongation 8; Hardness 8; Ductility 8.

Titanium Co., Ltd.—9, the reference number 134 appears twice—once followed by C and once by E since that term-role occurred in both links.

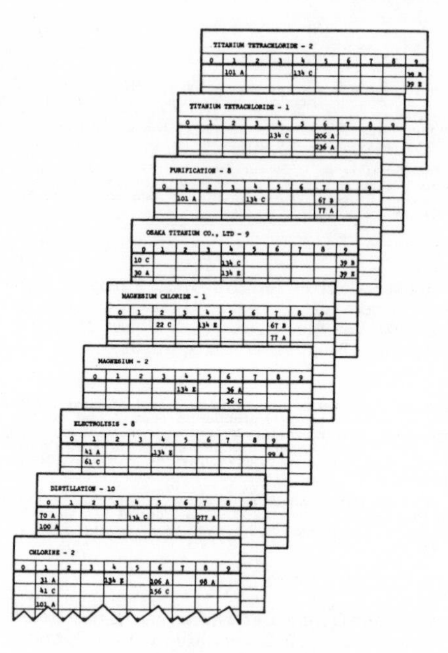

Figure 2

"Once a term-role card has been created for a keyword in a particular role, that term-role card is used for posting all accession numbers of abstracts for which that term-role is applicable. Each card shows all the accession numbers of articles that were indexed by the term-role written at the top. Accession numbers are always followed by a letter representing the link in which the term-role was used.

"Term-role cards are filed alphabetically, and the total file of these cards is the index.

"When a scientist or engineer wishes to use the index to retrieve references on a particular subject, the first step is to write a phrase or statement which best represents the information desired. The statement should be formed in terms as specific as the searcher's actual requirements. For instance, if the searcher is interested in the *use of electric arc furnaces for producing carbon steels,* the statement should not be generalized to read '*the use of furnaces for producing steels*.' The keywords selected during indexing are specific representations of the ideas expressed in each article and, in like manner, specific terms should be used in searching the index. In writing the search statement it is helpful mentally to picture the wording of a brief abstract statement of an imaginary article that the searcher would like to retrieve.

"The search statement is then converted to term-roles (indexed) in the same way that keywords and their roles are lifted out of the abstract statements when the articles are indexed. A searcher unfamiliar with the EJC system will need to refer to the definitions of the roles (Table 1) in order to apply the correct roles to the search keywords according to the context implied in the search statement.

"The following are a few examples of how the proper roles are selected. Consider the search statement, *the strength properties of beryllium alloys which are to be used in airframes*. The keyword *strength properties* should be searched in role 8 as the subject discussed; *beryllium alloys* is searched in role 9 as possessive object of strength properties; and *airframes* is searched in role 4 since its context is that of intended application. If the search statement is *the hydrogen reduction of molybdic oxide to produce molybdenum by means of a shaft furnace,* the cards that will be selected from the index to conduct the search are those representing the term-roles REDUCTION—8, MOLYBDIC OXIDE—1, MOLYBDENUM—2, and SHAFT FURNACES—10, since reduction is the subject under discussion, molybdic oxide is the starting material, molybdenum is the product, and shaft furnaces are used to accomplish the reduction process. *The effect of impurities on the ductility of steel castings* illustrates a search involving the cause-effect relationship. The term-role cards used in this search are IMPURITIES—6, DUCTILITY—7, CASTINGS—9, and STEEL—9.

"When proper search term-roles have been chosen, those term-role cards are pulled from the index. It is possible that a search might involve only one term-role as in a search for information on *uses of submerged-arc furnaces* (SUBMERGED-ARC FURNACES—10) or references on any *laminating processes* (LAMINATION—8). However, most searches require two or more term-role cards, as in a search for information on *preheating scrap* (PREHEATING—8, SCRAP—9) *or melting 4330 steel using a gas furnace* (MELTING—8,4330 STEEL—1, GAS FURNACES—10).

"When the search statement contains more than one term-role, the cards are visually examined to find accession number—link combinations which appear on all the term-role cards involved in the search. The process of finding these coincident accession numbers on term-role cards is called *concept coordination* and accomplishes the retrieval of accession numbers of pertinent abstracts. The links must agree as well as the accession numbers because, during the indexing step, keywords related to each other are grouped in the same link, while keywords not discussed in the same intellectual idea in the article are separated by using different links. If the index shows the accession number and link 201A on the card MELT-

ING—8 and 201B on the card 4330 STEEL—1, this indicates that abstract 201 identifies an article that discusses a process for melting something (not 4330 steel) and it also discusses some operation (but not melting) being performed on 4330 steel. Thus abstract number 201 would *not* be retrieved in answer to an inquiry about *melting 4330 steel.*

"When the searcher has obtained the accession numbers coincident to all term-role cards which satisfy the search statement, the abstract cards for those numbers are pulled from the abstract file. The searcher scans each abstract to gain further understanding of what each article is about, and requests, from the library, copies of those articles which are pertinent or finds the articles in his personal collection of journals."

The procedure discussed by Tippett and McCauley above is quite different from that usually followed by professional writers. Searching the literature stored in "standard" fashion in libraries or indexed by systems as yet incompatible with machine retrieval techniques involves a variety of procedures. None is particularly difficult—some are mutually exclusive, however—and the whole process takes longer when men rather than machines are involved. Standard searches are also less effective because reference card techniques, such as the Dewey Decimal and the Library of Congress systems are far less precise than that of the EJC outlined above.

An understanding of the standard card techniques for library research is almost a prerequisite for an understanding of the unique—and seemingly esoteric—techniques peculiar to specific disciplines. A few generalities pertain, not the least of which is the necessity of cultivating the reference librarian. He is a professional trained to know where information is to be found. Indeed, he will almost always be able to find information if given enough time and if told *exactly what information is sought.* Without a precise search statement or problem definition, however, his reference-card approach can only mine out great blocks of documents which must be laboriously read for relevant data.

The Library of Congress card system is used by the larger libraries because its alphabetical code accommodates many books in the same subject area without building up the involved decimals of the Dewey system. Books referenced by this system are divided into twenty classes by subject. The subject area has a preassigned letter. The basic classifications are

A	General works	M	Music
B	Philosophy, Religion	N	Fine arts
C	History, Auxiliary sciences	P	Language and literature
D	Foreign history and topography	Q	Science
E-F	American history	R	Medicine

G	Geography, Anthropology	S	Agriculture
H	Social science	T	Technology
J	Political science	U	Military science
K	Law	V	Naval science
L	Education	Z	Bibliography, Library science

Within the class, further subdivisions are shown by additional letters and numbers. In practice, one branch of classification H (Social science) is broken down to

H Social science
HF Commerce
HF 5500 Business organization and management
HF 5549 Personnel

All branches are similarly broken down; for example,

Q	Science	T	Technology
QA	Mathematics	TA	General and civil engineering
QB	Astronomy	TJ	Mechanical engineering
QC	Physics	TK	Electrical engineering
QD	Chemistry	TK	7800 Electronics
	etc.		etc.

The Dewey Decimal system, or a local modification of it, is used by the majority of libraries. Books referenced by this system are divided into ten classes by subject. The basic classifications are

000	General works	500	Pure science
100	Philosophy	600	Applied science
200	Religion	700	Arts and recreation
300	Social sciences	800	Literature
400	Linguistics	900	History

Each of the ten classes is subdivided into ten smaller parts. These subdivisions are broken into ten sub-subdivisions, and so on. The resulting number is the *classification number*. In practice, a random example from the 800's (Literature) breaks down as follows.

800 General literature
810 American literature
    811 American poetry
        811.3 American poetry—middle nineteenth century (1830–1861)
            811.34 Longfellow, Henry W.
820 English literature
830 German literature
    etc.

An additional code number called the *author number* is added *below* the classification number, and both are printed on the reference card and on the spine of the book. The author number is composed of the first letter or two of the author's last name followed by a number taken from a standard code called the "Cutter Table of Author Numbers." Thus, if someone named Oscar Zilch wrote a book *about* Longfellow, the classification number and the author number would appear as

811.34
Z 3

The only variation to the above system involves works of fiction. No call number is given; instead, the word "Fiction" is printed on the card and below that is printed the first letter of the author's last name. Thus a book called *The Martyred* by Richard *Kim* would be assigned a card showing

Fiction
Ki

At least three cards are used in cataloging all books except literature—one for the *author,* one for the *title,* and one or more cards for the *subject.* All cards are usually arranged by one alphabet, but there may be local exceptions when subject cards are filed separately.

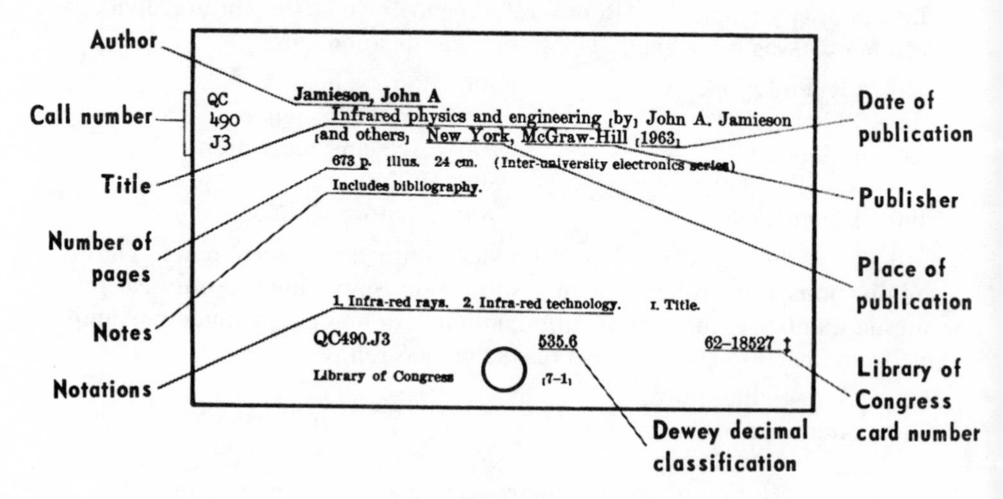

Figure 3

Figure 3 shows a typical author card. The card contains a great deal of information, but it is information that is helpful primarily to librarians.

The book is described physically, its numbers in the "standard" referencing systems are given, and complete publishing information is included. The relevance of the book to a specific problem, however, can only be partially determined from the data shown.

Nevertheless, the author card contains several items of value to a searcher. Specialization and the necessity for professional publication have forced individuals to become authorities in specific areas. If the author is such a specialist, his name and reputation are usually known to research scholars. Second, the title of a scholarly work is usually indicative of the contents. Indeed, an attention-seeking title which does not reveal the contents usually marks a book as being "popular" and perhaps lacking in depth as well as in seriousness. Third, the name of the publisher is significant. Publishers are jealous of their reputations, particularly in specialized areas. Fourth, the date of publication indicates whether recent information is included. Fifth, the number of pages usually indicates the depth of coverage. Although a short book on a general topic is likely to be superficial, a restrictive title and a lengthy coverage usually mean an exhaustive treatment. Sixth, the note "Bibliography" shows that the author has himself searched the literature and provided leads to additional information. Finally, as shown in Figure 4, the notations at the bottom of the card show where additional cards, called *added-entry* cards, are found. These added-entry cards show what subjects are covered in the book and how reference librarians have filed similar books. Added-entry cards have neither the depth nor the relevance, however, of keywords used in the EJC system.

Title and subject cards, as shown in Figure 5, contain the same data as the author card. Obviously, the number and location of copies are important information. Some libraries add symbols, such as an asterisk (*), before the call number to indicate that the book is oversize and housed in a separate collection. Others add a date below the call number to show that more than one edition has been published. All add "Mc" over the call number to indicate that the publication is on Microcard.

Cross-reference cards are used by most libraries to direct a searcher from subject headings which are not used to headings that are used: AEROPLANES, see AIRPLANES. *See also* cards are used to refer a searcher to additional subjects.

Perhaps the most effective method of searching a card catalog is to proceed from the specific to the general. A searcher should start with an author whenever the name and correct spelling are known. It should be remembered that an "author" can be a person, company, association, laboratory, or government agency. Next, the most definite subject should

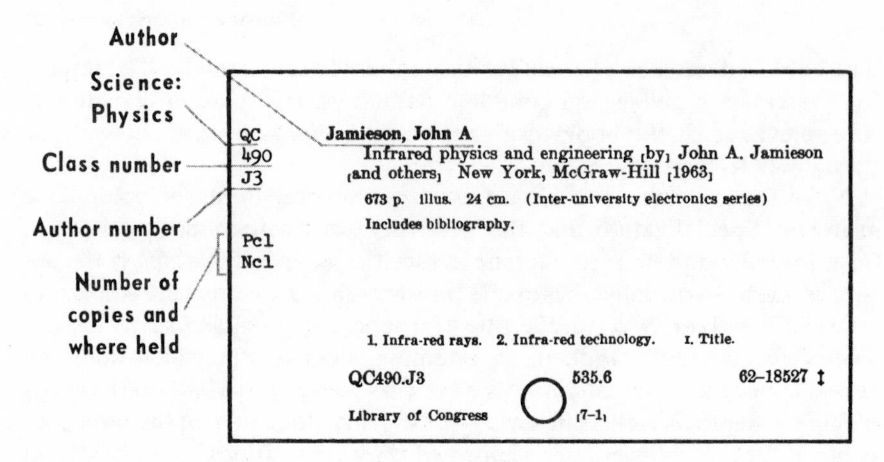

QC
490
J3

Pcl
Ncl

Jamieson, John A
Infrared physics and engineering [by] John A. Jamieson [and others] New York, McGraw-Hill [1963]
673 p. illus. 24 cm. (Inter-university electronics series)
Includes bibliography.

1. Infra-red rays. 2. Infra-red technology. I. Title.

QC490.J3 535.6 62–18527 ‡

Library of Congress [7-1]

Figure 4

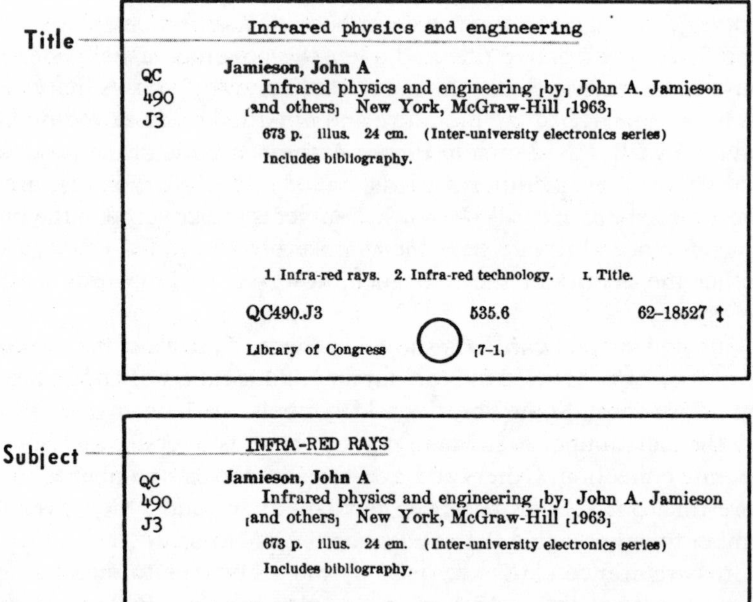

Infrared physics and engineering

QC
490
J3

Jamieson, John A
Infrared physics and engineering [by] John A. Jamieson [and others] New York, McGraw-Hill [1963]
673 p. illus. 24 cm. (Inter-university electronics series)
Includes bibliography.

1. Infra-red rays. 2. Infra-red technology. I. Title.

QC490.J3 535.6 62–18527 ‡

Library of Congress [7-1]

Subject

INFRA-RED RAYS

QC
490
J3

Jamieson, John A
Infrared physics and engineering [by] John A. Jamieson [and others] New York, McGraw-Hill [1963]
673 p. illus. 24 cm. (Inter-university electronics series)
Includes bibliography.

1. Infra-red rays. 2. Infra-red technology. I. Title.

QC490.J3 535.6 62–18527 ‡

Library of Congress [7-1]

Figure 5

be searched. CHEMISTRY, INORGANIC, not just CHEMISTRY; or ZIRCONIUM, not just METALS. If the definite subject and its synonyms list nothing, then the more general heading may be searched.

It is not presently possible to index periodical articles in sufficient detail in a card catalog. The number of articles is so great that only a system like that of the EJC which includes an abstracting stage *at the time of publication* could cope with the number of articles. Until such a system is widely adopted, a search of the periodical literature will require the use of specialized indexes.

Specialized indexes have existed since the eighteenth century, and new ventures appear annually. The American Chemical Society introduced *Chemical Titles* in 1961. This computer-produced and keyword-organized index gives current coverage of 650 of the 9000 periodicals subsequently abstracted by *Chemical Abstracts.*

It is indicative of the responsibility of professional societies for their disciplines that the American Chemical Society itself publishes fifteen first-rate journals in addition to *Chemical Abstracts.* The titles of those journals are listed below to indicate the complexities of indexing all areas of a specific discipline.

Analytical Chemistry
Biochemistry
Chemical and Engineering News
Chemical Reviews
Chemical Titles
Chemistry
Industrial and Engineering Chemistry
Inorganic Chemistry
Journal of Agricultural and Food Chemistry
Journal of the American Chemical Society
Journal of Chemical Documentation
Journal of Chemical and Engineering Data
Journal of Medicinal Chemistry
The Journal of Organic Chemistry
The Journal of Physical Chemistry

The majority of periodical indexes is useful only as a source for a bibliography of up-to-date material in specialized areas or in specific disciplines. *The Social Sciences and Humanities Index* is almost unique in attempting a synthesis. Because the literature of each discipline may best be organized in a framework appropriate to that discipline (subject determines organization and style, as shown in Chapter 1), indexes tend to differ from one another in arrangement.

An alphabetical listing, usually by author and subject, is the technique most frequently used to organize the general information found in popular and semipopular journals. *The Readers' Guide to Periodical Literature,* which indexes alphabetically by author, title, and subject most of the North American periodicals of large circulation, is a useful source for a bibliography of up-to-date material largely on a popular level. It is published twice a month (except in July and August) and is cumulated in three-, six-, nine-, and twelve-month intervals *for that year*. Bound volumes are published annually, and two-year cumulations are published in odd-numbered years. Abbreviations are explained at the beginning of each issue, and a random page takes the form illustrated in Figure 6.

Although most of the disciplines loosely termed "humanities" (e.g., art, English, history, languages, music, philosophy, religion, and speech) use the alphabetical system discussed above, the social sciences, earth sciences, life sciences, and applied sciences employ variations in organizing their indexes. In the social sciences the indexes *Psychological Abstracts* (1927 to date) and *Sociological Abstracts* (1953 to date) are basic reference works. The organization of each is by subject, and random pages take the form illustrated in Figures 7 and 8.

Within the social sciences of economics and government (or political science), indexes are organized by topic. Random pages from *Business Periodicals Index* (1958 to date) and *Public Affairs Information Service* (1915 to date) take the form shown in Figures 9 and 10.

Indexes of journal articles about the earth sciences (astronomy, chemistry, geology, meteorology, oceanography, physics, etc.) vary widely in organization. While the sample pages reproduced below indicate variations among these disciplines, interdisciplinary syntheses exist within the earth sciences, and all the earth sciences have a close interdisciplinary connection with the life sciences.

It is essential to read the prefatory information printed in indexes to scientific periodicals, because the organization is not in one alphabet. In *Chemical Abstracts,* for example, the abstracts are grouped in 74 sections within five areas as follows:

1. Physical including selected analytical and inorganic (Sections 1–15)
2. Applied chemistry (Sections 16–30)
3. Organic chemistry (Sections 31–44)
4. Macromolecular chemistry (Sections 45–55)
5. Biochemistry (Sections 56–74)

Indexes to journal articles about the life sciences (bacteriology, biology, botany, microbiology, zoology, etc.) are both interdisciplinary and

READERS' GUIDE

to periodical literature

AUTHOR AND SUBJECT INDEX

July 30—August 26, 1965

A and A distributors, Incorporated
A & A's new quarters greatly improving service. il Pub W 188:16-19 Ag 16 '65

AALL. See American association of law libraries

ABA. See American bar association

ALTA. See American library trustee association

ATS (advanced technology satellite) See Communications satellites

ABBEY, J. R.
Major Abbey's modern bookbindings. H. Nixon. il Craft Horiz 25:28-31 Jl '65

ABEEL, Erica
Daedalus at the Rollerdrome. Sat R 48:51-3 Ag 28 '65

ABEL, I. W.
Decision time again: is a steel strike coming? il por U S News 59:70-1 Ag 23 '65
Steel strike threat loses impact. por Bsns W p79-80 Ag 7 '65

ABERCROMBIE and Fitch company
Out to launch. J. Skow. il Sat Eve Post 238:20 Ag 28 '65

ABERNETHY, Roy
Mandatory inspection a sensible requirement; excerpts from testimony, July 1965. por U S News 59:101 Ag 9 '65

ABNORMALITIES (animals)
Growth of mouse mammary glands in vivo after monolayer culture. C. W. Daniel and K. B. DeOme. bibliog il Science 149:634-6 Ag 6 '65

ACADEMIC freedom
Academic revolution at St John's. F. Canavan. America 113:136-40 Ag 7 '65
How free is free? il Newsweek 66:73-4 Ag 30 '65

ACCELERATORS (electrons, etc)
Big accelerator: competition for AEC facility is stirring up communities throughout country. D. S. Greenberg. Science 149:730-1 Ag 13 '65
Nature of matter; purposes of high energy physics; excerpts. Science 147:1548-55 Mr 26 '65; Discussion. 149:584-6 Ag 6 '65

ACCIDENTS
See also
Explosions

Psychological aspects

Are you accident prone? reprint. J. Fetterman. il Sci Digest 58:53-6 Ag '65

ACETAZOLAMIDE
Deformity of forelimb in rats: association with high doses of acetazolamide. W. M. Layton, jr, and D. W. Hallesy. bibliog il Science 149:306-8 Jl 16 '65

ACHESON, Dean Gooderham
Advice to young academic propagandists. Reporter 33:30 Ag 12 '65

ACIDS
Soft and hard acids and bases. Sci Am 213:46 Ag '65

ACOCA, Miguel
June 14th: the young rebels. Life 59:56A Ag 13 '65

ACOUSTICAL materials
How to select acoustical materials. Arch Rec 138:185-6 Ag '65

ACOUSTICAL materials association
How to select acoustical materials. Arch Rec 138:185-6 Ag '65

ACOUSTICS, Architectural
See also
Acoustical materials

ADAMO, S. J.
Dilemma in the Catholic press. America 113:154-8 Ag 14 '65

ADAMS, Judith, and Dubrow, Heather
Something to talk about on campus. Mlle 61:312-15 Ag '65

ADAMS, Richard N.
Pattern of development in Latin America; with questions and answers. Ann Am Acad 360:1-10 Jl '65

ADAPTATION (biology)
Evolution of nest building. N. E. Collias. il Natur Hist 74:40-7 bibliog(p74) Ag '65

ADELSON, Joel. See Brodie, A. F. jt. auth.

ADENOSINE triphosphatase
Dynein: a protein with adenosine triphosphatase activity from cilia. I. R. Gibbons and A. J. Rowe. bibliog il Science 149:424-6 Jl 23 '65

ADENOSYLMETHIONINE
Adenosylmethionine elevation in leukemic white blood cells. R. J. Baldessarini and P. P. Carbone. bibliog il Science 149:644-5 Ag 6 '65

ADIRONDACK MOUNTAINS
Adirondack reader; ed. by P. F. Jamieson. Review
Liv Wildn 29:31-3 Spr '65. P. H. Oehser

ADIRONDACK trail. See Roads—New York (state)

ADJUSTMENT, Social
Dial F for fiction; plea for having life imitate art. S. Hixon. Seventeen 24:116 Ag '65
New girl in town; what to do when you move to a new school. R. Greer. il Seventeen 24:366+ Ag '65
See also
College students—Adjustment

ADLER, Myles
Turning the tables. U S Camera 28:50-1 Ag '65

ADMINISTRATIVE and political divisions
See also
Apportionment (election law)

ADRENAL hormones. See Hormones

ADULT education
See also
Service men, Discharged—Education
University extension

Library participation

See also
Libraries—Readers advisory service

ADULT music study. See Music—Instruction and study

ADVANCED technology satellite. See Communications satellites

ADVENTURE
One step to adventure. E. S. Hill. il Read Digest 87:129-32 S '65

ADVERTISING
Dream world of advertising. il Am Heritage 16:70-5 Ag '65
Naming names. il Time 86:70-1 Ag 20 '65

ADVERTISING agencies
Multi-agency marketing. G. Lazarus. Sat R 48:59-60 Ag 14 '65

ADVISORY service, Readers. See Libraries—Readers advisory service

AERIAL reconnaissance
Errant camera; air force above Pierrelatte, France. il Newsweek 66:35 Ag 2 '65
Radar imagery; used for mapping purposes by reconnaissance aircraft. J. L. Nelson. il Electr World 74:42-3+ Ag '65

AERO commander, Incorporated
Aero commander management shifts set. D. A. Brown. il Aviation W 83:20-1 Jl 19 '65

AERO spacelines, Incorporated
Super Guppy to make first flight Aug. 25. H. D. Watkins. il Aviation 83:42-3 Ag 23 '65

AERONAUTIC instruments
Pilot reaction to Sperry display assessed; windshield projection display. il Aviation W 83:115+ Ag 9 '65
Rainbow optical projector aids landing. P. J. Klass. il Aviation W 83:63-5 Ag 23 '65

AERONAUTICS, Military

Great Britain

See also
Great Britain—Royal air force

Figure 6

Psychological Abstracts

VOLUME 39 AUGUST 1965 NUMBER 4

EDITORIAL NOTE

The Galton Institute, Los Angeles, began publication of a quarterly journal, entitled *Perceptual Cognitive Development,* in February, 1965. It consists of bibliography keyword-in-context index, and an author list. The journal is "the first known computer-generated bibliographical index in the behavioral-biological-social science field scheduled for regular, continuing publication."

ERRATUM

The name of the author for abstract number 5063 should read: Handler, Leonard.

GENERAL

8912. **Barker, Roger G.** (U. Kansas) **Explorations in ecological psychology.** *American Psychologist,* 1965, **20**(1), 1-14.—The "great diversity of coupling between psychologists and psychological phenomena can be divided into 2 types which produce data of crucially different significance for the science of psychology." (1) Psychologists as Transducers—T Data: The psychologist transforms data and in effect is a translating machine. (2) Psychologists as Operators—O Data: The psychologist "achieves control which allows him to focus upon segments and processes of particular concern to him, via data that refer to events which he, in part, contrives." Psychologists "as operators and as transducers are not analogous, and . . . the data they produce have fundamentally different uses within science. A central problem of our science is the relation between ecological events (the distal stimuli) at the origin of E-O-E [environment-organism-environment] arcs and the succeeding events along these arcs." There "are a number of reasons for avoiding the role of transducer in psychological research. . . . The skills and personality attributes required of a successful transducer are different from those of a successful operator. . . . The techniques of the transducer are in many respects more difficult than those of the operator." —*S. J. Lachman.*

8913. **Devoto, Audrea. Prospettive della psicologia istituzionale.** [Prospectives in institutional psychology.] *Riv. Psicol. Soc.,* 1964, **31**(2-3), 177-188.—Institutional psychology is defined as the focal point of clinical criminology, penology, correctional sociology, and social psychiatry. Goffman's concept of "total institutions" is elaborated.—*L. L'Abate.*

8914. **Frey, Allan H.** (State Coll., Pa.) **Behavioral biophysics.** *Psychological Bulletin,* 1965, **63** (5), 322-337.—Electromagnetic energy is an important factor in the biophysical analysis of the properties and function of living systems. Due to technical advances in electronics, this energy is now being used as a research tool, both by study of its emission by living organisms and also by applying it to the organism. In this paper, the nature of the energy is sketched. Then, data on fingertip detection of color, neural emission of infrared energy, the use of electron paramagnetic resonance techniques to detect neural activity, brain impedance shifts and behavior, and the influence of UHF energy on behavior are considered. It is concluded that, though these areas are in the embryonic stage of development, most are potentially of great significance in the understanding of the nervous system and behavior. (2 p. ref.)—*Journal abstract.*

8915. **Gutman, Herbert. Structure and function.** *Genetic Psychology Monographs,* 1964, **70**(1), 3-56. —A definition of "structure" and "function" is followed by a discussion of the ways in which structure determines function and function determines structure. Based on a discussion of the genesis of structures, equipotentiality of structures, and equifinality of function, the author arrives at the conclusion of the "primacy of function" over structure. The results of the discussion are applied to the problems of "Machine vs. Organism," "Parts vs. Whole," and "Purpose in Structure and Function."—*Author abstract.*

8916. **Margoshes, Adam, & Litt, Sheldon.** (Boise Junior Coll.) **Psychology of the scientist: XII. Neglect of revolutionary ideas in psychology.** *Psychological Reports,* 1965, **16**(2), 621-624.—Radically new ideas typically meet with irrational opposition in all fields of thought, including science. As a result, progress is slowed down and some ideas need to be rediscovered before they become widely accepted. 2 examples are cited from the recent psychological literature: an extensive review of the role of muscle tension in personality theory failed to acknowledge Wilhelm Reich's major contributions; and a study of the personality of cancer patients failed to credit Reich's earlier work, although almost identical conclusions were reached.—*Journal abstract.*

8917. **Pflaum, John H.** (P.O. Box 1215, Madison, Wisc.) **The psychology of smoking.** *Psychology,* 1965, **2**(1), 44-58.—A review of the literature on smoking behavior and a theoretical position is presented by the author. The history of smoking, cultural differences, socio-economic and sex variables, the roles of anxiety and learning, role of the mass media and medical science, personality variables, and other aspects of smoking behavior are discussed. The author proposes that smoking behavior is part of an integrated pattern of learned behavior essentially motivated to establish stable expectancies in the social-psychological environment.—*Journal abstract.*

8918. **Platz, Arthur.** (VA Cent. NP Res. Lab., Perry Pt., Md.) **Psychology of the scientist: XI. Lotka's law and research visibility.** *Psychological Reports,* 1965, **16**(2), 566-568.—Previous research

Figure 7

0100 methodology and research technology

03 methodology (social science & behavioral)

B 4795 Becker, Howard S. (Stanford U, Calif), PROBLEMS IN THE PUBLICATION OF FIELD STUDIES, 267-284, Chpt in Reflections on Community Studies, A. J. Vidich et al, Eds, John Wiley & Sons, Inc, 1964, 349 pp, NP.
¶ Publishing the results of field studies may produce difficulties because of the possibility of harming those studied by revealing damaging information about them. Some authors suggest that this can be avoided by taking various precautions. In fact, there is an irreducible conflict between researcher & those studied, because a good study inevitably discloses deviations from rules or a disparity between a desired image & reality. A great danger is that the sci'st will censor his own report, either because he has been taken in or because he 'goes native', or he may censor his findings in deference to a bargain made with those studied or because he thinks publication would damage public instit's. The only solution to the dilemma of what to publish is dictated by individual conscience. But the researcher should aid those studied in assimilating the meaning of his report for them, by helping them to anticipate possible adverse reactions & to consider possible directions of change. (See SA 1117-B5335). AA

B 4796 Bergmann, Gustav (State U of Iowa, Ames), PURPOSE, FUNCTION, SCIENTIFIC EXPLANATION, Acta Sociol., 1961, 5, 4, 225-238.
¶ An analysis of the questions of whether it is in the role of the sci'st to deal with 'function' & 'purpose', whether he must use these terms if he can deal with them, & whether he may use these terms if he wants to. 2 meanings of the term 'function' & 3 derivative uses of the term 'purpose' are distinguished. The systematic use of 'purpose' in the soc disciplines is examined. It is concluded that, except when 'purpose' refers to human purposes, purpose has no place in sci. D. Cooperman

B 4797 Buck, Roger C. (Indiana U, Bloomington), REFLEXIVE PREDICTIONS, Philos. Sci., 1963, 30, 4, Oct, 359-368.
¶ An analysis of 'reflexive' (self-fulfilling) predictions, re 3 main issues: its precise definition; the methodological problems generated it in the soc sci's; & the suggestion that, since this type of prediction refers solely to HB, there is thus a 'philosophically signif' distinction between the natural & soc sci's. A discussion is presented of A. Grunbaum's attempt to negate the above thesis (originally proposed by R. K. Merton). It is seen that Grunbaum's rebuttal is ineffective. D. Cooperman

B 4798 Coleman, James S. (Johns Hopkins U, Baltimore, Md), INTRODUCTION TO MATHEMATICAL SOCIOLOGY, New York, NY: Free Press of Glencoe, 1964, xiv+554 pp, $9.95.
¶ A study providing a firm foundation for the Successful Use of Mathematics in Sociol, in 18 Chpts. Chpt (1) Uses of Mathematics in Sociology, surveys various applications of mathematics to sociol'al problems. (2) Problems of Quantitative Measurement in Sociology, studies special problems of measurement in sociol, beginning with the classical approach to measurement taken in the physical sci's. (3) Mathematics as a Language for Relations Between Variables, examines how, in mechanics & econ's, mathematics has provided a language for expressing the theoretical relations. (4) Mathematical Language for Relations Between Qualitative Attributes, describes the development of mathematical structure for expressing relations between qualitative attributes. (5) Relations Between Attributes: Over-Time Data, extends the model of (4) to its use with data from several points in time. (6) Multivariate Analysis, extends the model of (4) & (5) to the case of several independent variables. (7) Multiple-Level Systems and Emergent Propositions, applies this model to a problem where propositions are desired at both the level of individuals & groups. (8) One-Way Processes with a Continuous Independent Variable, applies this model to several sets of data where the independent variable is a continuous variable. (9) Social and Psychological Processes and their Equilibrium States, presents variations on the problem of (8). (10) The Poisson Process and its Contagious Relatives, introduces a general form of process applicable to individual & group behavior, of which the simplest case gives a Poisson distribution. (11) The Poisson Process and its Contagious Relatives: Equilibrium Models, presents an extension of (10) into processess with 2-directional flow. (12) Social and Psychological Organization of Attitudes, offers a model in which relations between att responses are seen as conditioning of elements in the individual. (13) Change and Response Uncertainty, combines the model of (12) with those of (4) & (5), to provide a model in which both change & unreliability of response occur. (14) Measures of Structural Characteristics, develops 5 measures of structural characteristics by use of models to mirror processes that might have produced the structures or that might be consequent upon them. (15) The Method of Residue, offers a new treatment of the old problem of distance & interaction. (16) The study of Logical Implications, describes logical implications following from some very simple assumptions about process together with variations in the structure of a community. (17) Diffusion in Incomplete Social Structures, offers an extension of diffusion models into structures which are not random or completely intermixed. (18) Tactics and Strategies in the Use of Mathematics, makes suggestions in the application of mathematics to sociol'al problems. AA

B 4799 Dahrendorf, Ralf (U of Tuebingen, Germany), SYMPOSIA ON POLITICAL BEHAVIOR, Amer. Sociol. R., 1964, 29, 5, Oct, 734-736.
¶ An evaluation of the claims pro & con the 'behavioral sci' trend among pol'al sci'ts. 3 criticisms not raised by others are leveled at the approach: (1) the behavioral approach tends to reduce pol to "no more than one segment of its historical substance"; (2) there is no reason to equate behavioralism with sci'fic rigor; & (3) more attention is devoted to dispute as to the proper orientation of the discipline than is directed toward specific problems. D. Cooperman

B 4800 Freymann, Moye W. & Herbert F. Lionberger (The Ford Foundation, New Delhi, India), A MODEL FOR FAMILY PLANNING ACTION-RESEARCH, 443-462,

Figure 8

Business Periodicals Index

August 1965

A.I.D. See Agency for international development
ABEL, Iorwith Wilbur
New temper in the steelworkers. Bsns W p43-4 Je 5 '65
Steel negotiators get down to brass tacks. por Bsns W p65-6 Je 19 '65
ABILITY, Creative. See Creative ability
ABILITY tests
Minimum loss function as determiner of optimal cutting scores [personnel selection tests] I. Guttman and N. S. Raju. Personnel Psych 18:179-85 Summer '65
Testing minority applicants for employment. H. C. Lockwood. Personnel J 44:356-60+ Jl-Ag '65
ABRASIVE belts
High speed abrasive belt grinder refinishes rubber and composition rolls. Bsns W p 150 My 22 '65
ABSENTEEISM (labor)
Busy social life of germs; psychologist develops a theory about microbes. T. J. Fleming. Personnel Mgt 47:99-101 Je '65
ACCELERATION principle (economics)
Capital appropriations and the accelerator. A. G. Hart. R Econ & Stat 47:123-36 My '65
ACCELERATORS (electrons, etc)
Race for atom smasher. Bsns W p 112 Je 26 '65
ACCELERATORS, Electron. See Accelerators (electrons, etc)
ACCIDENTS
Curbing a clear and present danger [injuries to people who barge through glass doors and panels] il Bsns W p84+ Jl 10 '65
Posters against the old Adam [call to keep safety in mind at home, work, play, and on roads] Economist 216:25 Jl 3 '65
See also
Collisions at sea
Drowning

Statistics

Accident hazards on the farm. Metropolitan Life Stat Bul 46:8-10 Ap '65
ACCOUNTANTS
See also
Accounting ethics
ACCOUNTANTS, Public
CPAs cite unsatisfactory IRS procedures in New York city area. J Taxation 23:56-7 Jl '65
Impact of electronic data processing on relations between banks and CPAs. M. B. T. Davies. J Account 120:60-2 Jl '65
Local bar attacks attorney-CPA in first court action on dual practice. J Taxation 23:53-4 Jl '65
See also
American institute of certified public accountants

Ethics

See Accounting ethics

Examinations

Income taxes on the CPA exam; what role do they play? R. M. Sommerfeld and S. Ritzwoller. J Taxation 23:54-5 Jl '65

Service

Casualty insurance survey for the smaller client. F. P. Walker. J Account 120:30-3 Jl '65
ACCOUNTING
Accounting flow chart. W. J. Pladies. J Account 120:45-54 Jl '65
Accounting for management, by E. L. Kohler; review. Fin Exec 33:6+ Jl '65
See also
Budget, Business
Cash flow
Cost accounting
Depreciation
Funds statements
National accounting
Overhead expense
Securities and exchange commission—Accounting practice and procedure
also subheads Accounting and Accounting aspects under various subjects, e.g. Corporations—Accounting; Electronic data processing—Accounting aspects

Computer aids

National cash [register co] to introduce 395-300 system via Wall Street journal [all-electronic data processing system for accounting] il Adv Age 36:3 Jl 5 '65

Mechanical aids

See also
Adding machines
Bookkeeping machines
ACCOUNTING as a profession
Professional organization and growth. D. F. Linowes. J Account 120:24-9 Jl '65
ACCOUNTING ethics
Professional ethics of the future. J. R. Ring. J Account 120:40-4 Jl '65
Review of professional conduct rule number four. L. Price. N Y Cert Pub Acct 35:514-16 Jl '65
ACCOUNTS receivable
Trend in collections (cont as) Trend in receivables (cont) Credit & Fin Mgt 67:23-4 Jl '65
ACRYLATES
Acrylic monomers and polymers; the prices are reduced sharply. Oil Paint & Drug Rep 187:4+ Je 21 '65
ACTORS and actresses
See also
Moving picture actors and actresses
ADDING machines
Keeping mistakes from computers [Addo, Swedish adding machine company] il Bsns W p 166+ Je 12 '65
ADDITIVES (petroleum products) See Petroleum products—Additives
ADHESIVES
Sprayable adhesive speeds bonding of insulation to ductwork. il Air Cond Heat & Refrig N 105:11 My 24 '65
ADMINISTRATION, Public
Strive for better planner-administrator relations. H. W. Starick. Pub Mgt 47:138-43 Je '65
See also
Agricultural administration
ADMINISTRATION of justice. See Justice, Administration of
ADMINISTRATIVE agencies
See also
Regulatory commissions
ADVERTISEMENTS
See also
Copyright—Advertisements

Books and booklets

New view on teen influences [brochure is good way for business to communicate to young buyers or prospects] Ptr Ink 290:5 My 28 '65

Printing

See also
Advertisements—Typography

Readership

Research shows ads boost purchases of brands 14 per cent; abstracts. D. Starch. Adv Age 36:4 My 10 '65; Ed & Pub 98:17 Je 12 '65

Samples

See also
Samples (merchandising)

Typography

Typographic scoreboard (cont) A. Lawson. Inland Ptr/Am Lith 155:87 Je '65
ADVERTISEMENTS, Classified
Association of newspaper classified advertising managers annual meeting, San Francisco. Ed & Pub 98:12+ Je 26; 42 Jl 3 '65
Court's ruling shows need for wariness [newspaper's right to classify ads] S. Finsness. Ed & Pub 98:32 Je 19 '65
Quality classified trend not indicated. S. Finsness. Ed & Pub 98:20 My 22 '65
Speedy sales ups volume in Detroit [free press] S. Finsness. Ed & Pub 98:24 Je 5 '65

Help wanted ads

Court upholds right to classify ads [case against Chicago tribune] Ed & Pub 98:20 Je 12 '65
Daily's right to classify its ads upheld by court. Adv Age 36:32 Je 7 '65

Figure 9

Public Affairs Information Service

ABBREVIATIONS
† Thomas, Robert C. and others, ed. Acronyms and initialisms dictionary: a guide to alphabetic designations, contractions, acronyms, initialisms and similar condensed appellations. 2d ed '65 767p $15—Gale
LC 64-8724

ABDULLAH, Mohammad
Kashmir, India and Pakistan. Sheikh Mohammad Abdullah. For Affairs 43:528-35 Ap '65

ABEL, I. W.
Meet the press: guest David J. McDonald, president, United steelworkers of America and I. W. Abel, secretary-treasurer, United steelworkers of America. ['65] 11p (v. 9, no. 5) 10c with s.a.e.—Merkle
Television and radio interview, Feb. 7, 1965.
Union leaders contending for the presidency of the United steelworkers of America.

ABILITY grouping in education
Jackson, Brian. Streaming: an education system in miniature. '64 ix+156+13p bibl tables chart (Inst. of community studies. Repts. 11) *21s; pa 10s—Routledge
Classification of primary school children into able, average and backward groups; Great Britain.

ABORTION
Contraception versus abortion [address]. H. Fabre. tables Eugenics R 57:21-5 Mr '65
The criminality of abortion in Korea [Republic]: to what extent can the criminal law serve as a means of social control? with specific concern for the attitudes of Koreans toward Korea's "imported" law relating to abortion. Pyong Choon Hahm and Byong Je Jon. J Criminal Law 56:18-26 Mr '65

Statistics

The birth and abortion rate in Czechoslovakia. Rudolf Urban. tables R Soviet Med Sciences 1:24-33 no 3 '64

ACCEPTANCES
Bankers acceptances used more widely [United States]. Joseph G. Kvasnicka. charts Federal Reserve Chicago p 9-16 My '65

ACCIDENT research
† Stratemeyer, Clara G. Accident research for better safety teaching. '64 32p bibl Single copies, payment with order $2—Nat. educ. assn., Nat. comm. on safety educ.
Findings drawn from selected research studies in safety education and the behavioral sciences, with ideas and suggestions for teachers in planning classroom applications.

ACCIDENTS, Industrial

Prevention

Accident prevention in Switzerland. Giacomo Bernasconi. il Free Labour World p 19-21 O '64

Conferences

† President's conference on occupational safety. Proceedings of the . . . June 23-25, 1965, Washington, D.C. ['65] viii+441p bibl il tables maps plan (Bul. 263) pa—Bureau of labor standards, Department of labor, Washington, D.C.

Statistics

* New York (state). Dept. of labor. Div. of research and statis. Injury rates in New York state industries, 1963. Lela Keogh and Herbert Prince. O '64 21+42p (processed) tables charts (Pubn. no. B-149) pa Free—80 Centre st., New York, N.Y. 10013

ACCIDENTS, Traffic

Claims and liability

The economic treatment of automobile injuries. Alfred F. Conard. Mich Law R 63: 279-326 D '64

Prevention

Automobile safety. Richard L. Worsnop. Editorial Research Repts p 403-18 Je 4 '65
Contents: Stress on safer cars for safer driving; Auto exhaust fumes and air pollution; Debate over what causes car accidents.
Congress vs. cars: pressure is increasing for new federal rules on auto, tire safety; Sens. [Abraham] Ribicoff, [Gaylord] Nelson urge dual brakes, padded dash; tires called inadequate; Detroit defends its products. Dan Cordtz. Wall St J 165:1+ Je 7 '65

Interstate cooperation

The case for compacts: the states must do more in the highway safety field, and interstate compacts can be the answer. Paul A. MacDonald. Highway User 29:19-21 D '64

ACCOUNTING
Embattled CPAs: they fret over rise in law suits, domination by big firms. Lee Silberman. Wall St J 165:10 My 24 '65
† Storey, Reed K. The search for accounting principles: today's problems in perspective. '64 65p pa $1.50—Am. inst. certified public accountants

Bonds, Revenue

† Munic. fin. officers assn. Accounting and legal aspects of revenue bond financing: Accounting viewpoint, by John B. Reid, jr.; Legal viewpoint, by Frank E. Curley. D 16 '64 8p (Special bul. 1964C) $1

Business

Small business

Small business looks at the C.A. [chartered accountant]: Financing, by R. A. Wildgoose; Budgeting and a standard cost plan, by W. F. Gibson; The organization structure, by D. D. Irwin. Can Chartered Accountant 86:47-9 Ja, 115-17 F, 190-3 Mr '65

Information processing systems

The accounting service bureau: one CPA firm's experience [Lennox and Lennox of Staten Island, N.Y.]: even the small CPA firm can make use of the short cuts provided by automation to handle large volumes of work—and expand its client services at the same time. John E. Lennox. J Accountancy 118:49-54 N '64

Public relations

The CPA in the public eye [feels the profession has isolated itself from public understanding and could gain much from a careful program of public exposure; address]. Osgood Nichols. J Accountancy 118: 33-6 D '64

Public utilities

Regulation and accounting: action to eliminate regulatory accounting differences is urged by the only certified public accountant ever to serve as a member of a federal regulatory commission. Lawrence J. O'Connor, jr. Federal Accountant 14:8-28 Fall '64

ACCOUNTING ethics
The CPA in the public eye [feels the profession has isolated itself from public understanding and could gain much from a careful program of public exposure; address]. Osgood Nichols. J Accountancy 118: 33-6 D '64

ACRONYMS. See Abbreviations

ACT of state
The Act of state doctrine after Sabbatino [Banco nacional de Cuba v. Sabbatino, involving nationalization of sugar company in Cuba, owned almost entirely by Americans]. William J. Bogaard. Mich Law R 63:528-42 Ja '65

Figure 10

Figure 11

segregated by discipline. The examples reproduced in Figure 12 indicate representative variations in organization.

Medical literature is indexed in a variety of ways. While many medical articles overlap allied disciplines in the life sciences and are printed in and indexed by journals and indexes mentioned above, the periodical literature of medicine is best indexed in the following:

1. *Abstracts of World Medicine and World Surgery* (1947 to date)
2. *Current List of Medical Literature* (1941 to date)
3. *Dental Abstracts* (1956 to date)
4. *Excerpta Medica* (1947 to date)
5. *Index Medicus*
6. *International Abstracts of Surgery* (1913 to date).
7. *International Medical Digest* (1920 to date).

Encyclopedias and other reference books are usually of value only as sources of general or historical information. Their contents, particularly in scientific areas, are usually dated. There are, however, yearbooks and

Figure 12

annual supplements to encyclopedias which are excellent sources for recent facts and statistics. The following is only a partial list.

1. *Americana Annual*
2. *Britannica Book of the Year*
3. *Collier's Year Book*
4. *New International Yearbook*
5. *Stateman's Yearbook*
6. *World Almanac and Book of Facts*
7. *Yearbook of the United Nations*
8. *Yearbook of World Affairs*

The following partial list of indexes to periodical literature covers about 90 percent of the recognized North American professional journals.

1. *Agricultural Index,* 1916–
2. American Economic Association. *Index of Economic Journals* 1961–

3. *Annual Library Index*. 1905–10
4. *Annual Magazine Subject Index*. 1907–
5. *Annual Literary Index*. 1892–1904.
6. *Applied Science and Technology Index*. 1958–
7. *Art Index*. 1929–
8. *Bibliographic Index*. 1938–
9. *Biography Index*. 1946–
10. *Biological Abstracts*. 1926–
11. *British Abstracts of Medical Science*. 1954– (now *International Abstracts of Biological Sciences*).
12. *Business Periodicals Index*. 1958–
13. *Catholic Periodical Index*. 1930–33, January 1939–
14. *Chemical Abstracts*. 1907–
15. *Current Contents;* Chemical, Pharmaco-Medical, and Life Sciences Editions. 1958–
16. *Christian Periodical Index*. 1959–
17. *Education Index*. 1929–
18. *Engineering Index*. 1884–
19. *Helminthological Abstracts*. 1932
20. *Historical Abstracts*. 1775–
21. *Index to Legal Periodicals*. 1908–
22. *Index to Legal Periodical Literature*. 1888–
23. *Index to Little Magazines*. 1948–
24. *Industrial Arts Index*. 1913–
25. *International Index*. 1907–15 (now called *Social Sciences and Humanities Index*).
26. *Library Literature*. 1921–
27. *Mathematical Reviews*. 1940–
28. *Mineralogical Abstracts*. 1959–
29. *Nineteenth Century Readers' Guide to Periodical Literature*. 1890–1899
30. *Nutrition Abstracts and Reviews*. 1931–
31. *Poole's Index to Periodical Literature*. 1802–81. Supplements, January 1882, January 1907.
32. *Psychological Abstracts*. 1927–
33. *Public Affairs Information Service*. 1915–
34. *Readers' Guide to Periodical Literature*. 1900–
35. *Review of Reviews* (Index to periodicals of 1890–1902).
36. *Sociological Abstracts*. 1953–

The structure of the indexes cited above is alphabetical, and the American Library Association has fixed rules for alphabetization. The following condensation discusses procedures used in all card catalog systems and in most indexing systems.

1. All entries, whether authors, titles, or subjects, are filed alphabetically, letter by letter to the end of the word, and word by word to the end of the entry, as in
   Cherokee Indians
   The Cherokee Strip
   Cherokees of the East
   . . .
   Brown, Herbert Ross
   Browne, Andrew Wyeth
2. The initial articles are ignored, as in
   Man against Space
   The Man Who Came to Dinner
   Les Misérables
3. Abbreviations are filed as though spelled out, as in
   Dr. (doctor)
   St. (saint)
   Mr. (mister)
   Mrs. (mistress)
4. Numerals are filed as though spelled in the way they are spoken, as in
   50,000 customers (*fifty* thousand)
   The Class of '67 (*sixty*-seven)
   100 million guinea pigs (*one* hundred million)
5. Entries beginning with the same word are filed as in
   France, Anatole (person)
   France – Description and Travel (place)
   The France I Knew (title)
6. Entries for the history of a country are arranged by dates; as in
   France – History
   France – History – Regency, 1715 – 1723
   France – History – 1789 – 1900
   France – History – Consulate and Empire, 1799 – 1815
7. Contractions are filed as though they did not have apostrophes; as in
   The boys' own boat
   A boy's treasure chest
   The boys who came back
8. Subjects with subdivisions indicated by a *dash* precede those indicated by a *comma;* as in
   Art – Dictionaries
   Art – History

Art—Russia
Art, Ancient
Art, Greek
Art, Roman

9. Cards for persons entered under their forenames, such as saints, kings, emperors, are filed before those whose surnames are the same; as in

   James, Saint, Apostle
   James II, king of Great Britain
   James V, king of Scotland
   James, Henry
   James, William

10. Cards for books *by* an author are filed before those for books *about* the author. The books about him have his name typed in *red;* as in

   Dickens, Charles
      Oliver Twist
   Dickens, Charles
      A Tale of Two Cities
      Dickens, Charles (in red)
   Chesterton, Gilbert Keith
      Charles Dickens

11. Initials are filed as though they were separate words; as in

   The A B C of Electronics
   The A B C of Fallout
   Ab, the Wolf Boy
   Abbott, Herschel

The Greek alphabet was occasionally used in indexing what are now historical collections of data. It is still used in organizing lists and tables, but such use is so rare that it seems an affectation. The sequence is as shown in Table 2.

The Russian language uses the Cyrillic alphabet, which is derived from the Greek. Several systems of transliteration are used in professional literature in the English language. Table 3 illustrates the form recommended by the American Institute of Biological Sciences until full agreement is reached.

Two documents published by the United States Patent Office index all new patents on a weekly and an annual basis. In addition, periodical indexes of some disciplines attempt to publish abstracts of patents which fall into specific subject areas.

The weekly publication is titled *Official Gazette of the United States Patent Office*. It is released every Tuesday (since January, 1872, when it

Table 2

Name of Letter	*Capital*	*Small*	*Latin and English Equivalent*
alpha	Α	α	a
beta	Β	β	b
gamma	Γ	γ	g (*or* n)
delta	Δ	δ	d
epsilon	Ε	ε	e
zeta	Ζ	ζ	z
eta	Η	η	ē
theta	Θ	θ	th (*or* t)
iota	Ι	ι	i
kappa	Κ	κ	c (*or* k)
lambda	Λ	λ	l
mu	Μ	μ	m
nu	Ν	ν	n
xi	Ξ	ξ	x
omicron	Ο	ο	o
pi	Π	π	p
rho	Ρ	ρ	r (*or* rh)
sigma	Σ	σ, ς	s
tau	Τ	τ	t
upsilon	Υ	υ	y (*or* u)
phi	Φ	φ	ph (*or* f)
chi	Χ	χ	ch
psi	Ψ	ψ	ps
omega	Ω	ω	ō

replaced the earlier *Patent Office Reports*), simultaneously with the weekly issue of new patents granted. The *Official Gazette* is sold, by annual subscription and in single copies, by the Superintendent of Documents, Washington, D. C. Among the specialized information it contains are indexes of patents and patentees, a claim and a selected figure of each patent granted on that day, lists of patents available for license or sale, and decisions rendered in patent cases by the courts and the Patent Office.

Since July 1962 the illustrations and claims have been arranged in order *according to the Patent Office classification of subject matter*. The Office indexes patentable processes, machines, manufactures, and compositions

Table 3

Draft Table for Modern Russian Letters

British Standards Institution—

American Standards Association Sectional Committee Z 39

Russian		*BSI*	*BSI-ASA/SC-Z39*	*Russian*		*BSI*	*BSI-ASA/SC-Z39*
1	А а	a	a	17	Р р	r	r
2	Б б	b	b	18	С с	s	s
3	В в	v	v	19	Т т	t	t[footnote (1)]
4	Г г	g	g	20	У у	u	u
5	Д д	d	d	21	Ф ф	f	f
6	Е е	e	e	22	Х х	kh	kh
7	Ж ж	zh	zh	23	Ц ц	ts	ts
8	З з	z	z	24	Ч ч	ch	ch
9	И и	i	i	25	Ш ш	sh	sh[footnote (2)]
10	Й й	ï	ï	26	Щ щ	shch	shch
11	К к	k	k	27	Ъ ъ	"	"
12	Л л	l	l	28	Ы ы	ȳ	y[footnote (3)]
13	М м	m	m	29	Ь ь	'	'
14	Н н	n	n	30	Э э	é	ē
15	О о	o	o	31	Ю ю	yu	yu
16	П п	p	p	32	Я я	ya	ya

(1) Use t• when т (No. 19) is followed by с (No. 18)
(2) Use sh• when ш (No. 25) is followed by ч (No. 24)
(3) ȳ (with a bar) may be used optionally when followed by а (No. 1) or у (No. 20)

of matter into 309 classes and 47,000 subclasses. Full explanation of the classification system is contained in *Manual of Classification.* This loose-leaf manual and subscriptions for substitute pages which are issued from time to time are also sold by the Superintendent of Documents. A sample page from the Official Gazette is reproduced in Figure 13.

The second important reference document issued by the Patent Office is the *Annual Index.* An annual index of the *Official Gazette,* it contains alphabetical indexes of the names of patentees and of the subject matter of the patents granted during the calendar year. Arrangement is in three numbers. The first indicates the class, the second indicates the subdivision, and the third (a seven-digit number) is the patent number. At present, the *Annual Index* is issued in two volumes, one for patents and one for trade marks. Both are sold by the Superintendent of Documents.

3,204,153
RELAXATION DIVIDER
William H. Tygart, Marietta, Ga., assignor to Lockheed Aircraft Corporation, Burbank, Calif.
Filed May 15, 1962, Ser. No. 194,807
11 Claims. (Cl. 317—148.5)

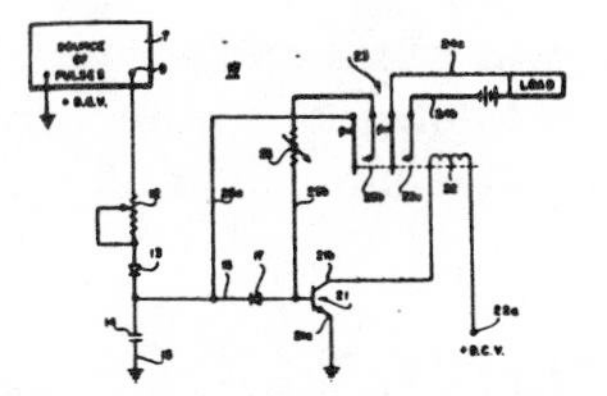

1. A relaxation divider comprising: a controlling circuit portion including a storage capacitor for storing input voltage pulses; a voltage reference device breaking down in the reverse direction upon receipt of a voltage of predetermined magnitude connected at one side to the discharge circuit of said capacitor for preventing premature discharge thereof; a switching circuit portion connected to the other side of said voltage reference device including means producing an output voltage pulse during the time said switching circuit portion receives a pulse from said controlling circuit portion; and means connecting a circuit in shunt around said voltage reference device during the time said switching circuit portion receives a pulse from said controlling circuit portion.

3,204,154
ELECTROMAGNETIC DOOR HOLDER AND STOP
Roy L. Crandell, Anaheim, Calif., assignor, by mesne assignments, to Yale & Towne, Inc., New York, N.Y., a company of Ohio
Filed Dec. 10, 1962, Ser. No. 243,400
5 Claims. (Cl. 317—159)

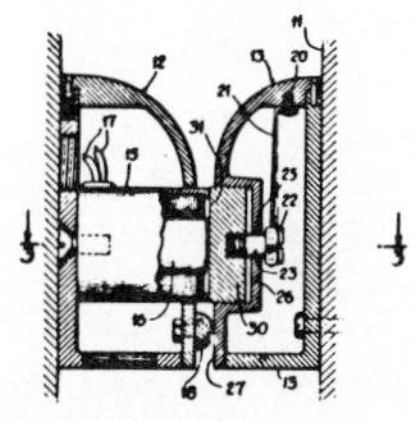

1. In a magnetic holder, a first casing adapted to be secured to a wall and having an electromagnetic core, a second casing adapted to be secured in position on a door to move into juxtaposed relation to the first casing when the door opens, an armature movably mounted on the second casing, yielding means normally pressing said armature to a predetermined inward position on its mounting relatively to said second casing, said armature being movable outwardly from said predetermined position, said electromagnetic core being effective when energized to pull said armature outwardly and against said core as the casings become juxtaposed, so as to hold the door open while a closing pressure applied to the door, as by a door closer, will stress said yielding means to hold the armature projected outwardly of the second casing.

3,204,155
MAGNETIC STRUCTURE HAVING A FIXED AND VARIABLE AIR GAP
Roger Charpentier, 27 bis Ave. des Lilas, Pau, Basses-Pyrenees, France
Filed July 31, 1961, Ser. No. 128,098
Claims priority, application France, July 29, 1960, 834,372, Patent 1,272,074; Dec. 22, 1960, 847,671
10 Claims. (Cl. 317—201)

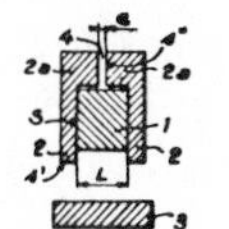

1. A magnet composed of an aluminum-nickel-cobalt-copper-iron base composition, and having a length L between $\sqrt{S}/2$ and $\sqrt{S}$, the magnet being combined with pole pieces arranged to provide a double non-magnetic gap of which one is fixed and the other variable, the dimensions of the fixed gap satisfying the following conditions:

$$\frac{L}{15} \leq e \leq \frac{L}{3}$$

$$\frac{5S}{12} \leq s'' \leq 3S$$

where

S is the area of contact of the magnet with the pole pieces,
L is the length of the magnet between the surfaces of contact with the pole pieces,
e is the length of the fixed gap,
s'' the surface area of each of those parts of the pole pieces which form the fixed air gap.

3,204,156
VENTED ELECTROLYTIC UNIT
Joseph A. Moresi, Jr., and Ralph D. Boisjolie, North Adams, Mass., assignors to Sprague Electric Company, North Adams, Mass., a corporation of Massachusetts
Filed May 1, 1961, Ser. No. 106,683
10 Claims. (Cl. 317—230)

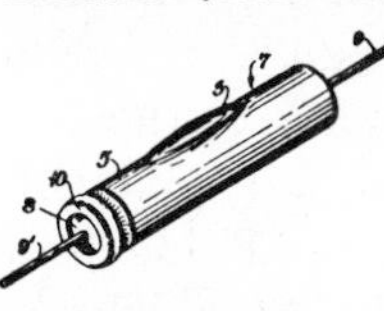

1. A vented can for an electrolytic unit comprising a lengthwise cut in and along the inner surface of the side wall of the can, said side wall having a thickness of at least 0.01 inch, and said cut leaving a wall thickness of approximately .004 inch.

3,204,157
CRYSTAL DIODE HEAT DISSIPATING MOUNTING
Allen R. Peterson, Dearborn, Mich., assignor to Welduction Corporation, Southfield, Mich., a corporation of Michigan
Filed Aug. 30, 1960, Ser. No. 52,934
13 Claims. (Cl. 317—234)

1. In combination, a unitary, hermetically sealed high-current crystal diode having an integral smooth frusto-conical base portion and a heat dissipating mounting therefor comprising a hollow metal body having a plurality of walls defining a cavity therewithin, one of said walls

Figure 13

Among the periodical indexes publishing abstracts of selected patents, *Chemical Abstracts* is the most significant. Since 1960, it has included between 20,000 and 30,000 patent abstracts annually. Classification is by author, patent number, and subject.

A writer searching the literature on patents will be most effective if he works either in the search rooms of the Patent Office or in one of the twenty-one depository libraries scattered among the fifty states. A list of these libraries follows:

Albany, N. Y., University of State of New York
Atlanta, Ga., Georgia Tech Library
Boston, Mass., Public Library
Buffalo, N. Y., Grosvenor Library
Chicago, Ill., Public Library
Cincinnati, Ohio, Public Library
Cleveland, Ohio, Public Library
Columbus, Ohio, Ohio State University Library
Detroit, Mich., Public Library
Kansas City, Mo., Linda Hall Library
Los Angeles, Calif., Public Library
Madison, Wis., State Historical Society of Wisconsin
Milwaukee, Wis., Public Library
Minneapolis, Minn., Public Library
Newark, N. J., Public Library
New York, N. Y., Public Library
Philadelphia, Pa., Franklin Institute
Pittsburgh, Pa., Carnegie Library
Providence, R. I., Public Library
St. Louis, Mo., Public Library
Toledo, Ohio, Public Library.

In its search rooms, the Patent Office maintains two sets of patents, one set in numerical order and one set arranged by subject matter, for the use of the public. Single sets of patents are maintained by the depositories.

It is time-consuming but possible to conduct a search by mail by buying back issues of the *Annual Index* and *Official Register* to determine the number of a specific patent. All patents identified by number may be purchased from the Patent Office at a cost of 25 cents each, postage free.

The only significant collection of foreign patents in the United States is housed in the Scientific Library of the Patent Office, where more than six million copies of foreign patents have been collected into bound volumes.

Translation services are available. The Scientific Library also makes available to the public its collection of over 35,000 scientific and technical books in various languages and its collection of over 40,000 bound volumes of periodicals devoted to science and technology.

New depositories will appear as rapidly as Public Law 89-182 is implemented. This law, known as the State Technical Services Act, was enacted in September 1965, to provide federal financial support to the several states that will establish and maintain technical service programs. The services are activities or programs designed to enable businesses, commerce, and industrial establishments to acquire and to use federal scientific and engineering information more effectively. The Act will work through designated agencies in each state, usually the state university. Other accredited universities and colleges will also be called upon to do work under this Act.

The Clearinghouse for Federal Scientific and Technical Information (U. S. Department of Commerce) issues on the fifteenth of each month a document titled *Government-Wide Index to Federal Research and Development Reports*. This single document contains the announcements made separately in the four specialized indexes *Nuclear Science Abstracts* (Atomic Energy Commission), *Scientific and Technical Aerospace Reports* (National Aeronautics and Space Administration), *Technical Abstracts Bulletin* (Defense Documentation Center), and *U. S. Government Research and Development Reports* (Department of Commerce Clearinghouse). Because it is a synthesis of the others, it is invaluable to a search of unclassified government literature. Because it also announces once-classified reports as they are downgraded and declassified, it is abreast of all documents that the federal government chooses to release. Its limitation is that it is simply an index (see Figure 14), while the specialized indexes mentioned contain abstracts (see Figure 15). Most larger libraries stock the *Government-Wide Index,* and it is also available from the Government Printing Office in single copies or on annual subscription.

Format of the *Index* is somewhat unusual because it is produced by computer manipulation of records prepared for other purposes. (The AEC uses two kinds of punched cards, NASA uses magnetic tape, and the other two agencies use punched tape.) Furthermore, contributing agencies use dissimilar rules for indexing and creating records that may be correlated by machine techniques. The importance of the *Index,* however, warrants adjustment to seeming eccentricities. Figure 14 reproduces the instructions for the use of the *Index,* together with a sample page.

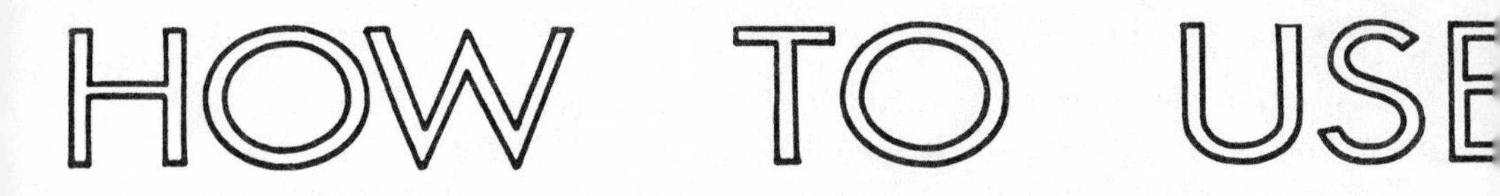

The access points to technical reports listed in this index include:

- Subject
- Personal Author
- Corporate Source
- Report Number
- Accession Number

Each index is arranged alphanumerically (that is, alphabetic data precedes numeric data). A typical entry is as follows:

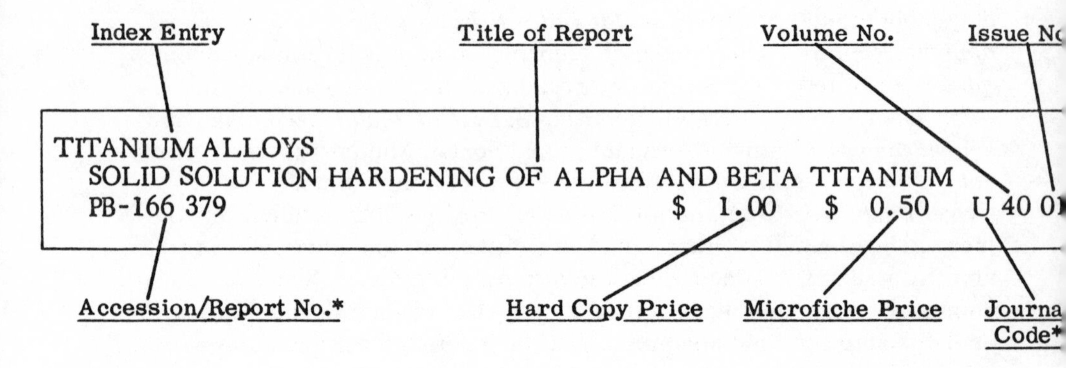

*The Accession/Report number is the number used by the announcing agency to uniquely identify the report; this number should be used in all correspondence with that agency.

**The code for the journal volumes and issues in which the reports are announced and abstracted are as follows:

U--U. S. Government Research and Development Reports, Clearinghouse for Federal Scientific and Technical Information.
T--Technical Abstract Bulletin, Defense Documentation Center of the Department of Defense.
N--Nuclear Science Abstracts, Atomic Energy Commission.
S--Scientific and Technical Aerospace Reports, National Aeronautics and Space Administration.

Figure 14

INDEXES

SUBJECT INDEX

Entries are placed in this index according to the subject terms assigned by the contributing agencies (this means "asterisked" descriptors for USGRDR entries, subject headings for STAR and NSA). The subject term appears at the top of each entry, flush left. The title of the report is indented below the subject term, except that for NSA entries the "modifier" which is assigned by AEC is used in place of the title. The "accession number line" follows the last line of the title. If two or more reports are indexed under the same subject term, they are listed in accession number sequence under the one term (separated by a blank line). If two or more reports under one subject heading have identically the same title, the title is printed once and the different accession number lines are printed below the title.

SAMPLES

```
HARDENING
   SOLID SOLUTION HARDENING OF ALPHA AND BETA TITANIUM.
   PB-166 379                                  $ 1.00  $ 0.50  U 40 01
   .
   .
   .
TITANIUM ALLOYS
   SOLID SOLUTION HARDENING OF ALPHA AND BETA TITANIUM.
   PB-166 379                                  $ 1.00  $ 0.50  U 40 01
```

PERSONAL AUTHOR INDEX

Entries are arranged in this index according to the personal author, or editor, or patent assignee, etc. The names are presented last name first. The title of the report is indented below the author's name; in all other respects, this index is prepared in the same manner as the Subject Index. While it is true for all of the indexes in this journal, it is most noticeable in the Personal Author Index that a sorting by computer does not always put entries in order the way a human being would do it. The requirements of a computer sort cause "JOHNS, J. J." to appear after "JOHNSON, F. X." (the comma sorts after the O in "son"), but "SMITH-JONES, A. B." will appear ahead of "SMITHERS, P. P." (the hyphen sorts ahead of the E in SMITHERS).

```
KESSLER, H. D.
   SOLID SOLUTION HARDENING OF ALPHA AND BETA TITANIUM.
   PB-166 379                                  $ 1.00  $ 0.50  U 40 01
   .
   .
   .
MCANDREW, J. B.
   SOLID SOLUTION HARDENING OF ALPHA AND BETA TITANIUM.
   PB-166 379                                  $ 1.00  $ 0.50  U 40 01
```

CORPORATE SOURCE INDEX

Entries are arranged in this index according to the corporate source of each report as cataloged by the contributing agencies. Some inconsistencies will be noticed during the first few months, but as the new corporate authority list prepared by the President's Council on Scientific and Technical Information (COSATI) comes into general use among the contributing agencies, it is expected that all entries for a given organization will come under a common heading in this index. In all other respects, this index is identical to the Personal Author Index.

```
IIT RESEARCH INST., CHICAGO, ILL.
   SOLID SOLUTION HARDENING OF ALPHA AND BETA TITANIUM.
   PB-166 379                                  $ 1.00  $ 0.50  U 40 01
```

REPORT NUMBER INDEX

Entries are arranged in this index alphanumerically by technical report and monitoring agency number. The accession number line appears indented under the report number. Only those reports which have been cataloged with report numbers by the contributing agencies are represented in this index.

```
AD-37 315
   PB-166 379                                  $ 1.00  $ 0.50  U 40 01
   .
   .
   .
IITRI-B-060-2
   PB-166 379                                  $ 1.00  $ 0.50  U 40 01
```

ACCESSION NUMBER INDEX

Entries are arranged in this index alphanumerically by the accession number assigned by the contributing agencies. The first line of each entry is the accession number line. All technical report numbers or monitoring agency numbers which have been cataloged by the contributing agencies (if any) appear indented below the accession number. Every report covered in this journal is represented in this index by at least one unique number.

```
PB-166 379                                     $ 1.00  $ 0.50  U 40 01
   AD-37 315
   IITRI-B-060-2
```

SUBJECT INDEX

Figure 14 Continued

The Clearinghouse publishes separately, at two-week intervals, the document titled *U. S. Government Research and Development Reports.* It is in two sections. The first announces reports released by the Defense Documentation Center (DDC) of the Department of Defense in its *Technical Abstracts Bulletin.* The second announces reports released by civilian agencies. Listing is by accession number under twenty-two major subject category headings. Both sections contain abstracts and complete bibliographic citation. The descriptors indicate the scope of the report, correlate with the *Government-Wide Index,* and are based upon the DDC *Thesaurus.* (Similar to and a source for the EJC *Thesaurus*). Figure 15 reproduces sample pages from the first or DDC section.

The Department of Commerce maintains field offices in the cities listed below for the purpose of providing ready access to the reports, publications, and services of the Business and Defense Services Administration, Bureau of International Commerce, Clearinghouse for Federal Scientific and Technical Information, Office of Business Economics, and the Bureau of the Census. Information on activities of the National Bureau of Standards, Patent Office, and the Area Redevelopment Administration is also available.

Field offices act as official sales agents of the Superintendent of Documents and stock a wide range of official Government publications relating to business. Each office maintains an extensive business reference library containing periodicals, directories, publications, and reports from official as well as private sources. Addresses of field offices and of federal regional technical report centers are listed on pp. 45-50.

DEPARTMENT FIELD OFFICES

Albuquerque, N. Mex., 87101, U.S. Courthouse. 247-0311.

Anchorage, Alaska, 99501, Room 306, Loussac-Sogn Building. Phone: BR 2-9611.

Atlanta, Ga., 30303, 75 Forsyth St., N.W. 526-6000.

Baltimore, Md., 21202, 305 U.S. Customhouse, Gay and Lombard Streets. Phone: PLaza 2-8460.

Birmingham, Ala., 35203, Title Bldg., 2030 Third Ave., North. Phone: 325-3131.

Boston, Mass., 02110, Room 230, 80 Federal Street. CApitol 3-2312.

Buffalo, N.Y., 14203, 504 Federal Building, 117 Ellicott Street. Phone: 842-3208.

Charleston, S.C., 29401, No. 4, North Atlantic Wharf. Phone: 722-6551.

TECHNICAL ABSTRACT BULLETIN

01. Aeronautics

AD-617 925 Div. 9/6, 1/2
CFSTI Prices: HC$4.00 MF$1.00

Vidya Div., Itek Corp., Palo Alto, Calif.
THEORETICAL AERODYNAMICS OF FLEXIBLE WINGS AT LOW SPEEDS. IV. EXPERIMENTAL PROGRAM AND COMPARISON WITH THEORY.
Final rept. for Feb 64-Feb 65,
by Jack A. Burnell and Jack N. Nielsen. 15 Feb 65,
127p Vidya rept. no. 172
Contract Nonr372800, Proj. 9064
Unclassified report

Descriptors: (*Parawings, Aerodynamic characteristics), (*Wings, Flexible structures), Subsonic characteristics, Slender bodies, Aerodynamic configurations, Aerodynamic loading, Lift, Drag, Flutter, Tests, Theory, Aspect ratio, Numerical analysis, Data, Graphics.

As Part IV of a general parawing study, low-speed measurements have been made of the aerodynamic characteristics of triangular one-lobed and two-lobed parawings of aspect ratios 1, 2, and 3 for comparison with the theory of Parts I and II of the study. It was found that this theory is adequate for determining the canopy shapes, the luffing boundaries, and the lift characteristics for parawings of low to moderate slackness ratio. The measured center of pressure positions are slightly forward of the theoretical position as a result, it is thought, of the presence of the leading-edge members. Also, the aerodynamic design of the leading-edge members is such that from 20 percent to 70 percent of the theoretical leading-edge suction is realized. The maximum lift-drag ratios appear to be adversely affected by trailing-edge flutter which in some cases is so intense as to preclude aerodynamic testing at angles of attack for maximum lift-drag ratio. A parawing of aspect ratio 8.5 with a taut trailing-edge cable exhibited no flutter, and when the tension was relieved the flutter was intense and the maximum lift-drag ratio actually increased. (Author)

AD-617 931 Div. 9/6
CFSTI Prices: HC$3.00 MF$0.75

Technion - Israel Inst. of Tech., Haifa. Dept. of Aeronautical Engineering.
APPLICATIONS OF A HIGH-FREQUENCY AERODYNAMIC THEORY TO THE CALCULATION OF FLUTTER AT SUPERSONIC SPEEDS.
Final rept.,
by M. Hanin and E. Nissim. 31 Dec 64, 60p
Rept. no. TAE-36
Grant AF EOAR63 66, Proj. 9781, Task 978101
AFOSR 65-1008
Unclassified report

Descriptors: (*Flutter, Supersonic airfoils), (*Supersonic flight, Flutter), Deformation, Aerodynamic loading, Ailerons, Numerical analysis, Aeroelasticity, Thickness, Israel, Theory.

Flutter of airfoil sections in bending, torsion, and control-surface rotation at supersonic flight speeds is analysed by employing an aerodynamic solution for high-frequency oscillations. The analysis includes effects of airfoil thickness and shape, and is valid at arbitrary supersonic Mach numbers and sufficiently high values of frequency variable. For flutter problems with two degrees of freedom (bending-torsion, bending-aileron, torsion-aileron) explicit expressions for calculating the flutter speed and frequency are obtained. The case of three degrees of freedom (bending-torsion-aileron flutter) is studied by solving the determinant equation numerically. Some calculated results are presented, showing the variation of flutter boundaries with the elastic and inertial properties of the airfoil, the flight conditions and the airfoil thickness. (Author)

AD-617 934 Div. 9/7, 12/1
CFSTI Prices: HC$3.00 MF$0.75

Electro-Optical Systems, Inc., Pasadena, Calif.
TURBULENT MIXING IN THE BASE FLOW REGION.
Research notes,
by Hartley H. King and M. Richard Denison.
May 65, 61p Rept. no. EOS-RN-24
Contract AF04 694 570, ARPA Order 254 62
Unclassified report

Descriptors: (*Reentry vehicles, Wake), (*Jet mixing flow, Turbulence), (*Base flow, Turbulence), Atmosphere entry, Hypersonic flight, Laminar flow, Viscosity, Enthalpy, Conical bodies, Wedges, Hypersonic flow, Integral equations.

An exploratory study is made in connection with the problem of predicting properties of the turbulent base flow region of a body at hypersonic speeds. An attempt is made to calculate the gross features of the base flow using a non-similar mixing model including the Chapman-Korst recompression condition. The expression for the eddy viscosity is based on a model which allows the turbulent equations to be transformed into incompressible laminar form. The empirical factor in the eddy viscosity expression is evaluated from data obtained from the near field (in the linear growth region) of jets exhausting into a quiescent region. The same expression for the eddy viscosity is then used in an attempt to estimate the growth of non-similar turbulent mixing layer using both turbulent and laminar initial profiles. The correlation of turbulent jet mixing data for speeds up to Mach 3 shows that the eddy viscosity is a very strong function of Mach number (or of density ratio across the jet, since the data are for adiabatic flow). If the eddy viscosity dependence on density ratio persists for density ratios typical of re-entry conditions, then it is found that at these conditions the growth rate of the turbulent shear layer is orders of magnitude slower approaching the rate of growth of the laminar mixing layer under the same conditions. (Author)

AD-617 935 Div. 1/4, 28/3
CFSTI Prices: HC$3.00 MF$0.75

Bolt, Beranek, and Newman, Inc., Cambridge, Mass.
SOME FACTORS INFLUENCING HUMAN RESPONSE TO AIRCRAFT NOISE: MASKING OF SPEECH AND VARIABILITY OF SUBJECTIVE JUDGMENTS.
Technical rept.,
by K. D. Kryter and C. E. Williams. Jun 65, 72p
Rept. no. BBN-1234
Contract FA64WA4951, Proj. 361 12H
FAA-ADS 42

Figure 15

AD-617 983 Div. 16/1, 16/3, 16/12

Karolinska Inst., Stockholm (Sweden).
UPTAKE OF L- AND D-ISOMERS OF CATECHOL-AMINES IN ADRENERGIC NERVE GRANULES,
by U. S. von Euler and F. Lishajko. 28 Jun 63, 6p
Grant AF EOAR62 14, Proj. 9777, Task 977701
AFOSR 65-0945
Unclassified report

Pub. in Acta Physiologica Scandinavica (Sweden) v60 p217-22 1964
(Copies available only to DDC users).

Descriptors: (*Nerve cells, Amines), (*Levarterenol, Molecular isomerism), (*Epinephrine, Molecular isomerism), (*Amines, Nerve cells), Adenosine phosphates, Absorption(Biological), Sweden.

The spontaneous depletion of endogenous noradrenaline (NA) from bovine splenic nerve granules is diminished or prevented by incubation with either L-NA or D-NA. However, the effect observed following the addition of L-NA is significantly greater than that produced by the D-isomer. No significant difference is observed between the uptake of the two isomers of adrenaline (A) in undepleted granules. The enhanced uptake of A or NA is a concentration of 10 micro-g/ml in the presence of adenosine triphosphate (ATP) similarly did not differ for the L- and D-isomers of the amine. However, the ATP-dependent uptake at an amine concentration of 1 micro-g/ml was 2-3 times as high for L-NA and L-A than for the D-isomers. It is concluded that the ATP-dependent uptake in nerve granules of NA and A at low amine concentrations is to some extent stereospecific, while the amine-uptake at higher concentrations with and without added ATP is not or only slightly dependent on the steric configuration. (Author)

AD-617 985 Div. 16/3

Johns Hopkins Univ., Baltimore, Md. Dept. of Biophysics.
A DEMONSTRATION OF CODING DEGENERACY FOR LEUCINE IN THE SYNTHESIS OF PROTEIN,
by Bernard Weisblum, Fabio Gonano, Gunter von Ehrenstein, and Seymour Benzer.
23 Dec 64, 8p
Contract AF49 638 1304
AFOSR 65-0956 Unclassified report

Research supported in part by National Institutes of Health, Bethesda, Md., and National Science Foundation, Washington, D. C.
Pub. in Proceedings of the National Academy of Sciences v53 n2 p328-34 Feb 1965 (Copies available only to DDC users).

Descriptors: (*Proteins, Biosynthesis), (*Genetics, Proteins), (*Amino acids, Genetics), (*Ribonucleic acids, Genetics), Nucleotides, Peptides, Hemoglobin, Ribosomes, Chromatographic analysis.

Different leucine sRNA's in E. coli are separable by countercurrent distribution. When these sRNA's are used as donors of labeled leucine during the synthesis of rabbit hemoglobin, in vitro, they distribute leucine differently into various peptides of the alpha-chain. One of the sRNA's introduces leucine only into a single position. The results indicate that there are at least two distinct codons for leucine in the amino acid code. (Author)

AD-617 990 Div. 16/3, 16/2

Leicester Univ. (England)
FINE CONTROL OF PHOSPHOPYRUVATE CARBOXY-LASE ACTIVITY IN ESCHERICHIA COLI,
by J. L. Cánovas and H. L. Kornberg. 7 Dec 64, 4p
Grants AF EOAR63 17, B SR652, Proj. 9777, Task 977701
AFOSR 65-0957
Unclassified report

Pub. in Biochimica et Biophysica Acta v96 p169-72 1965 (Copies available only to DDC users).

Descriptors: (*Eschirichia coli, Carboxy-lyases), (*Carboxy-lyases, Eschirichia coli), Bacteria, Metabolism, Enzymes, Coenzymes, Organic phosphorus compounds, Pyruvates, Adenosine phosphates, Bacterial extracts, Labeled substances, Great Britain.

Previous studies with mutants of Salmonella typhimurium and Escherichia coli, which fail to grow on glucose-salts media unless such media are supplemented with intermediates of the tricarboxylic acid cycle or direct precursors thereof, have shown that the carboxylation of phosphoenolpyruvate (CH2=C.(COOH).O.PO3H2 + CO2 yields HOOC.CH2.CO.COOH + Pi) is necessarily involved in the maintenance of the tricarboxylic acid cycle: this reaction serves to replenish the cycle with C4-compounds, as intermediates of that cycle are revoved in the course of biosynthesis. It was difficult to explain why CoASAc and citrate synthase should be so much more efficient a trapping agent for labelled oxaloacetate than was malate dehydrogenase and NADH2, even when the extracts were supplemented with crystalline malate dehydrogenase: this observation suggested that CoASAc played a role in this system additional to its function as a trapping agent. This suggestion was confirmed with a mutant AB1623 of E. coli K-12.

AD-617 991 Div. 16/3, 16/1, 16/4

Lund Univ. (Sweden).
A DETAILED METHODOLOGICAL DESCRIPTION OF THE FLUORESCENCE METHOD FOR THE CELLULAR DEMONSTRATION OF BIOGENIC MONOAMINES,
by Bengt Falck and Christer Owman. 1965, 25p
Grants AF EOAR64 5, PHS NB05236 01, Proj. 9777, Task 977701
AFOSR 65-0959
Unclassified report

Pub. in Acta Universitatis Lundensis, Sectio II Medica, Mathematica, Scientiae Rerum Naturalium n7 p1-23 1965 (Copies available only to DDC users).

Descriptors: (*Amines, Histological techniques), (*Histological techniques, Amines), (*Cells(Biology), Amines), (*Serotonin, Histological techniques), Phenols, Tissues(Biology), Fluorescence, Microscopy, Freeze drying, Sweden.

A methodological description gives detailed instructions for the preparation, freeze-drying, histochemical treatment, and sectioning of tissues for fluorescence microscopy of catecholamines, 5-hydroxytryptamine and their immediate precursors. (Author)

AD-618 000 Div. 16/1, 16/12

Massachusetts Inst. of Tech., Cambridge.
RESPONSES OF SINGLE DORSAL CORD CELLS TO PERIPHERAL CUTANEOUS UNMYELINATED FIBRES,
by Lorne M. Mendell and Patrick D. Wall. 1964, 6p
Contracts DA36 039AMC03200E, AF33 615 1747
Unclassified report

Pub. in Nature v206 n4979 p97-9 Apr 3 1965 (Copies available only to DDC users).

Descriptors: (*Nerve cells, Reflexes), (*Nerve fibers, Stimulation), (*Nerve impulses, Nerve fibers), Cats.

Charleston, W. Va., 25301, 3002 New Federal Office Building, 500 Quarrier Street. Phone: 343-6196.

Cheyenne, Wyo., 82001, 207 Majestic Bldg., 16th & Capitol Ave. Phone: 634-5920.

Chicago, Ill. 60604, 1486 New Federal Building, 219 South Dearborn Street. Phone: 828-4400.

Cincinnati, Ohio, 45202, 8028 Federal Office Building, 550 Main Street. Phone: 381-2200.

Cleveland, Ohio, 44101, 4th Floor, Federal Reserve Bank Bldg., East 6th St. & Superior Ave. 241-7900.

Dallas, Tex., 75202, Room 1200, 1114 Commerce Street. RIverside 9-3287.

Denver, Colo., 80202, 142 New Custom House, 19th & Stout Street. Phone: 297-3246.

Des Moines, Iowa, 50309, 1216 Paramount Building, 509 Grand Avenue. Phone: 284-4222.

Detroit, Mich., 48226, 445 Federal Bldg. 226-6088.

Greensboro, N. C., 27402, Room 412, U. S. Post Office Building. 275-9111.

Hartford, Conn., 06103, 18 Asylum St. Phone: 244-3530.

Honolulu, Hawaii, 96813, 202 International Savings Bldg., 1022 Bethel Street. Phone: 588667.

Houston, Tex., 77002, 5102 Federal Bldg., 515 Rusk Ave. CA 8-0611.

Jacksonville, Fla., 32202, 512 Greenleaf Building, 208 Laura Street. ELgin 4-7111.

Kansas City, Mo., 64106, Room 2011, 911 Walnut Street. BAltimore 1-7000.

Los Angeles, Calif., 90015, Room 450, Western Pacific Building, 1031 S. Broadway. 688-2833.

Memphis, Tenn., 38103, 345 Federal Office Building, 167 N. Main Street. Phone: 534-3214.

Miami, Fla., 33130, 1628 Federal Office Bldg., 51 S.W. First Avenue. Phone: 350-5267.

Milwaukee, Wis., 53203, Straus Bldg., 238 W. Wisconsin Avenue. Phone: BR 2-8600.

Minneapolis, Minn., 55401, Room 304, Federal Bldg., 110 South Fourth Street. Phone: 334-2133.

New Orleans, La., 70130, 909 Federal Office Building (South), 610 South Street. Phone: 527-6546.

New York, N. Y., 10001, 61st Fl., Empire State Bldg., 350 Fifth Avenue. LOngacre 3-3377.

Philadelphia, Pa., 19107, Jefferson Building, 1015 Chestnut Street. 597-2850.

Phoenix, Ariz., 85025, New Federal Bldg., 230 N. First Avenue. Phone: 261-3285.

Pittsburgh, Pa., 15219, Room 2201, Federal Building, 1000 Liberty Avenue. Phone: 644-2850.

Portland, Oreg., 97204, 217 Old U. S. Courthouse, 520 S. W. Morrison Street. 226-3361.

Reno, Nev., 89502, Room 2028, 300 Booth Street. Phone: 784-5203.

Richmond, Va., 23240, 2105 Federal Building, 400 North 8th Street. Phone: 649-3611.

St. Louis, Mo., 63103, 2511 Federal Building, 1520 Market Street. MAin 2-4243.

Salt Lake City, Utah, 84111, 3235 Federal Building, 125 South State Street. 524-5116.

San Francisco, Calif., 94102, Federal Building, Box 36013. 450 Golden Gate Avenue. Phone: 556-5864.

Santurce, Puerto Rico, 00907, Room 628, 605 Condado Avenue. Phone: 723-4640.

Savannah, Ga., 31402, 235 U. S. Courthouse and Post Office Bldg., 125-29 Bull Street. 232-4321.

Seattle, Wash., 98104, 809 Federal Office Bldg., 909 First Avenue. MUtual 2-3300.

FEDERAL REGIONAL TECHNICAL REPORT CENTERS

Each of the Federal Regional Technical Report Centers listed below contains a collection of USAEC, NASA, and DOD unclassified reports as well as reports of other U. S. Government agencies and provides reference, interlibrary loan, and reproduction services.

Carnegie Library of Pittsburgh
4400 Forbes St.
Pittsburgh, Pa. 15213

Columbia University
Engineering Library
Seeley W. Mudd Building
New York, N. Y. 10027

Georgia Institute of Technology
Price Gilbert Library
Atlanta, Ga. 30300

The John Crerar Library
35 West 33rd St.
Chicago, Ill. 60616

Library of Congress
Science and Technology Division
Washington, D. C. 20540

Linda Hall Library
5109 Cherry Street
Kansas City, Mo. 64100

Massachusetts Institute of Technology
M. I. T. Libraries
Cambridge, Mass. 02139

Southern Methodist University
Science Library
P. O. Box 1339
Dallas, Tex. 75222

University of California
General Library
Berkeley, Calif. 94704

University of Colorado
Boulder, Colo. 80301

University of Washington Library
Government Documents Center
Seattle, Wash. 98105

The Clearinghouse for Federal Scientific and Technical Information (U. S. Department of Commerce) releases twice a month a document titled *Technical Translation*. Available from the Government Printing Office either by annual subscription or by single copy, in its own words this document announces the availability of, and provides order information for, translated scientific and technical reports, periodicals, and books. Although most of the literature has been translated into English, some is in other Western European languages. The language of the translation, if other than English, is noted in the citation. Most of the material is of Soviet origin, but works from Eastern and Western European countries and from the Orient, especially Japan and Communist China, are also cited. In order to avoid duplication of effort, translations in process are also announced.

The journal is separated into five sections.

1. Bibliographic citations are divided into 22 major fields; they are further subdivided alphanumerically under 178 groups. References to sources of original text are given as they appear on the translations, with sufficient editing to assure uniformity in citations.
2. Translations in Process lists titles of translations being prepared by government and private sources. All Special Foreign Currency Science Information (SFCSI Program translations) are listed.
3. Translated or Abstracted Publications lists titles of approximately 190 technical and scientific periodicals regularly translated into English. Included is a cross-reference by English title.
4. Translated tables of contents of recent issues of monographs, journals, and journal sections; they are preceded by a list of the English titles of the publications from which they are taken.
5. Indexes; three indexes are arranged by author, journal, and number.

Each bibliographic entry contains the acronym or initials of the agency from which the translation is to be ordered; these are keyed to the complete list of ordering addresses listed in the pages headed "Directory of Sources."

Reference services for the location and identification of translated materials, completed or in process, are provided by the Clearinghouse, Special Libraries Association Translations Center, and the European Translations Center. Mailing addresses are listed in "Directory of Sources."

The Science Information Exchange operated by the Smithsonian Institution receives information on research in progress from the Clearinghouse of the Department of Commerce, the Defense Documentation Center, and other federal agencies. The Exchange compiles brief summaries of research projects *at the time they are initiated* and makes available information on what is being done in a particular subject field, by whom, where, and under whose support. This information is sent only to qualified individuals upon request to the Publications Division, Smithsonian Institution, Washington, D. C. 20560. Progress reports on research in progress are obtainable only from the sponsoring agency (such as the Defense Documentation Center of the Department of Defense) and only upon those projects whose contracts require such reports.

Information on classified research in progress is available only from the sponsoring agency to individuals and agencies meeting the requirements of the Department of Defense Instruction 5100.38 dated March 19, 1963.

Cataloguing and indexing are not very satisfactory for research that is not sponsored by a federal agency. A slight attempt is made by some of the professional journals, but most, such as *Publications of the Modern Language Association,* have abandoned their attempts to list academic research and dissertations in progress. Only completed and accepted dissertations are collated and indexed.

The doctoral dissertations submitted to some 140 cooperating institutions are abstracted and indexed in a monthly document titled *Dissertation Abstracts*. This document and the annual cumulative index titled *American Doctoral Dissertations* are in the reference rooms of most libraries and are sold by University Microfilms, Inc., 313 North First Street, Ann Arbor, Michigan. Because there are more than 140 institutions granting doctoral degrees, and because even the cooperating institutions do not report all of their dissertations, coverage is incomplete.

Indexing of both documents is under one or more subject headings as assigned by the Library of Congress from an examination of the abstract. Literary works are entered under Drama, Fiction, Poetry, etc.; musical compositions under Concertos, Symphonies, etc.; musical performances under Organ music, Piano music, etc. There is also an author index. Figure 16 reproduces sample pages from *Dissertation Abstracts* to illustrate length and style. Figure 17 reproduces a sample page from the author index of *American Doctoral Dissertations* to illustrate subdivisioning by institution.

Masters theses (in the sciences only) are indexed in a similar document titled *Masters Theses in the Pure and Applied Sciences*. On sale at the Thermophysical Properties Research Center, Purdue University, Lafayette, Indiana, this annual document has referenced some 3000 theses a year since 1956. Indexing is by subject and institution.

The United Nations Document Index, prepared by the Index Section of the Dag Hammarskjöld Library, United Nations, New York, lists and indexes all printed publications of the International Court of Justice and all documents and publications of the United Nations except restricted material and internal papers. This monthly index and the annual *Cumulative Checklist* are sold directly by the Sales Section, United Nations, New York, N. Y. 10017.

The *Index* and the *Checklist* are arranged alphabetically by subject, author, and title. As indicated on the sample page printed below, bibliographic information is compressed. Adequate explanation of abbreviations and references is contained in the introductory note to the subject

CENTRALIZED CONTROL OF FACULTY PROMOTION POLICIES AND PRACTICES IN LARGE STATE UNIVERSITIES

(Order No. 65-11,643)

Fred Luthans, Ph.D.
State University of Iowa, 1965

Chairman: Professor Henry H. Albers

Today's large universities present a tremendous challenge for effective academic administration. A search of the literature indicated that practicing university administrators and administrative scientists have not entirely met this challenge. There seems to be no consistent, systematic body of knowledge in academic administration.

This study utilizes the conceptual framework derived from general management knowledge to describe and analyze central controls of faculty promotion policies and practices. A confidential questionnaire survey of the president, academic vice-president, business college dean, one business department head, and three business professors from each of forty-six large state universities provided data for the study. Over 80 per cent of the 310 administrators and faculty members returned usable responses.

Almost all central administrators described well understood and accepted university-wide promotion policies and practices, but most of the decentral administrators and faculty members did not agree. The majority of the deans and department heads indicated independent promotion policies and practices at the decentral level. Moreover, only eight per cent of the faculty respondents felt their present promotion process was well accepted and contributed to high morale. While most of the central administrators felt they could presently evaluate faculty members' research and teaching, the professors again did not agree. The objective methods of evaluation reportedly used by central administrators were generally unknown to the professors.

The major investigation of the direct use of central control decisions over the promotion process uncovered two significant results. First, it was found that about half the central administrators seldom, if ever, reject decentral administrators' recommendations. Intra-organizational analysis and open-end responses indicated that the infrequent rejection was not necessarily the result of perfect accord between standards and performance or strict informal control. Rather, central administrators were viewed as a "rubberstamp" with "automatic approval." Second, the analysis of central control over research standards found that although research is a basic purpose of the university and the most widely recognized standard for promotion, over half the central administrators reported they sometimes approve for full professor faculty members who have no significant publication record. Furthermore, the publication records of the faculty respondents implied many more full professors who had no significant publication record prior to being promoted than central administration realized or cared to admit. In total, the study indicates that many central administrators are not applying university standards to their own promotion decisions and are not controlling the decentralized promotion decisions.

These results have several implications for academic administration. First, there seems to be a need for improved central promotion policies. To serve as effective guides for faculty performance, decentral policies and practices, and central control decisions, these policies should be understood and accepted by administration and faculty. In many universities the faculty members surveyed did not have this understanding or approval. Second, there seems to be a need for better utilization of objective methods of evaluation at all levels of the university. In particular, the central administrators are not performing an effective evaluative function if they make evaluations of faculty members based solely on recommendations from below. Finally, the study gives some evidence that many central administrators are not performing an effective control function. In many cases, central administrators do not verify that the university standards for promotion are in accord with decentral promotion recommendations and faculty performance.

Microfilm $3.00; Xerography $9.00. 198 pages.

NEGRO-WHITE DIFFERENCES IN THE PURCHASE OF AUTOMOBILES AND HOUSEHOLD DURABLE GOODS.

(Order No. 65-11,001)

Wayne Leroy Mock, Ph.D.
The University of Michigan, 1965

The purpose of this study is to contribute to the state of knowledge on the "Negro Market" through an empirical investigation of Negro-white differences in two specific economic behaviors: the purchase of automobiles and the purchase of household durable goods. Four major dependent variables are studied: the percentage of spending units purchasing automobiles, the net outlay on automobiles by purchasers, the percentage of spending units purchasing household durable goods, and the net outlay on household durable goods by purchasers.

A "separation-opportunity" schema is introduced to classify statements supporting a concept of the Negro market into those which contend that the Negro population is a separate market and those which point out the opportunity for sales to the Negro population. Negro-white differences in economic behavior are seen to be a critical element in assessing whether the Negro population is a separate market. A review of previous research on Negro-white differences in savings and consumer expenditures indicates that few studies have utilized a multi-variate approach to the analysis of Negro-white differences in economic behavior.

The research uses data pooled from the 1958, 1959, 1960, and 1963 "Survey of Consumer Finances" conducted by the Survey Research Center of The University of Michigan.

The analysis is organized around four basic questions concerning the purchase of automobiles and household durable goods:

(1) What are the actual Negro-white differences? (2) What are the differences between Negro and white spending units at the same income level? (3) What are the differences between "comparable" Negro and white spending units? (4) Do socio-economic and demographic factors differentially influence the purchasing by Negro and white spending units?

Findings on actual Negro-white differences indicate that a smaller percentage of Negroes purchase automobiles, a smaller percentage of Negroes purchase household durable goods, and the mean net outlays by Negro purchasers of automobiles and household durable goods are also smaller than those by whites. In addition a larger percentage of Negro automobile purchasers purchase used cars and a larger percentage of Negro purchasers of automobiles and household durable goods purchase on credit.

Controlling for income, Negro-white differences are still found in the values of the dependent variables. The only difference which is consistent in direction, however, is the percentage of Negro spending units purchasing automobiles, which is smaller at every level of disposable income.

To assess differences between comparable Negro and white spending units an iterative analysis of variance program

The items following each abstract are: the price of a microfilm copy; the price of a copy enlarged by the xerographic process to approximately 6 x 8½ inches; the number of pages in the manuscript. Please order copies by order number and author's name.

Figure 16

University of California, Los Angeles

KICIMAN, Mehmet Ozer. Optimum cost and weight design of structural members for air vehicles subjected to elevated temperatures.

REITER, Gordon Stamm. Dynamics of flexible gravity-gradient satellites.

SOLOMON, Louis Peter. Interaction of supersonic jets with cavities.

California Institute of Technology

VAN ATTA, Charles William. Spiral turbulence in circular Couette flow.

WU, Jain-Ming. A satellite theory and its applications.

University of Cincinnati

BOUDREAUX, Rodney. Boundary layer thermodynamic and hydrodynamic characteristics of a flowing liquid hydrogen tank, with and without internal heat generation.

University of Colorado

VINH, Nguyen Xuan. Geometrical studies of orbital transfer problems.

University of Connecticut

CHAU, Hancock. Binary diffusion of gases at high temperatures.

Cornell University

CRIMI, Peter. The effects of viscosity on the flow about an oscillating flat plate.

LEIBOVICH, Sidney. Sub-alfvenic flow past an airfoil for an aligned magnetic field.

University of Illinois

CARRUTHERS, George Robert. Experimental investigations of atomic nitrogen recombination.

DYNER, Harry Bernard. Strip film schlieren-interferometric observation of the effect of reflected shock-boundary layer interaction on reflected shock density in several gases.

MARTIN, James Thomas. Coupling and second-order effects of air drag and earth oblateness on satellite orbital elements.

STROM, Charles Ray. The method of characteristics for three-dimensional steady and unsteady reacting gas flows.

Iowa State University
of Science and Technology

BOCTOR, Magdy Luca. Hypersonic wedge flow of a dissociating and vibrationally relaxing diatomic gas.

FALKEN, Stephen Nathaniel. Some diffuse reflection problems in radiation aerodynamics.

SCHAFER, Robert Louis. Model-prototype studies of tillage implements.

SEVERSIKE, Leverne Kenneth. Wave propagation in a fully ionized plasma with uniform velocity.

University of Kansas

ABDUL-RAHIM, Abdul-Rahim, Abdul-Qader. Thermal stresses and stability of conical shells.

Massachusetts Institute of Technology

BIELAWA, Richard Lyle. Second order non-linear theory of the aeroelastic properties of helicopter rotor blades in forward flight.

BROCK, Larry Davis. Application of statistical estimation to navigation systems. (Classified)

BROWAND, Frederick Kent. Experimental investigation of the instability of an incompressible, separated, shear layer.

HALDEMAN, III, Charles Waldo. Unsteady velocity profiles in the traveling wave pump.

MARTIN, Frederick Harold. Closed-loop near-optimum steering for a class of space missions. (Classified)

MEIRY, Jacob Leon. Vestibular system in human dynamic spatial orientation.

SCOTT, Paul Brunson. Molecular beam velocity distribution measurements.

WILKINSON, Robert Hayden. Study of thermal gradients on instrument structural elements.

The University of Michigan

LENEMAN, Oscar Aszer Zelig. Stationary point processes and their application to random sampling of stochastic processes.

PHILLIPS, Richard Lang. The behavior of dynamic electric arcs.

New York University

CHIANG, Tom. Finite conductivity effect on magnetohydrodynamic flows.

SWENSON, Eva Valencia. Numerical computation of hypersonic flow past a two-dimensional blunt body.

Northwestern University

BOHRER, Lionel Curtis. Pressure distribution over an arbitrary airfoil section in ground proximity.

The Ohio State University

DALE, JR., Ralph Gibson. Flutter analysis of an airfoil exhibiting camber deformations.

PETRIE, Stuart Lee. Inviscid non-equilibrium flow of an expanded air plasma.

THOMAS, Richard Eugene. Similitude considerations in advanced aerothermodynamic facilities.

ZONARS, Demetrius. Theoretical and experimental study of equilibrium and non-equilibrium air flow through a Mach number 10 contoured nozzle.

The University of Oklahoma

HAYES, James Eugene. Structural design criteria by statistical methods.

Pennsylvania State University

CARSON, Bernard Hemphill. Experimental studies of the recombination of glow-discharged oxygen on catalytic spheres.

Polytechnic Institute of Brooklyn

BROOK, John Walter. A kinetic theory study of expanding gas glows.

FOX, Herbert. Laminar boundary layers with chemical reactions.

KORNREICH, Theodore Ronald. Analysis of interplanetary spacecraft missions.

MELNIK, Robert Edward. A systematic study of some singular conical flow problems.

ROSENBAUM, Harold. Turbulent compressible boundary layer on a flat plate with heat transfer and mass diffusion.

SFORZA, Pasquale Michael. Diffusive processes in fluid flow.

Princeton University

CURTISS, Howard Crosby. An analytical study of the dynamics of aircraft in unsteady flight.

GARVINE, Richard W. On high speed viscous flow near a flat plate leading edge.

GEORGE, JR., Albert Richard. Hypersonic flow over a wedge with upstream non-uniformities and variable wedge angle.

HEIE, H. Stability of the numerical solution of hyperbolic partial differential equations using the method of characteristics.

LASKER, Barry M. Boundary layers and dynamics for H II regions.

McINTYRE, John Edward. Optimal transfer of a thrust limited vehicle between coplaner circular orbits.

PAO, Young-ping. A uniformly valid asymptotic theory of rarified gas flows under the nearly free-molecule conditions.

SHERMAN, Martin Philip. Interactions between a partly-ionized laminar subsonic jet and a cool stagnant gas.

Purdue University

GASTON, Charles Arden. The singular case in trajectory optimization.

HOSACK, Grant Austin. On kinetic theory of fully ionized gases.

LEAR, William M. On the use of ultrastable oscillators and a Kalman filter to calibrate the earth's gravitational field.

MURTY, Sistla Sri R. Two-dimensional radiative transfer in attached shock layers.

SWEET, Arnold L. A stochastic model for predicting the subsidence of granular media.

YU, Chia-ping. Magneto-atmospheric waves in a horizontally stratified conducting medium.

Rensselaer Polytechnic Institute

DUFFY, Robert E. Experimental studies of non-equilibrium expanding flows.

GUNMAN, William J. Implosive behavior of magnetically-driven axisymmetric slug-type plasma sheets.

SUNDARAM, Thirukurungudi R. The flow of a dissociating gas past a convex corner.

Stanford University

BENSON, Arthur Searle. General instability and face wrinkling of sandwich plates: unified theory and applications.

BUSHNELL, David. Some problems in the theory of thin shells.

CHAPKIS, Robert Lynn. Newtonian hypersonic flow theory for power-law shock waves.

CHENG, Ping. Study of the flow of a radiating gas by a differential approximation.

CONTI, Raul Jorge. A theoretical study of non-equilibrium blunt-body flows.

DE VOTO, Ralph Steven. The transport properties of a partially ionized monatomic gas.

FANNELÖP, Torstein Kjell. Two-dimensional viscous hypersonic flow over simple blunt bodies including second-order effects.

FERNANDEZ-SINTES, Julio. Rotationally symmetric problems of highly elastic inflated thin membranes subjected to ring loads.

Figure 17

index (yellow pages). Complete explanation is contained in another document titled *Bibliographic Style Manual,* No. 8, Sales No. 63. I.16. This style manual is outstanding and is recommended as a complete discussion of bibliographic documentation. (See Chapter 5.) Figure 18 reproduces a sample page from the *United Nations Document Index.*

The Science and Technology Division of the Library of Congress, under sponsorship of the National Science Foundation, in 1963 released a guide to the world's indexing and abstracting services *in science and technology.* This document comprises 1855 titles originating in forty countries, and it is the major international index to journals of science and technology which publish abstracts. Because of its August 1962 deadline, it is slightly dated and is not all-inclusive. Nevertheless, because of world preoccupation with science and technology, almost all journals indexed not only continue but have also greatly expanded their services as entered.

The document is titled *A Guide to the World's Abstracting and Indexing Services in Science and Technology.* It was released as Report number 102 of the National Federation of Science Abstracting and Indexing Services, 324 East Capitol Street, Washington, D. C., and is available for sale at that office.

Services in the guide are first arranged by Universal Decimal Classification with numbers used from the *Abridged English Edition,* third edition, revised (London: British Standards Institution, 1961). Brief titles are listed under the Universal Decimal Classification numbers with each title followed by its numerical designation in the main list of titles. Complete bibliographical data are given in the second (main) list, arranged alphabetically by title and containing also cross references where needed. Alphabetizing is in accordance with *Filing Rules for the Dictionary Catalogs of the Library of Congress* (Washington, D.C.: Library of Congress, 1956). A country index listing short titles is provided in which an international section precedes the individual countries. An alphabetical subject index provides users with a ready approach to specific subjects.

As indicated in Figure 19, information for each entry is given in the following order: (a) title in the original language (Cyrillic and Oriental titles are transliterated); (b) title of section of journal containing service, if applicable; (c) name of issuing agency (sponsoring association, government organization, or professional society) and/or publisher and address; (d) frequency of publication (weekly, monthly, quarterly, etc.); (e) date service commenced; (f) average number of abstracts and/or references supplied yearly; (g) language sources covered (European languages, English language, Slavic languages, world coverage); (h) arrangement of entries (author, subject, geographical, Universal Decimal Classification, etc.); (i)

ABRAHAMSON, A.C., documents by
-- Israel: social welfare: training of personnel TAO/ISR/37

ABRAM, MORRIS B. (United States), documents by
-- religious discrimination: convention (draft) E/CN.4/Sub.2/L.361 & Rev.1

Abyssinia: See ETHIOPIA

AD HOC COMMITTEE OF EXPERTS ON LONG-RANGE PROGRAMME OF WORK IN THE FIELD OF POPULATION (E/CN.9/AC.5/-)
-- report E/CN.9/182/Add.1(text)

ADEN
-- & Federation of South Arabia
Documents
A/5800/Add.4 Special Cttee. Report on Aden
-- petitions A/5800/Add.4
-- self-government or independence
Reports
A/5800/Add.4 Special Cttee. Report on Aden
Meeting records
GA, 19th sess.: *A/PV.1290, 1294, 1296, 1298, 1299, 1302, 1303, 1305, 1306
-- UN presence (proposed)
Documents
A/5800/Add.4 Special Cttee. Report on Aden

ADVISORY COMMITTEE ON ADMINISTRATIVE AND BUDGETARY QUESTIONS
-- reports to GA, 19th sess.:
A/5842 Co-ordination among UN and specialized agencies: administrative and budgetary questions
A/5842 Technical assistance: Expanded Programme: administrative and operational services costs
A/5842 UN Special Fund: finances: overhead expenses

AFGHANISTAN, documents by
-- Asia and the Far East: economic co-operation E/CN.11/641 (Sales no.: 64.II.F.14)
-- women: advancement in developing countries E/CN.6/435/Add.1

AFRICA
See also Organization of African Unity and names of countries and territories
-- aerial surveys: training of personnel: centres (proposed)
See also Meeting of Experts on Regional Centres for Training in Photogrammetry and Airborne Geophysical Surveys and for Interpreting Aerial Surveys, Addis Ababa 1964
Reports
E/CN.14/CART/128 Report of the Meeting of Experts on Regional Centres for Training in Photogrammetry. Report

*Documents issued in provisional edition, subject to revision and to be re-issued at a later date in final form. NOT FOR GENERAL DISTRIBUTION, DEPOSIT OR SALE.

AFRICA (continued)
-- agricultural development
Documents
E/CN.14/297 ECA. Activities of the Joint ECA/FAO Agriculture Division
-- air transport
Reports
[E/CN.14/TRANS/26] African Air Transport Conference. Report
-- cartography: See Africa: aerial surveys: training of personnel: centres (proposed); Meeting of Experts on Regional Centres for Training in Photogrammetry and Airborne Geophysical Surveys and for Interpreting Aerial Surveys, Addis Ababa 1964
-- crime
Documents
E/CN.14/328 (E/CN.14/SODE/30/Rev.1) ECA Expert Group Meeting on Social Defence. Report
-- economic development: planning
Documents
E/CN.14/295/Add.1 ECA. Activities in economic development, planning and projections
-- industrial development
Documents
E/CN.14/298 ECA. Activities in industry
-- industrial development: training of personnel
Documents
E/CN.14/307 ECA. Training activities
-- industrial estates
See also Seminar on Industrial Estates (ECA), Addis Ababa, 1964
Documents
E/CN.14/IE/9 Industrial estates in Africa
-- international trade: transit trade E/CONF.46/AC.2/4
-- juvenile delinquency
Reports
E/CN.14/326 (E/CN.14/SWTA/36) First African Training Cource on Institutional Treatment of Juvenile Offenders. Report
E/CN.14/328 (E/CN.14/SODE/30/Rev.1) ECA Expert Group Meeting on Social Defence. Report
-- land-locked States: & international trade E/CONF.46/AC.2/4
-- locusts: control
Documents
E/CN.14/322 FAO. Locust control in Africa
-- nuclear-free zone (proposed)
Meeting records
GA, 19th sess.: *A/PV.1290, 1291, 1293, 1295, 1298, 1300, 1303, 1305, 1307
-- photogrammetry: See Africa: aerial surveys: training of personnel: centres (proposed)
-- social defence
Reports
E/CN.14/328 (E/CN.14/SODE/30/Rev.1) ECA Expert Group Meeting on Social Defence. Report

*Documents publiés en édition provisoire, sujets à revision, et dont le texte définitif paraîtra plus tard. DISTRIBUTION GENERALE, DEPOT ET VENTE INTERDITS.

Figure 18

indexes supplied (author, subject geographical, patent, etc.); (j) price per annum, when known; (k) field of interest or main subjects covered; (l) library of Congress call number for titles in the catalog collections of the Library of Congress (National Agricultural Library and National Library of Medicine call numbers have also been used). Currency abbreviations used are those used by the International Monetary Fund and set forth in *Currency Units of Various Countries and Areas* (Washington, D. C.: International Monetary Fund, 1962).

Locating the journals referred to by indexes often requires the use of a document called the *Union List of Serials in Libraries of the United States and Canada*. This document indicates which libraries house collections of journals and other publications of the various societies, institutions, and governments. Indexing to this document is described in the prefatory material.

"1. A serial not published by a society or a public office is entered under the first word, not an article, or the latest form of the title.

"2. A serial published by a society, but having a distinctive title, is entered under the title, with reference from the name of the society.

"3. The journals, transactions, proceedings, etc., of a society are entered under the first word, not an article, of the latest form of the name of the society.

"4. Learned societies and academies of Europe, other than English, with names beginning with an adjective denoting royal privilege are entered under the first word following the adjective (Kaiserlich, Königlich, Reale, Imperiale, etc.)."

Abbreviated titles in citations may be deciphered with the help of specialized lists of abbreviations. The *World List of Scientific Periodicals, Biological Abstracts, Chemical Abstracts,* and W. Artelt's *Periodica Medica* are particularly helpful.

In almost every discipline and profession individuals are the best source of current information. The men whose names recur on or in articles uncovered in a search of the literature usually know and are willing to share a great deal more information than can be discovered in library research. Indeed, most "authorities" in a field will agree to comment, *in their areas of competency,* for publication in the professional literature. Many have had second thoughts about or have updated the arguments of the very articles that placed them in authoritative positions. Having those arguments advanced, *with appropriate credit,* by others sometimes relieves these men of a reporting task. It also permits the airing of opinions which alone might seem too thin to merit separate publication.

556. **Drvna Industrija**
Section: Mi čitamo za vas
Institut za drvno industrijska istraživanja, Gajeva 5/VI, Zagreb, Yugoslavia
bimonthly; since 1951; 225 abstracts a year from world literature; 1,000 dinars.
wood industry

557. **Duodecim**
Section: Kokousselostuksia; Havaintoja Kirjallisuudesta
"Duodecim," Yrjönkatu 17, Helsinki, Finland
monthly (with irregular supplement); since 1885; 100 references a year from world literature.
medicine

558. **Dyna; Revista de la Asociación Nacional de Ingenieros Industriales**
Section: Documentación
"Dyna," Navarra, 1–1°, Bilbao, Spain
monthly; since 1935; 2,000 abstracts a year from English language, European, and Slavic literature; Universal decimal classification; 365 pesetas domestic, 500 pesetas foreign.
technology
TA4.D9

559. **E.E.N.T. Digest**
E.E.N.T. Digest, Inc., Box M. Winnetka, Ill.
monthly; since 1939; 500 abstracts a year from world literature; subject classification; $7.00.
ophthalmology, otolaryngology, laryngology, rhinology
(NLM) W1 E25

560. **E.R.A. Weekly Abstracts**
British Electrical and Allied Industries Research Association, Cleeve Road, Leatherhead, Surrey, England
weekly; since 1939; 1,500 abstracts a year from world journals; 25s. (members), £5 5s. (others).
electrical engineering

561. **ETR; Eisenbahntechnische Rundschau**
Section: Zeitschriftenschau
Carl Röhrig Verlag, Holzhofallee 33a, Darmstadt, Germany
monthly; since 1952; 900–1,000 abstracts a year from European and English language journal literature; DM 40.80.
railway engineering
TF3.E18

562. **L'Eau**
Section: Bibliographie
Comité Hygiène et Eau et l'Association Française pour l'Étude des Eaux, 9 Rue de Phalsbourg, Paris 17, France; published by Les Éditions de l'Eau, 23 Rue de Madrid, Paris 8, France
monthly; since 1908; about 900 annotations a year from world literature; subject classification; 28 NF domestic, 38 NF foreign.
water treatment
TD202.E38

563. **Eaux et Industries**
Section: Bibliographie
Association Française pour l'Étude des Eaux, 7 rue de Cirque, Paris 17, France
monthly; since 1950; about 3,500 references and abstracts a year of world literature; subject classification (A.F.E.E.); 10,000 fr. for members, and exchange basis.
water treatment

564. **Eisei Kôgyô Kyôkaishi. Journal of the Society of Domestic and Sanitary Engineering**
Sections: [Abstracts; Introduction to Important Current Periodical Articles]
The Society of Domestic and Sanitary Engineering, 1, 3-chome, Ginza-Nishi, Chuo-ku, Tokyo, Japan
monthly; since 1918; 40–50 abstracts a year from Western literature and 300–400 references a year to Western journals; 600 Yen domestic.
sanitary engineering, heating, air conditioning, refrigeration

565. **Eiyô to Shokuryô. Journal of Japanese Society of Food and Nutrition**
Section: Abstracts of Foreign Literature
Japanese Society of Food and Nutrition, c/o Department of Agricultural Chemistry, Faculty of Agriculture, University of Tokyo, Tokyo, Japan
4–5 issues per year; since 1947; 100 abstracts a year from Western literature; membership with journal 1,000 Yen per year.
food industries, nutrition

566. **Electrical Engineering Abstracts (Science Abstracts. Section B)**
Institution of Electrical Engineers, Savoy Place, London W.C. 2, England
monthly; since 1898; 6,500 abstracts a year of world literature; Universal decimal classification; monthly author, annual author and subject indexes; £5.
electrical engineering, electronics, telecommunications, data processing
Q1.S3

567. **The Electrical Review**
Section: New Patents
Dorset House, Stamford St., London S.E. 1, England
weekly; since 1872; about 5,000 references a year to patents; £4 domestic, £5 15s. foreign, $16 U.S. and Canada.
electrical engineering
TK1.E44

568. **Electronic Industries**
Section: International Electronic Sources
Chilton Company, Chestnut & 56th Sts., Philadelphia 39, Pa.
monthly; since 1942; 1,500 abstracts a year from English language, French, German, and Slavic journals; subject classification; annual subject index of Russian abstracts only; $10 domestic, $12 Canada, $18 other foreign countries.
electronics
TK7800.E438

569. **Electronic Technology**
Section: Abstracts and References (compiled by the Radio Research Station of the Department of Scientific and Industrial Research, State House, High Holborn, London, W.C.1, England)
Iliffe Electrical Publications Ltd., Dorset House, Stamford St., London S.E.1, England
monthly; since 1923; 4,000 abstracts a year from world literature; Universal decimal classification; annual author and subject index; £3. 7s. as separate, $9.50 as part of Electronic Technology.
Also published monthly in the Proceedings of the Institute of Radio Engineers, New York
electronics

570. **Electronics & Communications Abstracts**
The Multi-Science Publishing Co., 33 South Drive, Brentwood, Essex, England

Figure 19

There are many reasons why a man cannot or does not publish the results of his current work. First, of course, proprietary and security regulations raise artificial blocks to the flow of information. Second, there are not enough journals to print all the available information. (Six months after its inception the quarterly *Massachusetts Review* was receiving unsolicited articles at the rate of 50 a week, although it could print only 45 selected articles a year.) And, third, benevolent policies in the care and feeding of scientists have created ivory-tower atmospheres within which a research man does his research withdrawn from external pressures and reality. (The atmosphere—and delays—are created by pecking orders of secretaries, receptionists, and security personnel authorized to clear topics, approve agenda, and even to schedule publications). For these reasons, an interview is sometimes a welcome event to a professional. Although scientists may comment only on unclassified work, they are otherwise free. Only physicians and lawyers are prevented by the ethical rules of their profession from indulging in personal publicity.

Mechanical aids are often essential to interviewing, and a tape recorder and a Polaroid—Land camera should be used whenever possible. Many of the technical information divisions of larger companies will supply both visiting and staff writers with such equipment. At the Boeing Airplane Company, for example, it is not unusual to see a writer, charged with writing either a document for company use or an article for external publication, first describe a prototype while speaking into a tape recorder and then photograph (security permitting) selected areas with the camera. From the transcript of his recording he later mines the data needed to justify his writing, and with the photographs he later guides the illustrator in the preparation of exactly the sketches needed to reinforce the writing.

Staff writers for *Electromechanical Design* use prepared questions and tape recorders to interview scientists and designers. The procedure saves time in note taking, keeps the discussion relevant to the preplanned organization of the intended article, and speeds up the interview considerably. Yet the procedure is flexible enough to permit additional questions in areas requiring more information. Because the tape is a permanent record it permits writing up multiple articles later, at leisure, and in desirable surroundings.

However, the reticence and self-consciousness that most people feel when being interviewed may be compounded by the presence of recorders and cameras. When this is apparent, they should be removed, and the interviewer should then move, with good grace, to the use of a notebook. In most cases, a notebook is reassuring, for it indicates to the subject that he is not likely to be misquoted and that data in the final article will probably be accurate.

A formula for setting up and conducting an interview usually involves the following steps:

1. Request an interview well in advance and in writing.

The reasons for this are many. First, a person prominent enough to merit interviewing is probably too busy to grant an interview on short notice. Second, advance notice will give him an opportunity to brief himself and perhaps to collect supporting data. Third, a written request is a document which security advisors may discuss in advance and if necessary use to fix the type and depth of response permitted.

2. Give the person to be interviewed a choice of dates and times.

The person being interviewed grants the interview; therefore, the choice of time and place should be his.

3. Indicate how information gained in the interview will be used.

The person being interviewed is more likely to meet with a writer who intends to publish his findings in a reputable journal than a writer who is interested in documenting an article for one of the mass media. For this reason it is sometimes wise to name journals and editors and to forward any correspondence that commissions an article in advance. Further, and perhaps most important, the security limitations and the publicity ambitions of research laboratories vary widely, but no administrator is likely to fault an interview that will be reported in the professional literature.

4. Indicate the subject as specifically as possible and give illustrative questions.

Just as no editor will accept an article that discusses too broad a topic to make analysis in depth possible, few people will give time to discuss a general subject already adequately presented in an undergraduate textbook. The men being interviewed usually conduct specialized research; therefore, they should be treated as resource persons in their specialized areas.

5. Request permission to bring a camera and recorder if you wish them and if security permits.

This request will often save embarrassment if local regulations deny the use of recording equipment. Further, mention of recorders and cameras will often lead the person being interviewed to prepare data releases and visuals in advance. This is particularly important in camera work, because lead time permits the preparation and retouching of photographs that might not otherwise be released.

6. Brief yourself on both the subject and the person to be interviewed.

Most professionals are willing to talk with other professionals at an appropriate depth. Similarly, most will resist discussing background and general information that can be gained more readily elsewhere. Further, any person is likely to be more cordial and expansive to an interviewer who has secured background information about him as well as about his field. Indicating knowledge about a man's background is perhaps the quickest way for an interviewer to ingratiate himself.

7. Outline the form you would like your report to take.
A writer would probably have done this for his own use anyway; certainly an editor commissioning an article in advance would have demanded it. A person being interviewed can be guided in his observations if he knows the outline in advance. Further, his specialized knowledge sometimes prevents the writer from committing an error in logic or in fact that could be highly embarrassing.

8. Prepare a list of questions whose answers will fill your outline.
An interviewer assured of the information essential to the organization of his article can seek out illustrative details and specific data that might otherwise be irrelevant.

9. Arrive at an interview promptly.
Most persons resent giving time, and if the gift seems lightly regarded for any reason they will develop a block against an interviewer.

10. Give a copy of your prepared questions to the person being interviewed.
This technique not only puts a man at his ease, it reassures him that the interviewer has done his homework and is serious in his intent. Further, a pattern of questions leads to a pattern of answers. In effect, the man being interviewed writes the article by filling in the blanks of an outline.

11. Accede gracefully to the deletion of any questions.
A person who deletes questions has personal, professional, or proprietary reasons for doing so. Although he will usually be embarrassed for refusing to answer, he will be angered if he is nagged for not answering.

12. Maintain a candid, open, and interested attitude.
Most persons worthy of being interviewed chose their professions, and they are ready to defend their choice. They will be suspicious of interviewers who seem devious, and they will be angered by interviewers who seem bored.

13. Do not indicate boredom or impatience.
Usually a person who talks a long time supplying background or context for his answer to a question does so out of a fear of being misinterpreted or quoted out of context. If he is a professional, he will usually hedge his answers anyway.

14. Interrupt only when exact data, necessary for an article, are not given. Let the person being interviewed do the talking. No one likes to be interrupted. If a person has been elevated to the status of an authority

worth interviewing, he will dislike being heckled or cut down by the man who elevated him in the first place.

15. Add only those additional questions necessary to flesh out your report. Do not waste time in idle conversation.

The only important person at an interview is the man being interviewed. He is completely disinterested in the personality and opinions of the interviewer.

16. Thank the person interviewed, and promise him a copy of the finished report.

Always close an interview on a friendly note—you may want to come back. If a copy is promised, it should be delivered. Delivery in advance of publication helps to ensure against repudiation. When proprietary or security information is involved, clearance of the final article in advance will often be required.

Because any search of the literature will uncover dissociated data and random facts, rough correlation within the frame of reference established by the definition of a problem should be made during the data-collection phase. There are essentially three methods of correlating raw data: functional, physical, and chronological. The selection of a method or combination of methods will vary with the writer's intent.

Data are correlated by the functional method when the writer wishes to discuss an implement, an apparatus, a process, or a method. The functional technique forces him to write in the only frame of reference that can justify the existence of devices and systems—the accomplishment of work. (This is the only expository technique acceptable to the U. S. Patent Office.)

In the case of an implement, the writer should begin with an overview; that is, he should write a paragraph—or even a single sentence—which contains a statement of the primary function of the implement. The paragraph should state as succinctly as possible what the implement does and how each main functional part contributes to that mission. Such a statement puts the reader immediately in a functional frame of reference; he knows what something does and how it does it. Furthermore, the reader is conditioned by the sequential listing of main parts—with their function—to expect a similar sequence and functional linkage throughout the article.

For example, the picture tube in a television receiver is a common but fairly sophisticated component. Its function, however, can be broadly stated in a single sentence. Such a tube is "an electronic component which

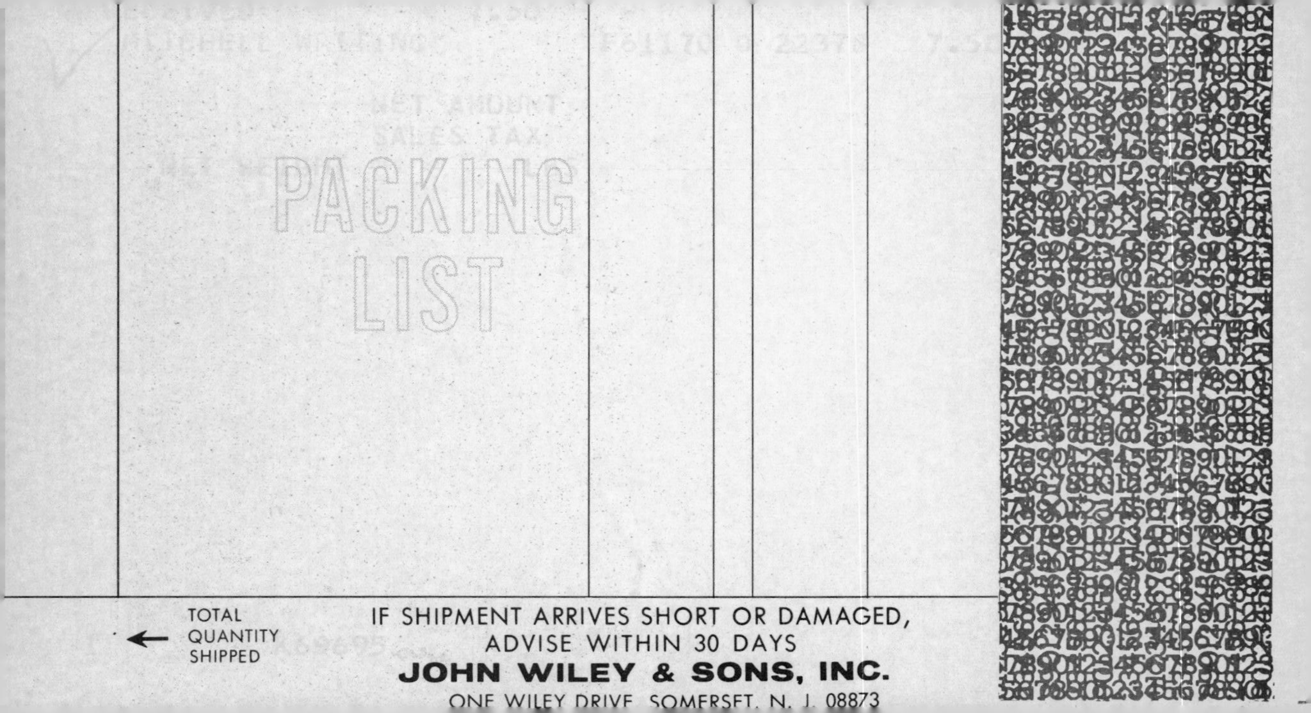
PACKING
LIST
TOTAL
QUANTITY
SHIPPED
IF SHIPMENT ARRIVES SHORT OR DAMAGED,
ADVISE WITHIN 30 DAYS
JOHN WILEY & SONS, INC.
ONE WILEY DRIVE SOMERSET, N. J. 08873

converts information from electrical terms to visual patterns by means of an electron beam which is successively created, accelerated, focused, deflected, and fluoresced." Such a statement permits the writer to correlate all his data under the five functional headings indicated by the series of verbs at the end of the sentence.

Analysis of the sentence shows techniques essential to this type of writing. First, the sentence begins with an adjective-noun sequence common to all organized sciences. In this case the family is "component" (as opposed to a general term such as "implement," "device," "system," etc., which would connote an entity or closed-loop situation). The branch is "electronic," which may itself be modified or subdivided by additional modifiers such as "thermionic," "magnetostatic," etc. Had the object under discussion been an entity like a screwdriver, a thermostat, a toaster, or even a flush toilet, the family term would have been successively "tool," "switch," "appliance," or "fixture." The branches of these families are normally "manual," "electrical," "mechanical," "thermal," "chemical," "hydraulic," etc., or any combination. The leading sentence for a discussion of each might begin: "A screwdriver is a manually operated tool . . . ," "A thermostat is an electromechanical switch . . . ," "A pop-up toaster is an electromechanical appliance . . . ," and "A flush toilet is a hydromechanical fixture . . . ." An almost limitless number of subbranches or adjectives can be added. The toaster is "spring-loaded," the toilet is "semiautomatic," and the thermostat may be "bimetallic." The procedure is analogous to all science; in the life sciences, for example, progressive subordination descends through phyla, classes, orders, families, and genera.

Stating the function of an implement requires selecting an exact verb for which the adjective-noun sequence serves as subject. Choosing such a verb requires thinking as the inventor thought. A cook created the toaster. Limited to the methodologies of baking, boiling, or broiling sliced materials, he chose the last. Accordingly, his electromechanical appliance "broils and positions." A plumber perfected the toilet. His hydromechanical fixture—in his jargon—"traps waste and seals off gases." An electrician needed the thermostat. His electromechanical switch "makes and breaks a circuit." In each case a *functional verb* describes the mission. Usually the verb or verbs can be applied even to the most complex form of the same invention. Thus the muzzle-loading cannon and the *Skysweeper* (the most sophisticated piece of fully automatic, radar-trained, rapid-fire ordnance in the United States arsenal) have the same function; they both simply "direct and propel." The latest typewriter still "prints and positions."

There remains only the necessity of telling the reader how each main functional part or subsystem contributes to the mission. This is best done with an expansion of the functional verbs introduced with the phrase "by means of." The toaster, for instance, performs its function—broiling and positioning—*by means of* "a resistance coil and a spring-loaded carriage."

The verbs themselves serve as a clue. The elaborate lands, grooves, choke, and radar-training gear of *Skysweeper* are all assumed by "direct and propel." Here, however, what might be called a *system* or *black-box concept* aids in the communication. A series of stages, systems, or black boxes usually work on a common element. Just as *Skysweeper's* projectile (itself the subject of a separate functional analysis) is the recipient of all the directing and propelling, so is the electron beam in the television picture tube that is mentioned earlier. This electronic *component* whose function is to "convert information from electrical terms to visual patterns by means of an electron beam which is successively created, accelerated, focused, deflected, and fluoresced" breaks into five main parts each denoted by a functional verb. Because each verb is functional and all are linked by the same element (the beam), the reader sees the component as having five parts. He is also led to expect a five-part discussion. The writer need only correlate his data within the five areas. Further, because each area has been introduced in terms of its function, the entire article may be developed in a functional frame of reference. This ensures both a sequitur in logic and a consistent pattern of organization.

The outline below reduces the functional technique of writing to absurdity. It is included, however, not only to show how a single sentence can contain all the information essential functionally to describe an implement, but also to show how placing main parts and subjects in their equivalent degrees of subordination creates an outline. A lengthy and definitive discussion of a typewriter could easily be made by expanding each numbered section of the outline to paragraph length. The organization of such a discussion would be functionally correct because the sentence is syntactically correct. The correlation of data to such an outline is relatively simple.

A Smith-Corona portable typewriter

1. is a manually operated, mechanical device for printing and positioning symbols on sheets of paper or suitably thin material by the impact, mechanically multiplied by a third- and first-class lever system, of raised reverse images of the symbols on the paper through a dye-impregnated ribbon and
2. consists of
   2.1 in general

2.1.1 a printing mechanism driven concurrently with
2.1.2 a positioning mechanism, and
2.2 in detail,
2.2.1 a printing mechanism consisting of
2.2.1.1 a reinforcing device,
2.2.1.1.1 of a material of suitable resistance, such as hard rubber, and
2.2.1.1.2 of a shape for convenient feeding, such as a cylinder, called a platen, and
2.2.1.2 an inking device composed of
2.2.1.2.1 a paste dye held in
2.2.1.2.2 a convenient textile material on synchronized drums, and
2.2.1.3 a selecting device, called a keyboard, whose individual keys activate finger-like levers which carry the reverse-image dyes, and
2.2.2 a positioning mechanism, consisting of
2.2.2.1 an automatic system to move the carriage of the platen under spring loading a short distance by
2.2.2.1.1 a space bar,
2.2.2.1.2 a backspace key, and
2.2.2.1.3 each type key, and
2.2.2.2 a manual system to move both carriage and platen a variable distance by
2.2.2.2.1 a knob on the platen for torque, and
2.2.2.2.2 a lever beside the platen for both torque and thrust.

The correlation of data essential to the discussion of a process may be done in a similar fashion. "Main steps" simply replace "main parts." These main steps break down into substeps, and the same analytic procedure is followed. The theme of function or mission is replaced by a superimposed time order. The organization is correct as long as the chronology is correct. The sequential arrangement of data is determined in a manner similar to that used in discussing an implement.

For example, the process for making home brew may be readily explained—and understood—if the writer will begin with an overview for an introductory paragraph. This overview should begin functionally and close with the sequence of steps. Consider the single sentence "Making home brew is a simple fermentation process that requires sprouting, mashing, and cooking wort, adding hops, fermenting the mixture, and straining, pasteurizing, and bottling the resulting liquid." This sentence contains the core information. The family and branch words "fermentation process" indicate the type of process, and the present participles (verbals ending in -ing) indicate the sequence of steps in which something specific is done. These verbals serve the same function as the verbals used in the functional analysis of the television tube discussed earlier: they fix the order of events, and they lead the reader to expect a multipart exposition of the topic.

There are eight verbals in the sentence quoted above. A writer need only discuss them in sequence to achieve understandable organization. He will achieve greater clarity if he develops each in a separate paragraph or section. An obvious linkage may be obtained if he begins each paragraph or section with "First," "Second," . . . "Finally," or if he repeats the verbal as the first word of the paragraph or section. The techniques may be combined, and use of the latter gives a reassuring parallelism; for example, the paragraphs (following the overview) might well be set up as follows:

"First, the sprouting of malt wort builds up the reducible starch content of the grain. It is achieved by . . . ."

"Second, the mashing of malt wort reduces particle size to accelerate enzymic action in fermentation. It is achieved by . . . ."

"Third, the cooking of the malt wort stops the sprouting action and further softens the particles. It is achieved by . . . ."

"Fourth, the adding of hops lends flavor and color to the liquid. It is achieved by . . . ." And so on.

Clearly, each of the above paragraphs may be expanded to contain whatever information the writer feels the reader should have. Furthermore, the arbitrary arrangement permits the writer to correlate his data as he collects it.

It should be noted that each of the paragraphs in the example above is concerned with *what, when,* and *why*. The last is a requirement of process writing; not only does the "why" communicate pertinent information, it also tells the reader or a technician how essential a step is and if it may be omitted.

The following example of a process description was written for a popular audience.

DEVELOPING BLACK AND WHITE ANSCO FILM NEGATIVES

"At its simplest, the development of a black and white Ansco film negative is a process of chemical reduction requiring two different chemical solutions—a *developer* and a *fixer*. The developing process follows the exposure process and precedes the washing process. The chemical sequence is as follows.

"When a picture is snapped, a measured amount of light enters the camera and strikes the film. This light reduces the silver salts in the emulsion coated on the film. The extent of the reduction varies with the amount of light striking each part of the emulsion.

"The *developer* reacts with salts reduced by light and changes them to metallic silver which shows up black. The degree of blackness depends upon the amount of previous reduction by light. Areas unreduced by light are unaffected by the *developer*.

"The *fixer* stops the action of the *developer* when a suitable contrast of dark and light has been reached. The *fixer* also changes salts unreduced by light to a soluble form which is washed away in a water bath process.

"A typical *developer* is an aqueous solution of pyrogallol, citric acid, potassium bromide, and ammonia. A typical *fixer* is an aqueous solution of sodium thiosulfate."[3]

This example is a good process description for several reasons. It opens with the reassuring phrase "At its simplest," which implies a general discussion. The first sentence establishes the family and branch as "chemical reduction." The number of solutions – two – is given, and their names are italicized to permit their use as a linking device throughout the rest of the description. Finally, the first paragraph places this process in time; that is, it follows one process and precedes another.

The second paragraph describes the preceding process because reductions made there are essential to the success of the process under discussion. The third paragraph discusses only the phase involving the first solution, the *developer*. The fourth paragraph discusses only the phase involving the *fixer*, but points ahead to the next process. The description closes with a general listing of chemicals.

Layout in process description is determined by analysis of the specific procedure. The writer knows he must present a linked narrative covering successive steps. As in the functional analysis, he can best do this by listing main steps in an opening paragraph. He can then discuss each step singly, since the reader is already aware of the sequence. Clearly, each step is a separate division or paragraph.

If steps are complex or lengthy, they should be introduced with subheadings. Substeps are then given separate paragraphs beneath each single

[3]From *The Everyday Reference Library*, edited by Lewis Copeland and Lawrence W. Lamm, published by J. G. Ferguson Publishing Co., copyright 1948 by J. G. Ferguson & Associates, copyright 1964 by J. G. Ferguson Publishing Co. Reprinted by permission.

subheading. Because readability increases with the number of subheadings, all but the simplest process descriptions should be so divided.

In general, art work reflects the same division as the written copy. The first figure is usually an overview illustrating the entire process. Subsequent figures illustrate specific steps. Correlation of illustrations is thus done in the same manner as correlation of data.

The techniques for physical descriptions are deceptively simple. In the standard procedure, the writer assumes a reference point—which he gives the reader—and moves from top to bottom, left to right, center to periphery, or in any meaningful direction. The procedure requires starting the reader at the stated reference point which he knows and then leading him to new and unknown points. Because printed words are involved, the reader can always fall back to the last known area if he becomes lost. In this type of description only spatial relationships are involved. Each relationship, however, may be subordinately developed to include data involving materials, surfaces, colors, and materials.

Whenever a comparison may be made to a familiar shape, a writer should organize and correlate his material to that frame of reference. The following description of the battlefield at Waterloo was written by Victor Hugo almost a century ago, but it is as sharp and accurate as any prose that has since been written on the subject. The frame of reference is a capital "A," and with the exception of the italicized phrase there is no irrelevant "literary" information.

"In order to get a clear idea of the Battle of Waterloo, we should imagine in our mind's eye a large capital letter 'A,' coming to a point at the top. The left leg of the 'A' is the road from the town called 'X.' The right leg of the 'A' is the road from the town called 'G.' The cross-bar of the 'A' is Mount Saint Jean. Wellington is here. General Jerome Bonaparte is located at the lower left leg of the 'A.' The right-hand lower leg is where Napoleon Bonaparte is located. A little below the point where the cross-bar of the 'A' cuts the right leg is the town called 'Q.' At this point the final battle word was spoken. *Here the lion is placed, the symbol of the supreme heroism of the Imperial Guard.*

The triangle contained at the top of the 'A' within the two legs and the cross-bar of the 'A' is the plateau of Mount Saint Jean. The struggle for this plateau was the whole of the battle. The wings of the two armies extended to the right and left of the two towns called 'X' and 'G.' Behind the point of the 'A,' behind the plateau of Mount Saint Jean, is a large forest. As to the plain itself, we must imagine a vast rolling country, each rolling hill commands the next; and these hills, rising toward Mount Saint Jean, are there bounded by a forest."

When an obvious comparison does not exist, more sophisticated techniques of physical description must be used. A writer can describe the hull of a dinghy—which has no straight lines—only if the intelligence of his audience permits the use of specialized language or if he can lead his audience to accept a hypothetical reference point. Mathematical or topological formulas may be used to describe the curved skew lines of the hull whenever a trained audience exists. Otherwise, a hypothetical reference line outside the boat must be established. Straight-line, plumb, and perpendicular measurements can be taken from this line. The appropriate degree of abstraction of a hypothetical frame of reference will depend upon the education and training of the intended audience.

The correlation of data to a physical description will vary with the frame of reference. The maneuvering and disposition of troops at Waterloo may be discussed with reference to the capital "A." Surfaces and materials of the dinghy hull may be discussed in terms of the lines and planes related to the hypothetical reference line. With mathematical formulas, however, a change in the frame of reference is necessary. Making such a change is most easily done with a visual aid. An illustration, summarizing the formulas and indicating through callouts the specific areas to be discussed, will serve to orient the reader. Data may be correlated in reference to the areas marked by callouts in the illustration.

Correlating data chronologically involves several techniques. Superimposing a time order on the functional analysis of a process has been discussed above. The techniques developed there are adequate unless simultaneity of steps or recycling is involved. In the case of simultaneous steps, adverbials such as "at the same time," "simultaneously," "concurrently," etc. are placed both in the overview (introductory paragraph) and in the specific step or steps involved. For example, the introductory sentence illustrated on p. 65 might be altered to end: ". . . and straining, pasteurizing, and simultaneously bottling the resulting liquid." This modification would require changing the final paragraph to begin: "Finally, the pasteurizing and bottling of the resulting liquid are done simultaneously to permit pasteurization of enzymes and sterilization of bottles in a single heating. These steps are achieved by . . . ."

Chronological correlation of steps when recycling is involved requires telling the reader how and when recycling is initiated and how and when it is stopped. The "when" is vital, for in a process description all things exist in order of time.

For example, part of the process for generating electricity with the aid of a pressurized water reactor involves recycling water at 2000 psi from the reactor core to the heat exchanger and back through the reactor core.

The main steps involving the forcing of the water into the core and through the heat exchanger would have been discussed in standard, separate paragraphs such as the following.

"Fifth, water, pressurized at 2000 psi in order to prevent the formation of steam within the reactor, is forced through the core to carry away the heat generated by fission and simultaneously to slow down the neutrons which are moving through the core. This last is achieved by the light hydrogen atoms in ordinary water which provide a cushioning effect which slows the neutrons to a speed at which the probability of additional fissioning is much greater. Without this slowing-down effect, or moderation, the chain reaction could not be maintained.

"Sixth, the pressurized water, having been forced through the core, next passes through the heat exchanger where it boils the cooler, lower-pressure secondary water, transforming it into steam."

The preceding paragraphs tell the reader how the pressurized water cycles through the tower once. Recycling is introduced *in a separate paragraph* beginning, perhaps, as follows.

"Seventh, the pressurized water, having given up much of its head as it passed through the heat exchanger, is drawn back to the pump and recycled to the core. Closed-loop piping . . . ."

A stylized drawing (see Figure 20) gives an overview of the entire process. Such illustrations communicate more effectively than words when size, shape, or sequence are involved. Indeed, two experiments (Air Force Institute of Technology Technical Report 65-1) have shown that verbal descriptions of size and shape add nothing to comprehension gained from drawings alone. Sequence, however, is most effectively communicated when both visual and verbal exposition are used.

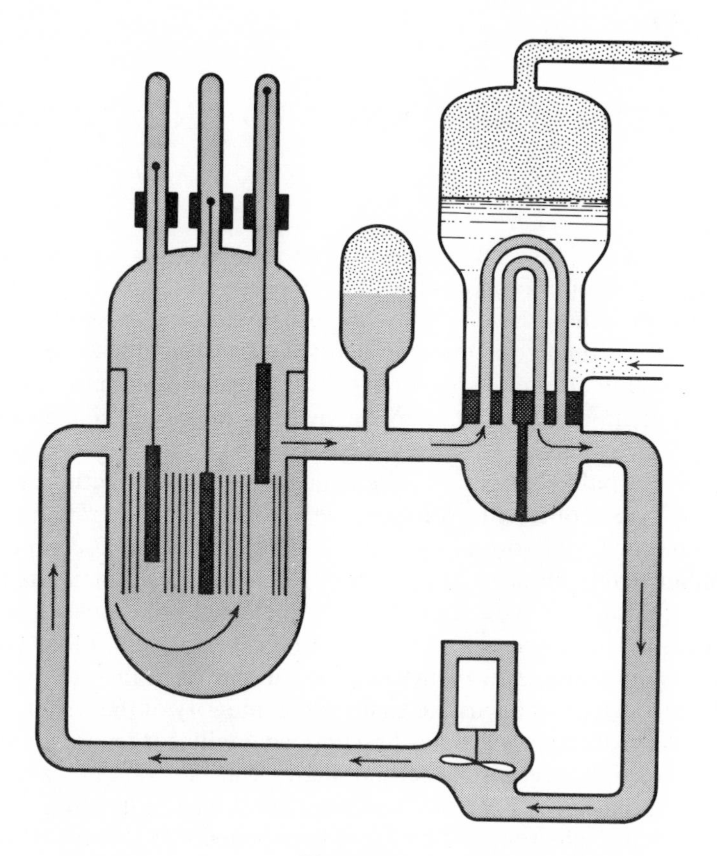

Figure 20

3
DATA SELECTION AND ARRANGEMENT

Determination of how much information to include in a scholarly article is always a matter of compromise. The integrity of the writer requires him to include data that may conflict with his findings, and the acuteness of his peers requires him to document in depth. Conversely, the expense of publication and the attention span of his audience require him to be brief. Yet selection of appropriate data may be the most important single step in the writing of an article.

Because a published article is examined by readers with training and points of view different from the writer's, its invulnerability can be achieved only by a rigorous and narrow statement of intent coupled with a monomaniacal adherence to that intent. This approach facilitates the selection of data, however, for only data relevant to the thesis need be included. If a writer states his thesis and reports only upon that thesis, he cannot be condemned for something he did not do. Thus, to play on the word, the integrity of an article ensures the integrity of the writer.

Because professional articles presuppose intelligent readers, most writers are aware that research methodologies will be examined as carefully as research findings. This situation requires a candid discussion of techniques, but the selection of data to explain the technique is fairly simple as long as the writer realizes that his role is to explain and not to justify. Similarly, an intelligent reader will not permit generalizing or conclusion-jumping; hence a writer need only select data essential to the logic of his conclusions. These conclusions must be solidly based, however.

The attention span of a professional audience is affected by both the depth and the style of an article. A lengthy, illiterate article with something to say will be read as long as it continues to say something. A lengthy, literate article with nothing to say will be thrown aside as soon as its shallowness is discovered. Because a shallow article can only be made long with the help of stylistic flaws such as padding, repetition, circularity, digressions, and redundancy, intelligent readers are hypersensitive to these aspects of style.

Happily, the expense of scientific publication has led editors to reject articles that are lengthy and inflated. This fact was one of the many findings reported in the NSF-supported study "Characteristics of Professional Scientific Journals" (available at the Clearinghouse for Federal Scientific and Technical Information on order number PB166 088N). Among the findings of that study pertinent to this chapter is the fact that average articles in university press journals are nearly twice as long (10, 000 words) as those published in society and commercial journals (under 6000 words). Significantly, university presses publish fewer research words per year than do physics journals, earth science journals, and chemistry journals which lead the field in that order. Equating this to cost, however, shows that physics journals have the lowest cost ($0.67 per 100, 000 research words) of any journal category. This is in contrast with miscellaneous society journals, which average $3.40 per 100,000 research words, and commercial journals, which have an average subscription rate of $3.94 per 100,000 research words and yet need to match this in advertising revenue to remain solvent.

Although the selection of data is limited by all the factors mentioned above, the arrangement of data is determined by the author's approach to the topic. "Standard" approaches are chronological, logical, physical-spatial, or fixed-formula, but combinations are always possible and are often preferable.

Data may be easily correlated in a chronological sequence. This type of correlation is achieved by viewing data only as they exist in order of time. However, selecting and arranging that data for chronological exposition is not as simple. For example, chronological correlation of raw data about an author might produce the following biography.

Born in log cabin, has fox for pet, sees father beat mother, goes to country school, has male pedant as teacher, told father is a drunkard, hears father called hero, has love affair with Jane Smith, learns to drink and smoke, enrolls in college, studies law, loses first case, is called liar, grows angry, knocks down constable, writes poetry in jail, marries Sue Brown, takes job with newspaper, writes novel, cannot sell novel, sells poetry, plants short story in newspaper, fired from paper, short story attracts patron, patron alcoholic, two novels published, wife leaves him, wife refuses divorce, patron leaves him, volume of poetry published, play based on novel produced, suffers from asthma, writes documentary TV script about dope, becomes addict, writes novel in sanatorium, run down by truck, wife claims body, wife publishes novel and sells movie rights, wife marries Italian nobleman who beats her.

The obvious faults of strict chronological order are the mixing of dissimilar events and the precluding of subordination. These faults can be corrected by selecting only those events pertinent to a writer's point of view and then showing how some of the events are causes and other effects; for example, a psychologist, given the chronological list of facts above, would select the pet fox, the father as hero, alcoholic, and wife-beater, the premarital love affair, the loss of temper, the loss of wife, the physical debilitation, and the drug addiction as significant data. He would keep these data in the same sequence, but he would link them to show a progression from cause to effect.

Writers with other points of view would select different data. A literary critic using the approach of the "new" criticism would ignore the facts chosen by the psychologist. His point of view—and, in this case, his technique—would find those facts irrelevant and unusable. Similarly, the points of view of a novelist or of a sociologist would require selection of different, although occasionally overlapping, data. Even a biographer will choose data most malleable to his thesis or to his style. The "snapshot" technique of Boswell's *Life of Johnson* is a striking example of selection and juxtaposition or arrangement. The "development-in-parallel" of Truman Capote's *In Cold Blood* is an example of the pairing of concrete facts to heighten both impact and communication.

Communication research at Northwestern University by the late Dr. Irving J. Lee showed that most individuals unconsciously manipulate data in the manner discussed above. In Lee's terms, all communicators *level* their data by selecting only the facts they feel to be important. Next, they *sharpen* their data by stressing specific preselected bits. Finally, they *assimilate* their facts into a story by twisting facts to relevancy within a rational framework.

Variations of truth are obviously created by this unconscious manipulation. The significant facts may be leveled out. The irrelevant or digressive facts may be sharpened. Concrete facts may be colored by the context of a nonexistent story or by being placed on an arbitrary framework. Equally important in Lee's findings, however, was that *the reader also levels, sharpens, and assimilates*. Errors are thus compounded—all unconsciously—by this two-stage, irrational approach.

Nevertheless, purely chronological articles are rare. Lee's findings have been well publicized, but most writers proceed with Olympian self-esteem, feeling that they can level, sharpen, and assimilate both rationally and accurately. In all events the clarity necessary for formulating and writing an article on a chronological framework is usually beyond the capacities—and perhaps the data—of the writer. What is often attempted is a compromise in which the chronological arrangement proceeds in three steps—past, present, and future. Such an arrangement carries the aura of chronology and permits three distinct assimilations. The only merit to this approach is that it increases the probability of accuracy.

Logical arrangement reflects patterns of thought and, essentially, there are three phases in the reasoning process: first, the mind stores up data based upon experience and the opinions of authorities; second, the mind draws conclusions from these data by *induction;* and, third, the mind applies principles to specific instances by *deduction*. Logical writing, designed to inform or to persuade, proceeds in the same fashion. The writer collects his data, fixes on a form of reasoning, examines his logic for fallacies, and then shows in writing just how he reasoned to reach his conclusions.

If he reached his conclusions through the use of *induction,* he used one—or any combination—of four processes. First, he probably *generalized;* that is, he drew a principle from any number of specific instances. He may once have seen a beagle with egg on its face. If he hears another observer say that he too has seen a hound eat an egg, he may infer from his own experience and from the experience of others that "all hound dogs suck eggs." He may easily detect the error of the principle he has drawn because the generalizing tests concern the samples. Were there enough samples, and were the samples representative? For example, few scientists quarreled with the findings of the Kinsey Report. Many, however, stated that the sample space should have included another 500,000 people. Others suggested that those who volunteered as subjects for Kinsey were not representative. Subjects willing to volunteer might be sexual athletes seeking recognition for their prowess. All subjects were North Americans, and one sociologist claimed all normal, "red-blooded American boys" will enlarge upon sexual adroitness.

Another phase of induction is *forming a hypothesis*. This process involves making tentative explanations for a set of facts. Each explanation is then tested as in any scientific investigation or problem-solving exercise, and all possible explanations are determined. From these, the best is chosen—the best being the simplest one that will fit all the facts.

An adequate example of the forming and testing of hypotheses involves eels. The Roman naturalist Pliny, who observed only female eels throughout Europe, hypothesized that eels were all of one sex. Others hypothesized that eels were born of morning dew in May or from hairs of a horse's tail. Finally, the Danish scientist, Johannes Schmidt, hypothesized that male eels did not like fresh water and that breeding must occur in salt water. When he discovered male and female eels breeding in the Sargasso

Sea, he proved that his hypothesis was correct and that the hypotheses of others were false.

Cause-and-effect reasoning is a treacherous technique often used with other kinds of induction. The technique involves finding causes of known effects. For example, there is an ancient joke about a magician performing on a troop ship. The magician would place an object on a table, cover it with a handkerchief, place a parrot on top, and fire a gun. Then he would remove parrot and handkerchief to show that the object had disappeared. During the final performance, a torpedo struck the ship at the precise moment the gun was fired. When he regained consciousness, the magician found himself floating on a hatch cover and the parrot saying, "I give up, what did you do with the ship?"

The main fallacy in the cause-and-effect reasoning above is called "after this; therefore, because of this." Classical Roman philosophers recognized it and fixed the fallacy as the *post hoc ergo propter hoc* argument. They realized that it is fallacious always to assume that because one event followed a first, the second event must therefore have been caused by the first. Indeed, we today brand as superstition something like: "After I walked under the ladder; therefore, because I walked under the ladder." For truth, cause, not time, is all-important. The parrot simply did not have enough evidence that the events he experienced were causally related, and any writer using cause-and-effect reasoning must furnish his reader such evidence.

Induction by analogy is another treacherous technique. It involves concluding that because two things are similar in some ways, they are also similar in other ways; for example, Sir Isaac Newton observed that combustible substances such as oil and camphor have refractive powers several times greater than their densities indicate. By analogy he reasoned that the diamond with its high refractive powers was combustible. He was lucky. Diamonds are combustible, but, as Brewster, a nineteenth-century physicist and biographer of Newton, pointed out, if Newton had reasoned by analogy that greenockite and octahedrite are combustible because they have high refractive powers he would have been dead wrong.

Because analogy is not logical proof, writers should employ it only for clarification and not for the logical arrangement of data. The popular literature of astronomy is full of specious articles reasoned by analogy. The United States space program has already shown that although Saturn has seasonal changes, atmosphere, etc. similar to earth's, it does not have a biology similar to earth's.

Although inductive methodologies are creative, *deduction* is not. Deduction involves the application of accepted rules to specific cases and the drawing of a conclusion.

Rule: All humans are bipeds.
Instance: Ducks are bipeds.
Conclusion: Therefore, ducks are human.

The mental gymnastic above is a syllogism and a false one at that. An example of a true syllogism is

Rule: All men are mortal.
Instance: Socrates is a man.
Conclusion: Therefore, Socrates is mortal.

There are three statements in the above—and in all—syllogisms. They contain three terms, each of which is used twice but not twice in the same statement ("mortal"). The *minor term* is the subject of the conclusion ("Socrates"). The *middle term* appears in both rule and instance ("men" and "man"). As long as this pattern exists, the syllogism is called *valid.* This does not mean it is *true.* A syllogism is only true when all elements of the rule and instance are correct and the conclusion contains the minor and major terms.

The difference between valid and true syllogisms is best illustrated by the diagrams in Figure 21. Using such diagrams is perhaps the easiest way to determine both truth and validity in syllogistic writing.

Just as there were common fallacies in inductive reasoning—incorrect or oversimplified hypotheses, after-this-therefore-because-of-this reasoning, false analogy, and so on—there are also common fallacies in deductive reasoning. The duck syllogism in Figure 21 is an example of an incorrectly drawn conclusion. Its incorrectness could be detected by both the absurdity of the conclusion and the use of the diagram.

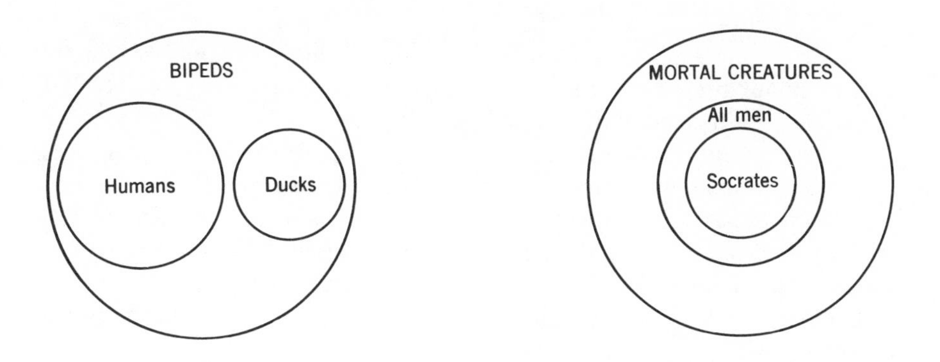

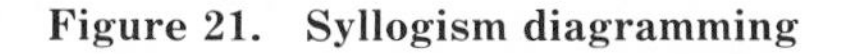
Figure 21. Syllogism diagramming

Basing conclusions on negative statements (or "premises" as rules and instances are often called) is usually a fallacious form of deductive reasoning, because if one premise is negative, the conclusion must be negative; or, if both premises are negative, no conclusion is possible.

Valid: All presidents have been men.
My wife is not a man.
Therefore my wife is not president.

Invalid: No woman has ever been president.
I am not a woman.
Therefore I am president.

Roman philosophers applied the term *non sequitur* ("it does not follow") to the fallacy of reaching a conclusion not warranted by the premises. Conclusions drawn from two negative premises are non sequiturs. So is the following:

Men who are religious should be honored.
I am religious.
Therefore I should be canonized.

Non sequiturs or inconsequent arguments also exist in truncated forms. For example, "The Navy does not need a supercarrier because the Air Force is responsible for the air defense of the United States."

Another fallacy in deductive reasoning involves shifting the meaning of terms between the major and minor premises. This makes the syllogism invalid (and usually untrue) because it puts a fourth term into the syllogism.

Man is the only creature equipped to reason.
A woman is not a man.
Therefore women are not equipped to reason.

In the example above, the meaning of "man" shifts from *mankind* in the rule or major premise to *male* in the instance or minor premise. The syllogism is therefore invalid; it might, however, be true.

Writers often arrange data in incomplete or abbreviated syllogisms. Such reductions imply three statements, but only one or two may be written. Such reduced syllogisms are called *enthymemes;* they are usually written to avoid laboring the obvious. For example, "Sailors who cannot swim are not allowed on ships; Admiral Brown cannot swim."

The omitted conclusion in the above example is obvious, and the writer avoids drawing it to heighten the effect of his writing.

A similar effect is achieved by omitting a minor premise, or both a minor premise and a conclusion. Consider the following.

I need a microscope capable of 300,000×. I shall buy yours. (Omitted minor premise: "Your microscope is capable of 300,000×.")

I will hire only men who have published. (Omitted minor premise: "You have not published." Omitted conclusion: "Therefore, I shall not hire you.")

Enthymemes may lend a sense of facility and confidence to a writer's style, but readers of professional articles are often suspicious of truncated logic. When logic is involved, even writers of "popular" articles should labor the obvious.

There remains something to be said about *begging the question, ignoring the question,* and *arguing in a circle.* The first amounts to stating without proof a proposition that requires proof. "That useless military requirement may be ignored" begs the question of whether or not the requirement is useless. Calling it useless it not a logical argument.

Ignoring the question amounts to diverting attention or shifting the argument. It is a courtroom cliché to hear the mother of an accused axe murderer say, "But he's a good boy, Judge," when the question is Did he commit the murder? Perhaps the most popular form of ignoring the question is the argument *ad hominem* ("to the man"). Politicians are singularly adept at this; rather than discuss an opponent's position on, say, a tax bill, they will attack his character, his ancestry, his religion, his wife, and occasionally his dog. The procedure is obviously unfair even in a profession not noted for clean tactics. It amounts to professional assassination when an article writer ignores a scientist's findings and snidely discusses the man and his qualifications.

Arguing in a circle involves proving an original proposition with a second proposition which can only be proved by proving the first. This hen-and-egg circularity is always inconclusive. It frequently goes undetected, however, because circular arguments tend to run on and obscure the absence of logic in a thicket of words.

There are other logically fallacious procedures that a writer should recognize and avoid. False analogies and hasty generalizations have already been discussed. Growing from the latter is the *implication of the total when only part is true.* A writer with integrity or a responsible scientist would add the words "some" or "part" to statements such as the following:

This report discusses the effect of an atomic bomb exploded in New York harbor.

Because this laboratory lacks an electron microscope, it can never do anything significant in microbiology.

Use of the *faulty dilemma* is a crafty but dishonest way to hide the fact that a rational course may exist between two extremes. For example, "The only choice left is either to wage a preventive war or to prepare a retaliatory attack against all possible aggressors. Because the latter would result in our own destruction, we have no alternative but to initiate a preventive war."

Avoiding a faulty dilemma does not justify use of the equally risky *argument for compromise*. A compromise is not necessarily correct or always practical; for example, in a report of a study of the science offerings at a major university, a writer (a) tried to define the terms "generalized" and "specialized"; (b) showed that science faculty are divided in opinion, one faction favoring generalized training and the other faction favoring specialized training; (c) concluded without any additional support that science training at that university should be partially generalized and partially specialized.

The trustees of the university recognized the report as an equivocation and took an extreme stand in favor of generalized training only.

The fact that the phrase "the exception that proves the rule" is a tired cliche indicates that this popular fallacy is easily recognized. Exceptions discredit rather than disprove rules. For example, "The fact that six nations including Liechtenstein don't owe us money only proves the rule that Uncle Sugar is being exploited by everyone." Substituting the word "tests" for "proves" allows the passage above to make sense. Indeed, the word "tests" was in the cliche as originally coined, but it was replaced by an over-zealous writer whose enthusiasm was contagious.

Loaded questions are as effective as *ad hominem* arguments in stopping further communication, and they are as illogical. A loaded question contains its own answer or additional assumptions. The question "When will we stop doing research in this blind alley?" assumes the research is leading up a blind alley. Such a beating-your-wife assumption will have to be answered before the question can be answered.

Citing an *authority out of his field* is a popular but illogical procedure. It is amusing to hear a politician trace his philosophy and lineage back to Washington, Jefferson, and Lincoln. It is insulting, however, for a writer to tell a reader that a physicist has this opinion about political science or that a doctor of medicine has that opinion about the stock market *unless* the individual cited is also an authority in those fields.

Glittering generalities are really a form of reverse name-calling. As a writing technique, this involves applying good labels to transfer approval to an idea. As generalities, they are usually illogical to begin with. When they are favorably loaded emotionally, as in advertising, they often amount to deception. The statement "Sound laboratory techniques are only possible with the use of ACME equipment" may sell equipment, but it can only be proved correct through further investigation.

The fallacy of *plain folks* stems from the cultural conditioning of North Americans involved with the democratic dream. It is a reverse form of snobbery presupposing both human perfectability and innate intelligence of the "common man." It is no more valid than the *authority-out-of-his-field* technique, but the "appeal to the people" and the "mandate from the people" is hugely effective. Usually it is answered by the equally illogical *catch phrase* or *quotable quote* such as Ibsen's generality, "A majority is always wrong."

Finally, there is the *bandwagon appeal.* One of the easiest ways for a writer to lose intelligent readers is for him to preface statements with "As everyone knows . . . ," "Authorities are agreed . . . ," "It is common knowledge that . . . ," and "No one disputes . . . ." However, conservative readers will often accept ideas so introduced without examining them. Indeed, many will believe whatever they are told their professional, social, or political group believes.

Most of the logical and illogical techniques discussed separately above are combined in use. All are important to a writer for a variety of reasons. Induction is essential to scientific thinking because it enables a scientist to arrive at generalizations and hypotheses, to see cause-and-effect relationships, and to recognize analogies. From this "knowledge" it is possible for the scientist to make predictions and to produce effects by controlling causes. The writer who retraces the process in words is simply making the logic of the scientist overt.

The procedures *currently in use in the professional disciplines* frequently provide fixed-formula arrangements that writers should follow. Often such arrangements are simply sequential accounts of laboratory or bench procedures. Whenever an article discusses "classical" procedures, fixed-formula arrangements are not only easier for a writer to follow, they are also easier for a conditioned audience to understand. Indeed, many editors insist upon a standardized topical arrangement even when atypical laboratory procedures introduce awkwardness into such predetermined organization. Thus, for the majority of research articles in professional journals, fixed-formula arrangements are the rule.

Among the physical sciences, the discipline of chemistry serves as a clear example. The American Chemical Society in its *Handbook for Authors* (1967) fixes the formula for articles submitted to its fourteen journals as follows:

Title
By-line and supplementary information
Abstract
Introduction
Experimental section
Results
Discussion
Acknowledgments
References

For the purpose of this discussion, the first four and last two sections may be regarded as "housekeeping." They are brief, they always appear in the order indicated, and they are essential for the reasons indicated earlier. The real body of the article will be contained in the sections labeled "Experimental," "Results," and "Discussion." These sections reflect the standard procedures currently in use in the professional discipline.

Examples of multiple, fixed-formula arrangements may be found in the social sciences. In the discipline of psychology, the depth and the purpose of the research determine the arrangement. A topical outline is followed when the purpose is the reporting—without evaluation—of tests. In this type of outline—and research—the tests are the topics:

Results of intelligence test
Results of Rorschach test
Results of TAT
Results of figure drawing
etc.

In the above the research had centered on tests, and so the orientation is to tests rather than to patients. In the formulation of cases, however, orientation shifts to the patient. Klopfer[1] postulates an arrangement which, although analytic rather than synthetic, constitutes a logical organization for ordering the discrete bits of information essential to psychiatrists.

Behavior during testing
Intellectual aspects of the personality
Affective aspects of the personality
Basic conflict areas
Adaptive and maladaptive techniques
Diagnostic indicators
Prognostic implications

[1]W. G. Klopfer, *The Clinical Report* (New York: Grune and Stratton, 1960).

Clinical psychologists tend to follow a theory or a school. The fixed-formula arrangement for a report or article about a clinician's work is fixed by the tenets of the theory or the school he follows.

A comprehensive, fixed-formula arrangement as postulated by Huber[2] is printed below. Its major limitation, to Huber, is a complex chronology.

1. *Intellectual functioning*
   Level of present functioning
   Level of capacity
   Reasons for failure to function up to capacity
2. *Conflicts*
   Major and minor conflicts
   People with whom conflicts are manifested
   Times and places where conflicts arise
   Etiology of conflicts
3. *Methods of handling conflicts*
   Manifestations of anxiety, symptoms, defense mechanisms
   Overt behavior
4. *Strengths and weaknesses in relation to goals*
   Needs and wishes, manifest and latent
   Strengths for pursuing them
   Weaknesses
5. *Recommendations*
   Therapy or no therapy, environmental change
   Form of therapy
   Predictions about therapy

Menninger[3] advocated the following chronological-topical approach, which he calls a "diagnostic synthesis."

I. The personality development and structure
   A. Heredo-congenital nucleus of the personality
   B. The general conditioning of childhood
   C. Special conditioning of childhood
   D. Data of the adolescent period relevant to present illness
   E. Maturity
II. The environment (present)

[2]J. T. Huber, *Report Writing in Psychology and Psychiatry* (New York: Harper & Brothers, 1961), p. 31.

[3]K. A. Menninger, *A Manual for Psychiatric Case Study* (New York: Grune and Stratton, 1952).

III. The maladjustment (precipitating causes, anxiety, insight, facade, defenses, focus of aggression, secondary gains, approach to treatment).

IV. Diagnostic summary
   - A. Personality type
   - B. Psychiatric syndrome
   - C. Medical, surgical, and dental complications
   - D. Unclassified symptomatic manifestations not included above
   - E. Sociological status, including economic situation
   - F. It is required in many places to add to the name-diagnosis of psychiatric illnesses the following:
     1. Brief description of syndrome, severity, and duration
     2. The precipitating stress
     3. Degree of predisposition recognized in the premorbid personality
     4. Degree of incapacity

Articles submitted to the journals published by the American Institute of Physics should follow the fixed-formula arrangement given in the Institute's *Style Manual* (Second Edition, 1959). The sequence is as follows:

1. Title
2. Author by-line
3. Abstract
4. Headings of text showing subordination by script, position, and numbering
5. Mathematical material
6. Tables
7. Figures
8. Footnotes

In those disciplines in which multiplicity of research or laboratory techniques makes standardization impossible, freedom from a fixed-formula organization is possible only in the test proper. The following outline is suggested by the American Institute of Biological Sciences in its *Style Guide for Biological Journals* (1964). Note that the phrase "in the natural sequence of its parts" in effect requires a tracing of procedures.

1. Title, by-line, running head, name and address for mailing
2. Abstract
3. Text, in the natural sequence of its parts
4. Acknowledgments
5. References to literature

6. Footnotes
7. Tables
8. Legends for figures
9. Figures

The only flexibility permitted by the AIBS *Manual* concerns the abstract. In the words of the *Manual,* "If the journal to which you submit the manuscript does not publish an abstract with the paper, include one on an unnumbered page for *Biological Abstracts.*"

Governmental agencies and facilities expect a fixed-formula organization from their writers. The preferred arrangement tends to follow the logic of the AIBS Manual in permitting freedom in the body of the paper while fixing the patterns or sequence of the remainder. The sequence is dissimilar, however, from that of the AIBS.

The National Aeronautics and Space Administration recommends the following fixed-formula organization in *NASA Publications Manual* (NASA SP-7013, 1964).

Summary
Introduction
Symbols (when a separate list is required)
Main body of paper
Concluding section
Appendixes (when needed)
References
Library-card abstract

The *layout of a book* is far more complex than the layout of an article. A book might contain the following, although few would have all these parts.

Front matter:
- Certificate of limited edition—an announcement of the number of copies printed in a limited edition
- Half-title (sometimes called the "bastard title")—the title of the book, standing alone on the page
- Imprimatur—the authorization to print
- Book card (or "face title")—a list of books by the same author or in the same series
- Frontispiece
- Title page
- Copyright
- Dedication

Preface
Acknowledgments
Table of contents
List of illustrations
List of figures, maps, charts, or tables
Introduction
Epigraph
Text
Appendix
Notes
Quotations
Bibliography
Glossary
Index
Colophon—a brief account of the book

Although the fixed formulas discussed above predetermine the sequence of main sections, a writer should predetermine the organization within those sections. A flexible system of outlining—such as the decimal system—is the most effective technique. The headings of this outline *in its final form* constitute the table of contents if the article is long enough to warrant such a table.

The mechanics of a decimal outline physically reflect organization and subordination (see pp. 64-65). For this reason, the system is recommended by the American Institute of Physics and is required by the Department of Defense. The latter agency is the tail that wags the dog, for its Mil-Specs require all research-and-development contractors as well as all military departments to use the decimal system. The flexibility of the system permits the Department to coordinate information by "weapons systems"; for example, one multisection manual can discuss the entire B-52 system. Because the manual's topic covers both the airplane with all its accessories and the weapons that it carries, a small library is required physically to house the documents. Correlation of documents is achieved by the decimal system, and the manual can easily be updated to include modifications to any part of the "weapon system" by the simple procedure of changing or withdrawing the document that contains the single subsection that discusses the modified part.

The decimal system of outlining is used in engineering reports and in articles concerned with the applied sciences. Major divisions (single-digit numbers) always begin on a fresh page and determine the left-hand margin. Their headings are typed in capital letters. In the subheadings the first letters of all significant words are capitalized; in the sub-subheadings only the first letter of the first word is capitalized. Subdivision headings are indented according to their degree of subordination and carry their own numbers (for example, 2.4, 2.4.1). The text always returns to the left margin. The decimal outline on p. 64-65 is an example of an outline without headings.

As many subordinate headings should be used as logic permits, because it has been shown that readability increases directly with the number of meaningful subdivisions.

Outlining an article in advance usually determines the sections and subsections. All writers should correlate and arrange their facts into meaningful groupings before attempting to write anything. These groupings form the major sections. Usually each divides logically into a series of subgroups of data. These become the subsections. Any data that do not belong in the subsections should either be omitted from the article or placed in a separate section.

Writing subheadings and headings to convey information is important. Principally, however, these headings should alert the reader as to what to expect and should indicate in which area the data are relevant. In cases in which headings are reproduced verbatim in the table of contents, they should indicate both organization and coherence within the article. Thus they serve as an outline of the report as well as a road map to the sections.

An outline may be checked for errors *before they appear*. The preceding passage discusses the special case of functional outlines, but the following general rules apply to all topical outlines.

- Use more than one point, because a subject cannot be divided into one.
- Use more than one subpoint for each main point, for the reason above. It is not always necessary to subdivide, but when it is done, it must be done *in parallel*.
- Limit the number of main points. More than 10 points usually indicates the presence of broader topics or categories.
- Check subordination. If an item does not belong under a point, put it where it belongs. If it does not fall logically anywhere in the body of the outline, put it in an appendix or in another *separate* article.
- Check coordination. A parallel element cannot be made subordinate, and vice versa.
- Check overlapping. An overlapping element should be separately developed, usually *on a higher level* of the outline.
- Check total effect. Taken together, the main points should give the overall design at a glance.

Categorically, there is no place for a digression in a formal technical article. As indicated above, the data that form a digression are usually detected by the use of an outline. When they exist, they should be published as a separate article or—rarely—in an appendix to an article to which they are somehow related. Even if included as an appendix, the data of a digression must be organized to form a coherent and logical article which could be published independently. Some publishers periodically issue chapbooks which contain just such articles.

In the case of informal articles, however, the level of usage and the insistence upon logical arrangement are relaxed. Digressions appear in such articles as illustrations and as anecdotes. Style and form approach that of the personal essay which, by definition, allows the personality of the author to intrude.

It is possible—and often profitable—to recast a formal technical article and to twist its logic to permit inclusion of whimsical asides. It is not possible, however, to recast a popular article upon a formal framework without deleting the digressions and deepening the remaining discussion.

Textbooks on journalism and the writing of feature articles discuss the techniques permissible and desirable in those genres. The present book is preoccupied with technical and professional writing.

4

ELEMENTS OF JOURNAL ARTICLES

All professional and technical writing must be factual and objective. It is not the role of such writing to entertain or, ordinarily, to persuade. Indeed, most readers and almost all editors of professional articles will react negatively to writing that has a gee-whiz tone or a hard-sell approach. Such readers will also object to the intrusion of a writer's personality.

This means that there are few personal pronouns in professional writing. The "I," "you," and "we" are deleted because only facts are important. Similarly, there are no exclamation points! If a fact is important, an intelligent reader will know it from the logic of the article, and he will be offended if the fact is illuminated with typographic fireworks. An exclamation point is acceptable as a factorial sign, but that is all.

The injection of personality and the use of personal pronouns are techniques common to persuasive writing but undesirable in informative exposition. Yet there are other techniques that are common to both types of writing. For example, the logical arrangements discussed earlier as being essential to an expository article are equally essential to a lawyer's argument or to a philosopher's reflections. The last two, however, are usually regarded as persuasive writings. Certainly a lawyer's argument, which follows the competitive techniques of debate, is more concerned with persuading than with informing a reader or jury.

Perhaps the only way to differentiate between informative and persuasive writing, and thus to determine appropriate stylistic techniques, is to determine the purpose of a specific piece of writing. Writers of both types present and explain facts. The informative writer, however, does so objectively. He exposes, as it were, data for consideration. The persuasive writer, on the other hand, selects his data subjectively and arranges it in a pseudo-logic designed to lead the reader into accepting the writer's evaluation. The use of pseudo-logic permits the introduction of human elements, of problems and emotions with which the reader can identify on a personal basis, which are more likely to persuade a reader to respond to a pattern of thought than are objective presentations of facts and figures. Because this approach is easily recognized—often unconsciously—the writer of professional and technical articles must avoid those techniques which signal the use of human elements. It is primarily for this reason that personality and personal pronouns have no place in informative or expository writing.

Fixed-formula organization tends to keep article writing informative and objective. When the organization of parts also follows a pattern or formula, the aura of objectivity is increased. This chapter discusses a formula approach to organizing specific parts of an article. It is interesting to note that writers who object to such writing-by-formula as suppressing their own identity are in effect complaining that they are not allowed use of the human elements which belong to persuasive rather than to informative writing.

All articles are forwarded to editors with a letter of transmittal. The format of correspondence has changed little since the eighteenth century. The layout used then is in use today in spite of the occasionally attractive departures taken by advertising copywriters conducting direct-mail campaigns.

The sequence of paragraphs within letters is essentially static, and most textbooks on business correspondence illustrate the sequence expected in letters of recommendation, application, complaint, and adjustment as well as transmittal. The present book is concerned only with the subclass of letters of transmittal appropriate to professional articles. The sequence and content of paragraphs has been fixed by custom to the following:

1. Identification of the letter and of the article.
2. Organization of the article or of the research upon which it is based.
3. Findings or conclusions of the article.
4. Statement of economics (if appropriate).

Two examples illustrate the formula. The first letter transmits an unsolicited article. The second letter transmits an article requested—or commissioned—in advance. As indicated in this letter, solicited articles frequently advance an argument that an editor wishes to present in the interest of balance and objectivity. To be professional, such an article must concern itself with a logical presentation of demonstrable facts. It should not deal in polemics or *ad hominem* arguments which, in the long run, tend to tar the writer with the brush of terms he applies to someone else.

Titles for professional and technical articles—as opposed to papers—must indicate content. Brevity and succinctness are desirable, but at no

27 Olympic Avenue
Ann Arbor, Michigan
June 10, 1966

r. John H. Hicks, Editor
he Massachusetts Review
niversity of Massachusetts
mherst, Massachusetts

ear Dr. Hicks:

The enclosed, unsolicited article entitled "Thoreau in Our ime" is submitted for possible publication in The Massachusetts eview.

You will note that the article is arranged chronologically y topic; i. e., the section entitled "Passive Resistance" begins ith John Marshall and ends with Martin Luther King. It is followed y a section, entitled "Secularization," which begins with Increase ather and ends with Martin Buber. The arrangement is thus thematic, nd the themes are presented in historical perspective.

The final sections of the article constitute a summary of the rguments that liberalism in North America is derivative from horeau. The last paragraph presents the conclusions which may be rawn from the article. While this paragraph is essential to the nity of the article in that position, it has been so written as to ermit separate reprinting either as an abstract or as a part of republication publicity.

Should you decide to publish this article, please consider his letter a formal release subject to the terms and rate scale nnounced on page three of the January, 1966, issue of The assachusetts Review.

Yours very truly,

David A. Porter

ncl.

327 Eames Avenue
Madison, Wisconsin
June 15, 1966

Mr. A. Stanley Higgins, Editor
The STWP Review
Westinghouse Research Laboratories
Pittsburgh, Pennsylvania 15230

Dear Mr. Higgins:

I enclose the article entitled "Data Management: Ten Tired Techniques" requested in your letter of June 12. As requested, the article answers the questions posed and allegations made by J. R. Hunsberger in his article "Data Management: New Approaches to Old Problems" (Review, June 1966).

The article follows Hunsberger's original format: ten types of data-management problems are isolated and defined, and solutions to the problems are given. Unlike the Hunsberger article, the solutions in this article are drawn from the history of the art and are not based upon applications of new hardware to typical problems.

The introduction identifies the article as an answer to Hunsberger, and the summary shows that six of his ten "new approaches" were known and in use at the turn of the century. The obvious conclusion is drawn that if Hunsberger had "managed his data," his search of the literature would have shown that none of his approaches was "new."

The fee of $75 offered in your letter of June 12 and payable upon publication will be quite satisfactory.

Yours very truly,

Harry P. Wagner
President, Wagner Associates

Encl.

time should aspects of style be permitted to inhibit communication. Further, aspects of tone such as cuteness, wit, and straining for effect should be avoided at all cost. Trick titles that will "hook the reader" are discussed in textbooks on journalism. Titles that will cue the reader, inform the research scientist, and guide the reference librarian are the concern of this book.

The difference between titles for professional articles and titles for all other types of writing is great. The following fail as titles for articles, but they were adequate for papers delivered at a recent meeting of the Society of Automotive Engineers.

1. The High Cost of Goofing
2. Management, Men, and Machines
3. An Overall Viewpoint of Systems Analysis
4. Value Analysis and Its Applications
5. Data Handling in Complex Systems
6. What's Ahead in Nuclear Energy

Title 1 neither cues the reader nor informs the researcher, nor guides the reference librarian. It contains no information; hence, it fails as an article title. It succeeded as a paper title, however, because "goofing" is not only an amusing word but it was also, at the time of delivery, a popular one. It was coined by teen-agers around 1960 and was understood by their parents by 1962 – when the paper was given. (By that time it was a stale or square word to teen-agers, and it embarrassed them then as much as it would a user today.)

Title 2 is alliterative, gives a librarian three words for cross reference, but fails to cue a reader of articles. The glibness which makes it an acceptable title for a paper spoils it as a title for an article whose reader seeks to be informed rather than to be entertained.

Title 3 fails by being too general for a presentation of article length. Systems analysis is too complex to permit an "overall viewpoint" in a single article or even in a single volume.

Titles 4 and 5 fail as article titles for the same reason Title 3 failed. They would attract an audience for an oral presentation, however, because their generality implies a predigested and simplified discourse on a popular level. Promise of a popular-level presentation will always attract an untrained audience and often professionals *from another field.*

Title 6 connotes a crystal-ball approach that few take seriously and that no professional would take. (Even if he did, only his enemies would read such an article.) The same title, however, would be effective in a nonprofessional publication such as the Sunday supplement to a popular newspaper.

There are two effective techniques to use in writing titles for professional articles. The first is to modify the problem definition, and the second is to organize the key words or uniterms into a meaningful phrase.

Problem definition has been discussed as an aid to searching the literature and to guiding research. It has been further discussed as an integral part of an article. In each discussion the statement was made that a good definition anticipates or predicts the organization of the article. A good title will retain the predictive aspects of the definition.

By reviewing the problems of semantics, generics, syntactics, and viewpoint already solved for the definition, it is an easy matter to write a title that is predictive of the article and that communicates to the reader. Problems of *semantics* (synonyms and homographs) can be resolved with a thesaurus which will indicate whether a specific discipline uses the word "imperfection" or the word "defect." Problems of *generics* (hierarchical word families) can be resolved by being as specific as the article. (Not "batteries," not "nickel batteries," but "nickel-cadmium batteries.") Problems of *syntax* (word order) can be resolved by the writer when he rereads his article to determine *what he really did say*. Finally, problems of *viewpoint* (individual orientation) seldom arise because a professional article is written for publication in a journal of a specific discipline to be read by those trained in that discipline. While this does not permit wholesale use of highly specialized jargon, it does permit the use of concrete terms parochial to a discipline.

The problem definition discussed earlier was refined as follows:

1. To determine a safe insecticide for houseflies.
2. To determine the effect of dichlorodiphenyltrichloroethane upon humans and houseflies.
3. To determine the physiological effect of skin penetration, inhalation, and ingestion of DDT in a 5-percent, wettable powder upon humans and domestic animals.

Definition 3 was fixed upon as a guide to both the research and the search of the literature. If, after rereading his article, the writer feels that the article actually discusses work falling within that definition, he may modify the definition to form the title. Word sequence in his modification will be guided by the characteristics of keywords and uniterms.

Because "physiological effects" is both the general area discussed and the specific preoccupation of the article, that term should be used first. In that position, it will guide reference librarians and cue readers. Further, in archaic recall systems which require filing under the first significant word in the title, placing the word "physiological" first in the title will ensure recovery of the article. (The title of this book was revised to use "writing" as the first word.)

Because "skin penetration," "inhalation," and "ingestion" are *generic* aspects of physiology, they should be included both to limit the predictive aspects of the title and to specify the areas of physiology actually discussed. Their use guides the librarian in cross-referencing and tells the reader what topics are discussed and (by omission) what topics are not discussed.

Because the article discusses relationships (i.e., the effect of DDT upon humans), the terms and their relationships to each other should be given. The eleven roles established by the Engineers Joint Council indicate the possible relationships of terms in a meaningful phrase. Frequently, this can be simplified by answering the implied question "What, to what, with what?" In this case the answer is "DDT to animals by spraying."

Synthesizing the area, subareas, terms, and relationships produces a title such as "Physiological Effects Caused by Skin Penetration, Inhalation, and Ingestion of 5 percent, Wettable-Powder DDT by Humans and Domestic Animals." This is lengthy, but it communicates. The only protest will come from a journal editor who must find space for the title in the table of contents as well as—in larger print—at the top of the article. The title is effective because it accomplishes its threefold mission: it cues the reader, it informs the research scholar, and it guides the reference librarian.

Artificial requirements for titles exist in some journals. Limitations to ten words or even to ninety characters and spaces are not unusual. Fortunately, machine-retrieval techniques which originally necessitated the ninety-character requirement have become so sophisticated that there is now no mechanical limitation upon title length. Indeed, the more keywords and uniterms a writer can place in a title, the more likely it is that a machine will retrieve the article. Conversely, the fewer keywords, the less chance there is that a machine will be programmed correctly for retrieval. In effect, a good title to an article can mean the difference between recognition and oblivion.

"It is unethical to include in the by-line the name of any person who was not actually engaged in the reported research. This breach of ethics is usually committed by an institutional superior who insists that his name appear."

Style Guide for Biological Journals

The statement printed above pertains to the best of possible worlds. Administrative and professional jealousy, empire building, and acceptance of the publish-or-perish theory are but a few of the snakes in Eden. Coupled with viperous industrial, contract, institutional, and foundation customs and regulations, they make a snake pit of the world of professional writing. Tangles over by-lines have sometimes grown so complex that only litigation was able to quell the hysteria.

In a rational and ethical situation the rules for by-lines and author credits are simple: the name of the author and the name of the institution at which the research was conducted are given. When multiple authors are involved, *the name of the person who did the research is listed first.* After his name come the names of professionals *in order of their contribution.* If this sequence is difficult to determine, an alphabetical pecking order is usually acceptable. Given mature professionals as authors, bylines can be fun, as indicated in an apocryphal article in physics allegedly written by the living scientists Alpha, Bethe, and Gamow.

When multiple institutions are involved, only one should be listed unless work was done at others. In that event, every institution making a contribution should be listed. It is unethical, however, to change institutions and give credit to the new one when work was done at the previous one.

In the applied sciences administrative seniority often determines the "principal investigator" and even a name or two after his. The name of the research person who actually did the work is buried in the boiler plate. Often, this is not the only injustice. Articles released by industries, national laboratories, and politicians are usually written or edited by technical writers who translate into English the raw data furnished them by the research specialist. Because these writers never receive a by-line, the Society of Technical Writers and Publishers is actively seeking a modification of federal specifications which govern articles released by national laboratories. It is unlikely that private industry, with a long tradition of "ghosts," will ever grant the by-lines. It is almost certain that politicians will never give credit to the actual author of their articles and speeches.

The terms of grants made by foundations, societies, and institutions usually require that credit be given or that mention be made of the role of the agency. Unless otherwise stipulated, the statement is usually printed as a *footnote to the title* (see p. 332). When stipulation is made, it usually requires that the statement be part of the prefatory material and appear either in the acknowledgments or introduction. Most authors are so willing to acknowledge foundation support that the name-dropping of agencies has become a popular ploy in professional writing. Indeed, a parlay of support by multiple agencies is usually noted with pride. The listing of multiple support is usually alphabetical by agency, but chronological sequences are occasionally printed.

Requests for reprints and correspondence about an article or its copyright are addressed to the first-named author. If his institutional address changes prior to publication, the new address should be noted in a footnote. Addresses of junior authors are not given; that of a coauthor is given only if his institution supported the work.

In the rare case when the appendix to a paper has been prepared by someone other than the author of the report or article, the name of the author of the appendix should be added below the heading of the appendix. His role should also be noted in the introduction. In the case when the data in the appendix are as important as those in the article, the author of the appendix should be listed as a coauthor of the article.

A list of keywords, links, and roles is required by many journal editors. The varied and often mutually exclusive indexing techniques discussed in Chapter 2 cause confusion in the retrieval of information. A writer can avoid heightening this confusion if he will employ three techniques. First, he can write titles, such as those discussed in the preceding section, which communicate. Second, he can anticipate semantic shift and change in viewpoint by following the procedures discussed in the following section. Third, he can employ key words or uniterms found in either the thesaurus of the discipline involved or in the specialized dictionaries of that discipline.

One positive attribute of a thesaurus is its organization. Synonomous terms used by specialists with different interests and different viewpoints are grouped together, and the specialized area of each term is indicated. With the aid of a thesaurus, a writer trained in one discipline can find the synonomous but different terms used in another discipline. These are the terms he must use when he translates information from one discipline to another. A physical chemist, for example, must know in advance whether a journal of chemistry or a journal of physics will publish his article. Not only are the notations and symbols different, the words themselves are different.

Only the applied sciences have correlated their specialized dictionaries into a single thesaurus. This 10,500-word document, discussed in Chapter 2, merges the eighteen vocabularies of the professional engineering societies, the Defense Documentation Center, and the National Aeronautics and Space Administration. Because publications of these societies and agencies require either the writer or the editor to supply interdisciplinary key words and uniterms as well as abstracts with each article, the EJC *Thesaurus* coordinates the terms and ensures that they are understood. Writers submitting articles to journals outside the applied sciences, however, must translate their words with the aid of specialized dictionaries.

The use of links and roles in the recall of information was also discussed in Chapter 2. Writers submitting articles to journals subscribing to the EJC system should follow the figures and tables illustrating that section.

While usage varies slightly among the journals of applied science, the keywords, links, and roles appropriate to a given article are usually submitted on a separate sheet immediately before or after the abstract. The *Style Manual for Biological Journals* states, "Select as many as 10 words essential for indexing your paper and place them at the bottom of the abstract."

Some journals require that keywords or index terms be italicized or noted in the abstract. The "Information for Authors" printed in *IEEE Transactions of EWS* contains the following.

Each index term should be a single word, number, or symbol or a brief phrase which gives a clue or index to a substantial information concept of the paper. Index terms should be selected from the following list or carefully chosen from the abstract. Those which appear in the abstract should be underlined. Those terms which represent major topics in the paper should be preceded by an asterisk.

AEROSPACE ENGINEERING
AGRICULTURAL ENGINEERING
ANALYZING
ARCHITECTURAL ENGINEERING
BIOMECHANICS
CERAMIC ENGINEERING
CHEMICAL ENGINEERING
CIVIL ENGINEERING
COMPUTERS
COOPERATIVE EDUCATION
CURRICULUM
DEMONSTRATION
DESIGN
EDUCATIONAL METHODS
ELECTRICAL ENGINEERING
ENGINEERING ECONOMY
ENGINEERING SCHOOLS
ENROLLMENTS
ETHICS
EVENING EDUCATION
FACULTY
FEDERAL GOVERNMENT
GOVERNMENT CONTRACTS
GRADES
GRADUATE STUDY
GRAPHICS
HUMANITIES
INDUSTRIAL ENGINEERING
INSTRUMENTATION
LABORATORIES
LIBRARIES
MANAGEMENT
MATHEMATICS
MECHANICAL ENGINEERING
MECHANICS
MINING ENGINEERING
MODELS
NUCLEAR ENGINEERING
PHYSICS
PROGRAMMED INSTRUCTION
PROGRAMS (COMPUTER)
SIMULATION
SOCIAL SCIENCES
STRUCTURES
STUDENTS
TECHNICAL INSTITUTE
TEXTILE ENGINEERING

Table 4 below indicates the system used by the EJC *Thesaurus* to show shades of differences among synonyms. The meanings of the reference notes are as follows.

- BT (broader term)—the term following this note is more general than the term looked up, but both are important.
- NT (narrower term)—the term following this is less general than the term looked up, but both are important.
- USE—the term following is more acceptable than the term looked up, and the term looked up is not likely to gain importance.
- UF (use for)—the term following is less acceptable than the term looked up and is not likely to gain importance.

- RT (related term)—the term following is sometimes related to the term looked up but not in all instances.

In general, the USE/UF relationship is employed if the overlap in meaning between the two terms is nearly exact but the more specific term is not expected to be a useful retrieval word. The BT/NT designation is reserved for the case of the more specific word being useful.

The fifth and final reference note found in the *Thesaurus* is a SCOPE NOTE, i. e., a qualifier. This is used to limit a term which is seriously ambiguous in meaning. It is clued into the system by parentheses. Examples-follow.

- Accelerator (excludes automobile accelerator)
- Optical measurements (measurements of optical properties, quantities, or conditions)
- Order (sequence)

Extremely ambiguous or excessively broad terms are affixed with the scope note "use more specific term if possible." These terms are then followed by the alternate, more specific terms, listed as related terms.

Table 4. EVC *Thesaurus* Sample

OPTICS

OPTICAL MASERS
  USE LASERS
OPTICAL MATERIALS +
  NT OPTICAL GLASS =

  RT FIBER OPTICS
     LENSES
     MIRRORS
     OPTICAL INSTRUMENTS
     OPTICS
OPTICAL MEASUREMENTS
  (MEASUREMENTS OF OPTICAL
  PROPERTIES, QUANTITIES
  OR CONDITIONS)
  NT ELECTROPHOTOMETRY
     FLAME PHOTOMETRY
     OPTOMETRY
     PHOTOMETRY
  BT MEASUREMENT
  RT CHEMICAL ANALYSIS
     CHROMATICITY COORDINATES
etc.

OPTICAL MICROSCOPES
  RT ELECTRON MICROSCOPES
OPTICAL PROPERTIES
  (INCLUDES PROPERTIES OF
  VISIBLE, INFRARED, AND
  ULTRAVIOLET
  ELECTROMAGNETIC WAVES
  AND THEIR EFFECTS)
  UF OPTICAL TRANSMITTANCE
  BT PHYSICAL PROPERTIES
  RT ALBEDO
      BIREFRINGENCE
      BRIGHTNESS
      DICHROISM
      DIFFRACTION
      DISPERSION (WAVE)
      ELECTROMAGNETIC
        ABSORPTION
      GLOSS
      INTERFERENCE
etc.
OPTICAL TRANSMITTANCE
  USE OPTICAL PROPERTIES
    TRANSMITTANCE
OPTICS
  (USE MORE SPECIFIC TERM
  IF POSSIBLE)
  RT ANGLE OF INCIDENCE
      ASTRONOMY
      BEAMS (RADIATION)
      DISPERSION
      DISPERSION (WAVE)
      ELECTRON OPTICS
      FARADAY EFFECT
      FIBER OPTICS
      FOCAL LENGTH
      FOCUSING
      LIGHT (ILLUMINATION)
      MAGNETO OPTICS
      MAGNIFICATION
      MIRRORS
etc.

Almost all professional journals require author-written abstracts. This is a healthy situation, for the abstract is the most important section of an article. Certainly, it is the section most read.

Basically, there are two types of abstract. The *descriptive abstract* discusses the article itself and is written for the use of research personnel and reference librarians. The *informative abstract* discusses the work and its findings. It is written for professionals in a discipline who do not require the full contents of a paper because they are capable of understanding a condensation. Only rarely are the two types combined in the same section.

It is not unusual, however, for editors of journals of applied science to require both types of abstract *in the same report.* The following examples, reproduced by permission of the Hamilton-Standard Division of the United Aircraft Corporation, are from the same report and illustrate the difference between types of abstract.

DESCRIPTIVE ABSTRACT

This report describes the flight testing of four Integral Oil Controls installed on two propeller designs, the 24260/2J17H3-8W and the 34G60/C7021C-8. This equipment was flight tested on an Air Force KC-97G airplane equipped with Pratt and Whitney Aircraft R4360-59B engines.

INFORMATIVE ABSTRACT

The flight testing of four Integral Oil Controls was conducted on an Air Force KC-97G airplane equipped with Pratt and Whitney Rf360-59B engines. The first tests were conducted with four specially prepared and instrumented Integral Oil Controls installed with four 24260/2J17H3-8W propellers. The tests were then repeated using four 34G60/C7021C-8 propellers with the same controls. The flight tests were designed to simulate a Strategic Air Command mission, and certain additional operational conditions were introduced to provide a full range of control data.

The results of these tests may be summarized as follows:

1. There was no significant difference in engine torque between the 34G60 and the 24260 installations.

2. There was no consistent correlation between prominent control vibration frequencies and pressure pulsation frequencies. Pressure pulsation frequencies were predominantly two times engine shaft speed (2N); control amplitudes at this frequency were generally minor.

3. The pulsating and steady pressures in the 34G60 installation were considerably higher than those in the 24260 installation.

4. The frequency of pressure pulsation for all test configurations varied in the range of 55 to 90 cycles per second. During takeoff, the frequency increased with the increasing high pitch pressure and appeared independent of either engine speed or control vibratory motions.

Analysis of the examples printed above shows the characteristics of each type of abstract. The *descriptive abstract* states what the report contains. The standard opening "This report describes ..." permits the writer to catalog the precise areas discussed. In a sense, these areas are key words or uniterms. They serve to guide a reference librarian who wishes to cross-reference the article, if key words and uniterms have not been required. The rule of thumb in writing a *descriptive abstract* is to *talk about the article.*

The great majority of professional journals require *informative abstracts.* This type of abstract talks about the research and its findings (as opposed to what the article "describes"). In the example printed above, the te‚ts are described and the results are summarized. Although the listing of results in a vertical, numbered sequence is not acceptable practice in some journals, it is standard procedure in others.

An *informative abstract* must be complete and self-explanatory. It is more than an expansion of the title and a condensation of the article; it often takes the place of the article; for example, the abstracts of doctoral theses and dissertations are separately printed, as indicated in Chapter 2. The abstracts of papers to be read at symposia and professional meetings are often collected and released prior to the meeting. Perhaps most important, author-written abstracts are usually collected and reprinted in abstract journals. Figure 12 is an example of author-written *informative abstracts* reprinted in *Nuclear Science Abstracts.* Figure 11, however, is a montage showing sample collections of abstracts of articles in the earth sciences. Some collections require *informative abstracts* and others require *descriptive abstracts.* Although *Chemical Abstracts* prints only *informative abstracts* (which are required for articles submitted for publication in any journal published by the American Chemical Society), annotated bibliographies such as those compiled for *Economic Geology* will be descriptive in nature (see Figure 11) and print only *descriptive abstracts.* The reason, of course, is the intended audience. Research scholars and reference librarians use bibliographies and expect *descriptive abstracts.* Professionals capable of understanding condensations of articles read collations such as *Chemical Abstracts.* They expect informative abstracts which discuss findings and results as opposed to contents and chapter headings.

In the applied sciences, in which reports rather than articles are often the accepted form of communication, an atypical pattern of abstracting pertains. Because the federal government sponsors a lion's share of research in applied fields—and requires reports upon that research—its patterns are followed in most industrial and in-plant situations. They are also followed by the Department of Commerce Clearinghouse for Federal Scientific and Technical Information. The discussion below is from the Armed Services Technical Information Agency (now the Defense Documentation Center) publication entitled *ASTIA Guidelines for Cataloging and Abstracting,* 1962.

ABSTRACTING

The primary reason for abstracting reports in the ASTIA collection is to supply the reader with information which will assist him in determining whether or not he needs the report. Although the abstract should be informative in style, it is not intended as a replacement for the report itself. The abstracts are reproduced in the *Technical Abstract Bulletin* for announcement purposes. It is intended that the information will be stored in an electronic computer so that at some time in the future complete bibliographies may be prepared which will give the descriptive cataloguing, the abstract, and the descriptors.

A. Abstract Elements

1. Reports of Monographic Nature. An adequate abstract for this type of report should include four items: what was sought, how it was sought, what was learned, and what can be concluded therefrom. Such components might logically be termed the purpose, the method, the results, and the conclusion.

a. *Purpose.* A beginning statement for an abstract of a technical report should inform the reader of the reason for the investigation. The abstractor should not, however, merely rephrase the title of the report. If the title is sufficiently informative, the abstractor can begin immediately to tell how the research was accomplished.

b. *Method.* The reader should be told briefly what was done in the investigation, how it was done, with what materials or data, and under what circumstances. This is most frequently the phase where the abstractor must weigh his words and be brief.

c. *Results.* What was learned from an investigation is probably the most important material to include in an abstract. Frequently, however, there are too many specific results for inclusion. To prepare a fully informative abstract, the abstractor should reflect that which was found to be new in the field under investigation. That is what the reader will want to know.

d. *Conclusions*. The abstractor should include a statement as to what may be concluded from an investigation, what it meant, and how it may be of value or interest to investigators in similar fields of endeavor.

2. Reports of Serial Nature. The first progress report on a contract should give the purpose of the investigation being made. Succeeding reports, however, should not repeat the purpose unless the scope of the problem changes. The items comprising the abstract for a progress report are essentially the same as those for a technical report, except that conclusions frequently cannot be made for the progress report. The progress report, by its nature, records progress made during a specific interval of time; pertinent conclusions can be made only in a final report or in a technical report on a single phase of an investigation.

3. Evaluation or Qualification Tests. Reports on the testing of items of interest to agencies of the Department of Defense can be abstracted by summarizing (1) the results of the tests, and (2) the recommendations as to disposition of the items.

4. Bibliographies. A bibliography can be abstracted by citing the scope of the bibliography and giving the time span covered by the entries and their method of arrangement, i.e., whether by topic or alphabetical by author.

B. Security Classification of Abstracts

The security classification of an abstract for a classified report should be clearly indicated in parentheses at the end of the abstract.

Frequently the abstract may have a lower classification than the overall report. Bibliographic and reference services are aided whenever this is possible. Originating offices are urged to prepare an abstract having the lowest security classification, which will permit the inclusion of the information elements of abstracting listed herein.

C. Machine Language for Abstracts

In order that abstracts may be machine-stored for ASTIA purposes, the originators of the reports are requested, insofar as possible, to write the abstracts in machine language. For the present, there is capability for only fifteen symbols other than ordinary letters and numerals. These are the period, comma, semicolon, colon, apostrophe, hyphen, percentage sign, parentheses, dollar sign, item number sign (#), ampersand, plus sign, asterisk (*), and the slant sign or virgule (/).

An introductory section—with or without heading—is essential to any article longer than a note. It not only introduces the reader to the work under discussion but also to the style and tone of the author. For this reason, it should neither be pompous and puffed up with the whole efflatus of acknowledgments nor make elaborate claims for the need or significance of the work. It is still possible to find articles beginning "This article fills a long-felt need for...." Nothing, however, has quite equaled the recent opening gambit "There are five living economists of importance, and we feel...." A writer who feels compelled to use such an opening should recall that an offensive or supercilious tone in the beginning of an article will dissuade most readers from reading further.

There are essentially six areas an introduction must cover if it is to prepare the reader for the material and to relate the work to the field.[1]

1. The status of the problem prior to the present research.
2. The purpose of the investigation precisely defined.
3. The conditions under which the work was done and the procedure, if unusual.
4. The scope of the present work and its connection with the general problem.
5. Recognition of similar work on the subject.
6. Significance of the material treated.

The *status statement* not only states the problem or topic clearly but also indicates how the present state of the art was reached. The review should assume an intelligent reader. No elaborate tracing of history or defining of terms need be done. In effect, only as much background as is necessary to an understanding of the problem should be given.

The *purpose statement* should orient the reader as to why the work was done, what the goal was, and how the goal differed from that sought by other work.

The *method statement* again assumes an intelligent reader capable of recognizing the usual techniques of a discipline. He wants, in this section, to learn only the variables or conditions under which work proceeded. Unusual methodology should be called to his attention, but full descriptions should be placed in the appropriate section.

The *scope statement* should cover both the work and the article. A reader should be told at the outset of any limitations or omissions. He will be angered if he reads the entire article without finding something he had hoped to find. He must be told the exact phases of the general problem

[1]Scientific and Technical Information Division, National Aeronautics and Space Adminstration, *NASA Publications Manual,* NASA SP-7013 (Washington, D. C., 1964), p. 22.

covered by the article. He should be told whether the work was experimental or theoretical.

The *recognition statement* equates the work with other work in the field. Any evaluations or comparisons should be objective and essential to understanding; they should be made only to show the significance of the present work.

The *significance statement* should warn the reader of flaws in sampling or other techniques. It is important that a writer point out faults in his own work; if he fails to do so, others will delight in doing it for him.

Most writers will need to condense their introductions. There is a tendency to write essays upon each of the six areas described above. This wordiness not only delays the reader, it also convinces him that the writer does not regard him as a professional correlative.

Examples of succinct introductions are printed on pp. 253,260, and 310.

Article 1, Section 8 of the Federal Constitution guarantees freedom of the press. Professional journals qualify for federal protection, but their articles are frequently emasculated prior to printing by reviewers concerned with security, rights in data, proprietary and patent regulations. Security regulations are absolute. They fix practice in the collection, preparation, storage, and dissemination of classified information. They are too complex to discuss here. Writers working in sensitive areas may request from the Department of Defense through the Superintendent of Documents, Government Printing Office, a summary (DD-441 and attachments) of the nine federal statutes and executive orders relevant to classified information. This summary, together with those of other agencies such as the Atomic Energy Commission, should be strictly observed. In the realm of security it is interesting to note that journals, such as *Mechanical Engineering,* which cover developments close to classified and patentable areas enjoy a wider foreign than domestic circulation.

Writers and research persons can easily recognize classified documents —as opposed to information. All contain the following statement taken from the Department of Defense *Industrial Security Manual.*

"This document contains information affecting the national defense of the United States within the meaning of the Espionage Laws, Title 18, U.S.C., Sec. 793 and 794. Its transmission or the revelation of its contents in any manner to an unauthorized person is prohibited by law."[2]

[2]Missiles and Space Division, Lockheed Aircraft Corporation, *Style Manual,* LMSD-5000 (Sunnyvale, California: Lockheed Aircraft Corporation, December, 1958), pp. 3–4. Reproduced by permission.

Exclusive possession of data, know-how, or trade secrets constitutes a tremendous competitive advantage to a company operating in a free-enterprise system. For this reason, information that might otherwise find its way into professional journals is suppressed by the company. For example, in proposals and bids for contract work for the federal government, proprietary data need not be furnished unless specifically negotiated for. Even when a company wins a contract from the government, its rights in data remain—although copying by competitors is legal if an item or part can be reproduced by reverse engineering. An article which discusses new items or parts is seldom released by a company unless the company enjoys a clear-cut patent position.

Documents that contain rights in data carry either of the following notices on the title page.

NOTICE

"These data shall not be disclosed outside the government or be duplicated, used or disclosed in whole or in part for any purpose other than to evaluate the proposal, provided that if a contract is awarded to this offeror as a result of or in connection with the submission of such data, this legend shall be of no force or effect, and the Government shall have the right to use the data for any purpose except as otherwise provided in the contract. This restriction does not limit the Government's right to use information contained in such data if it is obtained from another source."

NOTICE

"Furnished under United States Government Contract (place contract number here) and only those portions hereof which are marked (for example, by circling, underscoring, or otherwise) and indicated as being subject to this legend shall not be released outside the Government (except to foreign governments, subject to these same limitations) nor be disclosed, used, or duplicated for procurement or manufacturing purposes, except as otherwise authorized by contract, without the permission of (name of the originating company). This legend shall be marked on any reproduction hereof in whole or in part."[3]

No process, machine, manufacture, or composition of matter may be patented if it has been described in a printed publication in this or a for-

[3]Lockheed, *Style Manual,* p. 5.

eign country more than one year prior to the date of application for patent. For this reason, writers of professional articles have an obligation to industry, to private designers, and indeed to any source of patentable information. Further, if writers discuss their own patentable research findings, they have an obligation to themselves. It is unwise to rush into print before filing for patent.

A writer who serves as a consultant is sometimes at a disadvantage; he knows more than he ethically can discuss. Another writer not so encumbered can make an educated guess and publish a thin but creditable article.

A word about *shop right* is necessary here. Shop right is an employer's right or license to use any invention made by an employee during working hours or with the employer's tools or materials. A consultant is legally an employee. If he did appropriate an idea and patent it in his own name, the patent would be reassigned to his employer. Because shop right sometimes extends to work done on federal or foundation grants, patents growing out of such work often must be assigned to the sponsor. A writer who reports the results of sponsored research should determine if he is reporting patentable information and if the sponsor wants it reported. The federal government, which created the Patent Office to promote the distribution of knowledge in "science and the useful arts," usually allows information to be reported so that no single person may patent it.

Proprietary information is a term loosely applied to any information unique to a company, laboratory, or individual. Rights in data and unpatented but novel processes, machines, manufactures, or compositions of matter constitute proprietary interests. Writers should respect the proprietary interests of others. Failure to do so will cause them to be regarded as spies.

Industrial espionage is an incredibly sophisticated and lucrative business. The security officers of federal laboratories admit that they fear industrial espionage from within more than they fear international espionage. As a result, security officers have become "editors" in many of the applied fields of science and technology.

Documents which contain proprietary information carry a statement similar to the following on the title page.

NOTICE

"The information and design disclosed herein were originated by and are the property of Lockheed Aircraft Corporation. Lockheed reserves all patent, proprietary, design, manufacturing, reproduction, use, and sales rights thereto, and to any article disclosed therein, except to the extent rights are expressly granted to others. The foregoing does not apply to vendor proprietary parts."[4]

The word "proprietary" is also applied to *trade names*. When a writer wishes to name an item, he is free, as a private citizen, to use a trade name subject to trademark regulations. When a writer is associated with an agency or federal facility, however, he cannot put the agency or facility in the position of seeming to endorse the item by using the trade name. As the following passage indicates, even private citizens should not play fast and loose with trade names. "Reciprocal courtesy and a desire to stay out of trouble requires caution of anyone who uses trademarks belonging to other companies. Some companies bristle at misuse of their trademarks; they retain legal counsel just to warn or sue anyone who misuses the trademarks. He who uses some trademarks as common nouns will soon get an unpleasant legal education."[5]

Historically, the Bureau of Standards has been a prime target for litigation arising from the use of proprietary or trade names. Profiting from the experience of its sister agency, NASA included the following statement in its *Publications Manual*.

PROPRIETARY IDENTIFICATIONS

"As a responsible agency operating in the public interest, NASA considers it improper to advertise, endorse, or criticize commercial products in its formal publications available to the general public. Cases in which trade names and similar identifications cannot be replaced by adequately descriptive generic terms should be extremely rare. Evaluations of proprietary products in a report may introduce delays in dissemination of the results to allow review by the manufacturers involved or may require special limitations on distribution of the report. Regardless of how convenient authors may find the use of commercial identifications, attention should in every case be given to their possible deletion for the following reasons:

"1. Use of commercial identifications serves, in effect, to publicize a product and could be interpreted by competitors as endorsements by NASA of a manufacturer's product.

"2. Evaluations that show a commercial product in an unfavorable light

[4]Lockheed, *Style Manual, p. 5*.

[5]R. Hays, *Principles of Technical Writing* (Reading, Mass.: Addison-Wesley Publishing Co., 1965), p. 160.

may be objectionable to a manufacturer and could involve NASA in troublesome controversy."

The majority of writers wishes copyright protection for articles, and the majority of professional journals give that protection. The "boiler plate" printed at the bottom of the table of contents of a journal states whether or not the articles in that issue are protected by copyright.

Theses and *dissertations* should be copyrighted. Individuals may secure their own copyrights by following the instructions contained in Copyright Office Bulletin No. 14, *Copyright Law of the United States of America,* Government Printing Office ($0.20). A simpler procedure is indicated in the following passage from the University of Massachusetts Graduate School's *Instructions for Thesis Typing.*

"As a general rule, all students are urged to copyright their dissertations. Dissertations filed without copyright go into public domain once they appear on the library shelves. Students in the College of Agriculture whose investigations were supported wholly or in part by Federal funds are not permitted to copyright the results of their investigations.

"Students who wish to copyright should insert a page immediately following the title page containing a copyright notice.The copyright notice should give the full legal name of the author and the date of publication on microfilm.

"It should appear on the page as follows:

Copyright by
John Arthur Brown
1966

The dissertation should be forwarded with either a cashier's check or a postal money order made out to *University Microfilm, Inc.* in the amount of the microfilm fee calculated as follows:

Microfilming (Includes publication of abstract in *Dissertation Abstracts.*)	$20.00
Copyright fee	5.00
Cost of two copyright copies	

(University Microfilms will file these two copies with the Library of Congress. Cost is $0.025 per page. Two copies are required.)"

In the event that a writer wishes to release an article containing rights in data, patentable information, or proprietary information or if he wishes to copyright the article, he should place the appropriate statement(s) between the title page and the table of contents if any.

A section devoted to the statement of the problem is optional in most journals. Usually the introduction section contains the essential information.

The techniques for defining a problem have been discussed earlier. In the case when the style guide of a specific journal requires restatement of the problem as a specific section in an article, emphasis should move from the hypothetical and theoretical to the specific and applied. The writer who originally defined a problem as "to determine the physiological effect of skin penetration, inhalation, and ingestion of DDT in a 5-percent wettable powder upon humans and domestic animals" should now clarify problems of generics and of word order. In so doing he should also clarify those aspects of methodology and scope that might be ambiguous or misleading.

For example, the phrase "physiological effects" would perhaps be limited in scope to heart, liver, and central nervous system. "Skin penetration" might be limited in method and scope to adsorption, subcutaneous injection, and abrasion. "Domestic animals" might be limited in scope to three yearling Portland China pigs and five adult male cats.

Atypical samples and situations which must be reviewed later in "Discussion of Results" should be mentioned here. For example, the six humans who ingested 0.5 cc of 5-percent DDT might range in age from undergraduates to a department head—and the department head might already be ulcerous. Similarly, methodology should also receive limited discussion but should not be explained in this section. Structurally, the section should be organized to lead logically to a discussion of research technique.

The object of the "experimental" section is to present so detailed a discussion of methodology that others could repeat the work. The presentation may be correlated functionally, physically, or chronologically with the techniques discussed in Chapter 3.

The major preoccupation of this section should be with accuracy, and the generalized phraseology common to most process descriptions should be made specific; for example, citing the quantity of a specific material is not enough. Where that material was obtained, how its degree of purity was established and maintained, and even what it cost are essential information.

The description of apparatus must be made clear. Standard terms for standard equipment should be used, and illustrations that show the entire construct should be furnished. In specialized and nonstandard cases, both specific and trade names should be given. Sources and perhaps list prices should be mentioned.

The following example[6] discusses a single laboratory device. Discussion of a more complex device or series of devices would be correspondingly longer. (The footnoting in the article is standard in the field of chemistry.)

IMPROVED CARBON DIOXIDE PRESSURE REGULATOR

by Carmine DiPietro, Warren A. Sassaman, and Charles Merritt, Jr.

"In the determination of nitrogen by the micro-Dumas method, the pressure of carbon dioxide delivered to the combustion tube is ordinarily regulated by use of a column of water[3] or mercury.[1,3,5] Since the water head is somewhat cumbersome, it has been replaced mainly by devices using mercury. One type consists of an open capillary T-tube extending a predetermined depth below the surface of mercury in a cylinder to give the desired gas pressure and allowing the CO_2 to escape. This results in the scattering of fine particles of mercury, thereby constituting a health hazard.[1] Hershberg and Wellwood[1] substituted a modification using a filter paper valve which allows the CO^2 to escape but retains the mercury at the desired level. In other modifications, a porous fritted glass disk was substituted for the filter paper,[2] and to keep traces of mercury vapor from contaminating the laboratory, activated charcoal impregnated with iodine[4] was placed between two porous disks.[5]

"The mercury-type regulators are compact and the desired pressure can readily be adjusted. However, the oxides which form in the mercury eventually fill the pores of the filter paper or fritted disk, preventing the escape of excess CO_2 and thus forcing the stopper out of the CO_2 generator which is usually a Dewar flask containing Dry Ice. It then becomes necessary to disengage the regulator, change the filter paper or clean the disk, reassemble the apparatus, and allow time for the air to be flushed out of the system. Furthermore, there is no assurance that the same situation will not recur within a short time.

"To avoid disruptions of the nitrogen determinationa and subsequent delays due to the outlet becoming plugged, a simple and more dependable regulator has been designed. A schematic diagram of the device is shown in [the accompanying drawing]. It consists of a U-shaped manometer and a wide bubble tube partially filled with mercury. The bubble tube was a

[6]C. DiPietro et al., "Improved Carbon Dioxide Pressure Regulator." Reprinted from the *Microchemical Journal,* Vol. IV, Issue No. 1. Copyright 1960 by Interscience Publishers, Inc., New York.

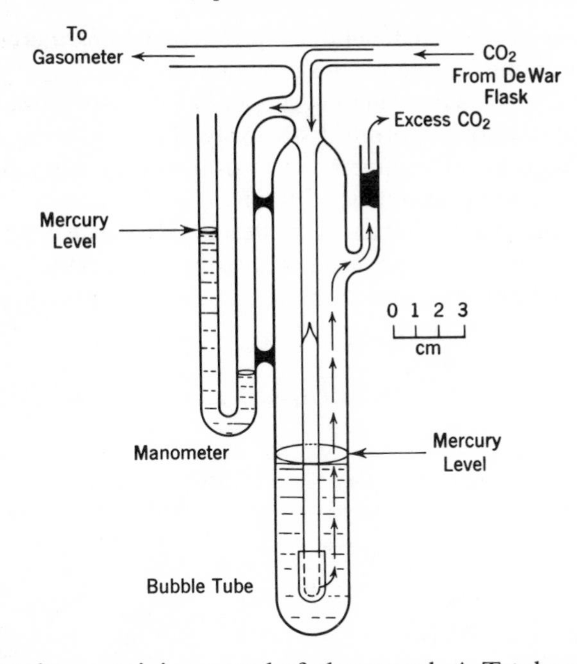

small outlet tube containing a wad of glass wool. A T-tube with the third arm "A," made to extend almost to the bottom of the bubble tube, is closed at the lower end. This arm is slotted about 12 mm from the end and a piece of rubber tubing with a crosswise slit, approximately 8 mm long, is fitted over the gas arm so that the openings coincide. At the upper end, the third arm is then sealed into the bubble tube. Excell CO_2 escapes from the system as small bubbles through the slit rubber opening, leaving the apparatus via the side arm where the glass plug prevents the passage of particulate mercury to the work area. The amount of mercury vapor being carried by the excell CO_2 into the work area has been found to be negligible* and no health hazard is involved.

"Final pressure CO_2 in the system is determined by the mercury height above the rubber valve in the bubble tube and may be measured by means of the manometer. For the apparatus dimensions shown in [the diagram], pressure, equivalent to about 3 in. of mercury, is available. For convenience in making the initial assembly of the device to the micro-Dumas train, it is desirable to fill the bubble tube with a smaller amount of mercury than is finally required to obtain the desired pressure. Then, after the

*A General Electric Co. Mercury Detector, Type B, Cat. No. 5142355G2, using selenium coated paper was set up directly in front of the side arm outlet. There was no apparent discoloration of the paper after 12 hours, indicating no appreciable transfer of mercury vapor to the laboratory. Only very slight discoloration occurred after 72 hours.

apparatus is in position, the height is adjusted by adding mercury through the side arm until the desired level is attained.

"Preliminary designs omitting the rubber valve or substituting a plastic tube with a slit did not perform satisfactorily. The rubber valve is necessary to allow CO_2 gas to escape at an even rate, thus maintaining nearly constant pressure in the system at all times.

"Although this regulator has been designed for use with the micro-Dumas Apparatus, it is also used with other apparatus requiring the delivery of a laboratory-generated gas at constant pressure."

REFERENCES

1. Hershberg, E. B., and Wellwood, G. W., *Ind. Eng. Chem., Anal. Ed.*, 9, 303 (1937).
2. Stehr, E., *Ind. Eng. Chem., Anal. Ed.*, 18, 513 (1946).
3. Steyermark, Al., *Quantitative Organic Microanalysis,* Blakiston, New York, 1951, P. 60.
4. Stock, A., *Angew, Chem.*, 47, 64, 184 (1934).
5. Zimmerman, W., *Microchèmie,* 31, 42 (1943).

Pioneering Research Division
Quartermaster Research & Engineering Center
U.S. Army
Natick, Massachusetts
Received November 20, 1959

The experimental section should also furnish a description of how the work was done. This is more than a process description. Citations must be given for any established or derivative procedures. The mensuration techniques and their degree of accuracy must be shown, and the extent to which the procedure follows another already reported in the literature must be indicated.

Whenever the article discusses work that is theoretical rather than experimental, all formulas, equations, and background data are included in this section. Derivations of formulas and equations are placed in this section if they are not too lengthy. In the event their length or their inclusion constitutes a digression, derivations may be placed in the appendix.

Warnings must accompany discussions of hazardous procedures. Legal adjudications and insurance coverages vary widely in laboratory situations. One point is clear, however. If a technician harms himself, others, or equipment while following a written procedure, he will seek reprisal. The writer will escape if he can show negligence on the technician's part, but negligence can only be shown if the discussion is clear and cautions and warnings have been employed. Courts usually adjudicate against ambiguous discussions on the grounds that *a writer is responsible for his own words.*

An objective presentation of the results must be given. Objectivity is never easy to attain; it is particularly difficult to achieve when the writer discusses his own work. Emotional involvement with something that has taken time or that might affect reputation can lead a writer unwittingly to color or even to "rig" his results.

The pioneering work in communication theory of the late Dr. Irving Lee at Northwestern University has shown that both writers and readers unconsciously level, sharpen, and assimilate data. While the resulting distortion may communicate with greater impact and even with reassuring logic, it usually has drifted away from truth and is only a distortion. Lee's work showed the following: (a) Data are "leveled" when either writer or reader feels that a set of facts is neither significant nor relevant. (b) Data are "sharpened" when either writer or reader feels that another set of facts is keenly important and should be stressed. (c) Data are "assimilated" when either writer or reader feels that the data should tell a story; that is, they should form a meaningful whole, lead to a conclusion, or prove a point. Lee's conclusion was that it is possible for either writer or reader unconsciously to "level" out the important data, to "sharpen" the unimportant data, and to "assimilate" a conclusion from an entirely subjective interpretation of "facts." A writer can be more objective, perhaps, if he appreciates Lee's findings and overcompensates by reporting everything without interpretation. To reverse a canon of journalism, "the unrelated fact is always worth mentioning."

The Results section, therefore, should be written without interpretation. All analysis belongs in the following section, Discussion of Results (see below). The Results section should be a flat statement of what each part of the research discussed in the preceding—Experimental—section showed. The statement will be assimilated—that is, it will form a whole—if it reflects the logic that binds the Experimental section together. For this reason, it is often easiest to write up results as a sequitur to the research pattern already described to the reader.

The organization of the Results section is illustrated in Chapter 5 by an article that discusses five experiments. Each experiment has its own Results section, because each experiment sought different information. Organization thus reflects methodology.

Content of the Results section, naturally, will vary. In all cases, however, data should be introduced in as concise a manner as clarity will permit. A conscious attempt to be brief is healthy; it inhibits the desire to interpret or to comment.

A tabular form should be used in the Results section whenever possible. This form not only compresses data but it also permits readers to use the data for a variety of private purposes. This aspect is particularly important in articles submitted to journals, such as *Electromechanical Design,* whose orientation is interdisciplinary. Notice may be taken, without comment, of specific and significant aspects of tables, graphs, and other illustrations. Care, of course, should be used that results and numerical values so noted agree precisely with the tabular presentation.

When research produces peripheral data outside the main purpose and of possible interest to only a few readers, that data may be omitted to save space. Its omission and availability should be indicated by a statement such as the following.

> "The basic results are presented in the graphs of temperature as a function of dynamic pressure. Comprehensive tables of the test data are available upon request. A request form for this purpose is bound in the back of this report."[7]

Techniques for organizing tables and other illustrative materials are given later in this chapter. Note should be made here, however, that many journals, particularly those in the field of chemistry, require that all numerical data be reported in the metric system.

Just as the section on Results grows organically out of the Experimental section, so the Discussion of Results section should reflect the organization of the Results section. The linkage of these three sections may be either functional, physical, or chronological. At all costs, however, the succession must be logical. In the event that the journal to which the article is to be submitted requires a Conclusion section or a Recommendation section, the logic of the Discussion section may anticipate them and show that the conclusions and recommendations are warranted.

Content in the Discussion section will vary usually with the results. The results have been stated earlier; restatement is unnecessary, but evaluation is vital. A writer should defend or attack his findings as if his worst enemy had written them. This will make the defense factual, hence objective or "professional." In all events it will preclude elaborate claims that the work is the most significant thing since Galileo. Similarly, the attack will steal the thunder of others who might want to comment negatively.

This does not mean that a writer should be humble; persons writing up their own research seldom are. It simply means that the tone should not

[7]Scientific and Technical Information Division, National Aeronautics and Space Administration, *NASA Publications Manual,* NASA SP-7013 (Washington, D. C., 1964), p. 31.

be apologetic. Alibis have no place in the Discussion section; the problem was solved or it was not. Partial solutions permit partial claims, but the Discussion must show limits in findings, samples, or conclusions. The tentative language of announcements in journals of medicine should be used when it is necessary to qualify the significance of data.

Style in the Discussion of Results should be almost clinical. If possible, data should be explained according to the laws of chemistry, physics, or the discipline involved. The symbols and language of mathematics/statistics should be used. This language will ensure that discussion is more than repetition of the detailed data on graphs and tables. Comparisons, relations, conclusions, and generalizations are more meaningful when given a mathematical basis.

Comparisons with the work of other investigators should be made. Citations and credits are necessary here, particularly if the discussion is pejorative. Fair use of the work of others is ethically essential, however. When data from the literature are given, the purpose of the research that produced those data for another should be mentioned to show the pertinence and relative accuracy of the work.

Application of findings to other research areas should be indicated. Regardless of the discipline, theoretical and practical implications should be shown. The method of the application may be suggested or discussed, but the drawing of conclusions and the solving of other problems constitute a digression. A separate article should be written in these cases; only projected applications should be made in the parent article.

Projection should not be in the form of a promised article, because redirection of work activities often leaves these promises unfulfilled. Reference to work in progress should be made in guarded terms to avoid inquiries that might be embarrassing.

Controversial work does not justify the use of controversial language. When there is a difference of opinion among scientists or in the literature, fair comment in the light of present data may be made. *Ad hominem* arguments do not constitute fair comment. Data and ideas are important; men and their motives are not. Thus, when the data under discussion apply to a controversial area, a writer should make the application (winning both friends and enemies) but leave the conclusion to others or to another article. The latter is the preferable procedure because a logical reorientation of data is usually required, and additional data, often the work of others, may be necessary for reinforcement.

The solution of the problem defined in the opening section of the article remains the goal and the theme of the article. Thus the Discussion section must reach some sort of conclusions. The problem has either been resolved or it has not. If not, a partial solution has perhaps been furnished, and recommendations may be made as to a possible path to the solution. If the work failed or if the method and data are ineffective, a positive statement to that effect should be made. The discovery of procedures that will not work is almost as important as the discovery of those that will. In logic and in research, definition by negation is a standard technique.

Conclusions should be qualified, however, when there is a gap in the data, reservation on the part of the writer, or the possibility of criticism of technique. One essential form of qualification for a conclusion is the precise definition of the limits within which the conclusion applies. The range of validity of a conclusion should be stated objectively, and *no additional claims should be made.*

Those journals that require a formal section headed "Conclusions" are seeking numbered itemizations of what has been concluded. Argument and justification have no place in this section; they belong in the preceding Discussion section.

The content of the Conclusions section is limited by the Results and the Discussion sections. *Conclusions may be drawn only from the data presented in the article.* There is a high probability that the sun will rise at a predictable time on a specific day, but unless an article proves the statement with data and supporting references, that conclusion cannot be drawn. In effect, a conclusion may only be stated *after* it has been defended. Certainly, no new material may be presented in this section.

Like the Abstract, the Conclusions section should be self-contained because it is sometimes published separately. Further, specific conclusions are often quoted verbatim in other documents. This situation makes it essential to phrase conclusions so that limits and qualifications are apparent.

When no specific conclusion can be drawn, a brief statement of what has been accomplished should be made. Such a statement should close with a restatement of the present status of the problem. Although this procedure gives the reader no tangible or specific findings, it does leave him with a sense of logical and thematic unity.

Probably the best way to write a Conclusions section is to do so only after rereading the Statement of the Problem and the Discussion of Results. This will usually limit a writer to reality; at least it will dampen his enthusiasm.

Although very few journals require a section devoted exclusively to recommendations, almost all will print articles that contain such a section. Further, few articles are organized to lead logically to recommendations, but editors who feel that additional positive statements will increase the effect of an article will accept a recommendation section that suggests applications in other areas or fields.

A writer who chooses to add a Recommendations section should be sure of three things. First, all recommendations must be logically consistent with the article. Second, all recommendations must be based upon data contained in the article. And, third, all recommendations must be professional rather than tactical or political.

In practice, particular types of articles lend themselves to a Recommendations section. Work in the life sciences, for example, is often tentative. When a professional publishes an article about an advance in the field, he usually limits his conclusions carefully. From these limitations and qualifications grow recommendations as to how the conclusions might be expanded. This is both a logical and a professional procedure, for the recommendations are upon stated conclusions and thus serve further to alert the reader to the limited applicability of those conclusions.

In the applied sciences the standard reporting sequence requires a Recommendations section; for example, through specification MIL-M-005474C, the Department of Defense requires that reports contain the following nine sections in the sequence shown.

Cover
Title page
Letter of transmittal
Abstract
Table of contents
Text
Conclusions
Recommendations
Appendix

No variations from this list and sequence may be made. In the event that work reported upon failed or was inconclusive, the appropriate entry in a Department of Defense Report would be "Recommendations: None."

When a similar situation occurs in work being discussed in article form for a journal that requires a Recommendations section, a statement that no recommendation can be drawn from the data should be made. This statement should be followed by the general recommendation of whether or not the work should be continued. If this recommendation is for continuing, specific recommendations should indicate the direction the work might take. There is nothing unprofessional about this. The writer who has done the work is usually the person best informed upon the topic. No one is better qualified than he to make such a recommendation.

Gratuitous recommendations should never be made. Such recommendations are seldom justified by the text, they are usually non sequiturs in logic, and they will be viewed with suspicion by a reader who feels that an article should advance the state of the art and not the career of the writer. It was possible for Cicero to end every Senate speech, regardless of content, with "Carthage must be destroyed." Cicero was then acting as a politician talking to other politicians. He wrote logically, objectively, and well when he had other audiences.

Articles in the fields of psychology and psychiatry pose a different problem. An article in these fields is often a statement of *opinion*. Although the work, like most scientific work, has a goal of prediction and control, formulations cannot always be closed with *quod erat demonstrandum*. When a subject's total functioning is involved, an article is usually too thin in content to merit the prediction of future performance and the recommendation of future therapy. For this reason, the Recommendations section of an article in the earth sciences is most likely to contain positive and unqualified statements. The same section of an article in the life sciences will usually make limited recommendations.

Finally, the Recommendations section of articles in the social sciences will usually be hedged with the elaborate impediments and jargon of the professional fields. Partly to avoid this, the American Psychological Association recommends the following five sections as meeting the requirements for articles submitted to the professional journals of the field.

Problem
Method
Results
Discussion
Summary

As a general rule, anything that might seem a digression in the Text or Discussion section of an article is placed in the Appendix. Because most journal editors realize that digressive material dulls the effect of an article, most journals apply the rule and permit the use of an Appendix section.

Content in the Appendix is usually subdivided with appropriate headings and subheadings. Possible content includes the derivation of equations, multiple diagrams, figures containing whole families of curves, foldout diagrams, before/after photographs, and even price lists.

In the applied sciences and in areas in which an interdisciplinary application is possible the Appendix is often the longest section in the article.

Although it is possible to argue that all data are important and have other applications, by limiting space arbitrarily, some editors resist the desire of writers to use everything and to throw nothing away.

Organization of the Appendix is random; it reflects the paste pot more than any form of logic. The determining factors seem to be what a writer feels is important and what an editor will agree to print. Referencing in the Text or Discussion sections as well as cross referencing in the subsections of the Appendix may lend a spurious unity. Such linkage is essential, however, and writers should justify the inclusion of the material in the Appendix by showing how that material expands the applications of the research.

Accurate documentation is essential to professional writing. The legal requirements of copyright, the moral aspects of plagiarism, and the professional responsibility of giving credit (and responsibility) to others all contribute in making documentation essential. There is no need to review the evolution of documentary systems or to recount anecdotes about research charlatans—some of whom were magnificent rogues. It is only necessary to accept the existence of archaic, mutually exclusive customs of documentation so ingrained in every profession and discipline that standardization of techniques can come only through revelation. Until that time, a writer must accept the customs and usages of separate disciplines and journals as an act of faith.

Because of the wide variety of styles involved, the documentation practices of the various disciplines have been anthologized in Chapter 5 of this book. A person writing an article *in a specific discipline* will find the appropriate style in that chapter *segregated by discipline.* If he intends his article to be interdisciplinary, however, he should examine recent issues of the journal to which he intends to submit his article. There he will find examples of the documentation techniques *appropriate to that journal.*

The present section discusses the "logic" of documentation and attempts partially to indicate the variations illustrated in Chapter 5.

There are two types of footnote. One supplements or clarifies the text by explaining it; the other validates the text by citing an authority. In the jargon of scholarship the first is a "footnote" proper, and the second is a "citation" or "reference note." Because of the wide stylistic variations among disciplines and even among journals of the same discipline, the two must be discussed separately.

The footnote proper—hereinafter referred to as a "content" note—is used for data that are not sufficiently relevant to the text to be incorporated in it. The note must bear upon the subject of the article, and it must be significant in itself. It must also be self-contained. Because its peripheral relevance makes a content note a digression, content notes longer than 150 words are placed in an appendix. If a generality applies to content notes, it is that they are overused.

Typical uses for content notes are

- For editorial comment, as in

  [1]Name dropping is a basic ploy in footnotemanship. If two men say essentially the same thing, always quote the more prominent.
- For an explanation of a technical or jargon word as in

  [2]A "therblig" is the term for an identifiable part of a repeated work-act. Term devised by F. B. Gilbreth—an inversion of his name.
- For the conversion of unfamiliar monies, weights, and measures into modern American usage. The terms must be unfamiliar; otherwise an intelligent reader will be offended by an explanation of the obvious. For example, the rate for the pound sterling and the deutsche mark is seldom cited.

  [3]In July 1966 the *pengo* was listed at $0.0025.

  [4]A *hectare* is equal to 2.7 acres.
- For the presentation of a passage, in a foreign language, that has been loosely translated into the text.
- For a fair presentation of the other side of a minor controversy from the side presented in the text.

  [5]But Mike Jacobs saw the play in a different light. His considered opinion from the dugout was "We was robbed."
- For the illustration of a complicated sequence such as a corporate structure or a geneological tree.
- For the illustration in tabular form of statistics difficult to write out.
- For the derivation of equations when the derivation is not so lengthy as to warrant placing it in the appendix.

Citation or reference notes are a professional, legal, and moral requirement. The code of research requires giving credit where it is due. (First corollary: The code requires escaping responsibility by footnoting whenever points are in doubt or in error.) The federal courts usually accept footnotes or releases as a legal protection against copyright suits except when wholesale copying and exploitation violating "fair use" are involved. The standards of morality are so widely accepted that readers will be disturbed by recognizable plagiarism even when the writer is not.

Essentially, citations answer a reader's question "Where did he get that from?" The writer is obligated to answer.[8] Answers must be given in the following cases:

[8]Answers must be accurate. A recent PhD in history was ungowned when it was discovered post facto that his dissertation cites over a hundred nonexistent sources.

- For all direct quotations not commonly known.
- For all paraphrases, equations, and quotations from secondary sources.
- For all statistical data not derived by the author.
- For all new concepts from nonstandard sources; i.e., comments in letters, interviews, and lectures.
- For all statements of unverifiable opinion not originating with the author.
- For all illustrations, tables, photographs, or maps not the work of the author or of his staff.
- For all original material cited by another but unavailable to the writer. In this case a compound footnote is used as follows.

[1]O. H. Ammann, "Tentative Report of Bridge Engineer on Hudson River Bridge at New York Between Fort Washington and Fort Lee," Port of New York Authority, 1926. Cited by Walter J. Miller and Leo E. A. Saidla, eds., *Engineers as Writers* (New York: D. Van Nostrand Company, Inc., 1953), p. 237.

- For all industrial, corporate, and institutional publicity releases even though not copyrighted. (Source and verifiability are as important in professional writing as literary rights.) The ascending administrative sequence should be given – if known – when a section within a division within a corporation is the author. For example,

[2]Management Development Section, Engineering Department, Hamilton-Standard Division, United Aircraft Corporation, *Relocation of Newly Hired Personnel,* HSIR-103, May, 1961.

Citations of classified publications may be made when the titles of those publications are themselves unclassified. Discretion must be used here, however, for no classified information requiring a citation may be released in an unclassified article. Further, if the logic of the article makes apparent the classified content of the cited publication, security has been compromised as effectively as if the publication had been quoted directly. Finally, because most classified publications are available only to qualified readers, no useful purpose is served by the citation.

When highly dangerous procedures, items, or materials are discussed, the reader should be alerted and warned either in the text or in a footnote.

Footnotes within illustrations and tables differ from footnotes to those illustrations and tables. Thus symbols and letters may be used within the table, and the whole table will be footnoted with a number in superscript at the end of the title. Single and double asterisks and daggers are the symbols most commonly used, but usage varies among disciplines and journals.

Because the positioning of a table may change the normal sequence of exponential numbers used as footnotes, the journals of the biological sciences use only symbols and lower case letters in footnoting tables.

Figure 22, reproduced from p. 56 of the *Style Manual for Biological Journals,* illustrates the use of symbols and letters for footnotes both to and within tabular matter.

Mathematical notations, equations, and formulas (use the anglicized spelling in mathematics journals) receive citation footnotes whenever they are not the work of the writer. Because the use of superscript numbers for footnotes might be confused with exponential use, symbols are generally used. A suggested list of mathematical symbols appears on 308. A useful discussion of style in the preparation of mathematical articles is entitled "The Preparation and Typing of Mathematical Manuscripts." It may be obtained by writing to Mathematical Manuscripts, Bell Telephone Laboratories, Inc., 463 West Street, New York, New York 10014.

The physical positioning, content, and arrangement of footnotes also varies with discipline and journal. When back issues of the specific journal for which an article is intended are not available for imitation, the *Style Sheet* of the Modern Language Association may be followed with some factor of safety. If the *Style Sheet* fails to cover a point, the University of Chicago Press *Manual of Style* or *Webster's New International Dictionary* should be followed.

Abbreviations within footnotes have been standardized only in the *Union List of Serials in Libraries of the United States and Canada* (New York: H. W. Wilson Co.). While some journals use variations (see, for example, the section "Rules Peculiar to the *Hispanic American Historical Review*" in Chapter 5), a reader searching a library for a reference will find that the library follows both the pattern and the sequence of the Wilson publication. Listed below are the rules governing the *Union List of Serials*.

GENERAL CHARACTERISTICS OF THE SECOND EDITION CONTINUED IN THE THIRD EDITION*

1. A serial not published by a society or a public office is entered under the first word, not an article, ofthe title.
2. A serial published by a society, but having a distinctive title, is entered under the title with reference from the name of the society.

*From *Union List of Serials in Libraries of the United States and Canada,* third edition (New York: H. H. Wilson Co., (1966).

3. Learned societies and academies of Europe, other than English, with names beginning with an adjective denoting royal privilege are entered

TABLE 5. *Occurrence of grasses of the tribe Hordeae in certain counties of Arizona**

County	*Number of species of the genus*		
	Hordeum	*Elymus*†	*Lolium*
Apache	2	7	0
Graham	1	7	1
Cochise	3	8	2

* Compiled from Gould, Frank W. 1951. Grasses of the Southwestern United States. Univ. Ariz. Bull. 22:1-343.

† Includes those species often segregated in the genus *Agropyron.*

TABLE 6. *Temperature characteristics of various homeothermic animals*[a]

Animal	*Rectal temperature, C*			*Critical air temperature,*[b] *C*		*Presence or absence of temperature regulating mechanism*[c]			*Thermo neutrality zone,*[d] *C*
	Normal	*Min*	*Max*	*Low*	*High*	*Sweating*	*Shivering*	*Panting*	
Man	37.0	22	44	17 to 22	32	+	+	0	23 to 34.5
Cat	37.2–39	17	42		32.2	0	+	+	10 to 30
Cow, dairy	38–39		42.8	–40	21–27	0		+	4.4 to 15.6
Dog	39.0	17	42.8	–40	29	0	+	+	–40 to 30
Elephant	35.9–36.7					0	+		
Guinea pig	38.5–39.9	21		–15	29.5			+	30 to 31
Chicken	40–42	25–27	45	–34	32.2	0	+	+	16 to 35

[a] Adapted from original table. Spector, W. S. (Ed.). 1956. *In* Handbook of biological data. W. B. Saunders Co., Philadelphia.

[b] Air temperature at which the first indication of change in rectal temperature occurs in the animal.

[c] Symbols: + = mechanism present; 0 = mechanism absent.

[d] Range of air temperature over which the metabolic rate is lowest and constant in the animal.

Figure 22

under the first word following the adjective. These adjectives, Kaiserlich, Koniglich, Reale, Imperiale, etc., are abbreviated to K., R., I., etc., and are disregarded in the arrangement.

4. Colleges and universities having a geographical designation are entered under the name of the city, state or country contained in the title.

5. Observatories, botanical and zoological gardens, etc., not having a distinctive name, are entered under the name of the place in which they are located, unless affiliated with a university, in which case they are entered under the name of the university.

6. References have been made from earlier forms of a title and/or name of issuing body to latest form known, and in general whenever a reference might facilitate the use of the list.

7. Volume numbers and dates are inclusive in all cases and the dates are for the period covered by the serial, not those of publication.

8. A library's symbol is made up of a combination of letters for state, city and library. A symbol alone indicates the library holds the complete set.

9. A hyphen (-) between volume numbers or dates signifies "from and including the former to and including the latter."

10. A plus sign (+) indicates that the serial is currently received and that the set is complete from the last date or volume number given.

11. Parallels (‖) indicate that publication ceased with the preceding date or volume.

12. Brackets ([]) indicate that the volumes or years so enclosed are not complete.

Permissions or releases are required whenever longer passages from copyrighted works are quoted. Interpretations of the meaning of "longer passages" vary from ten or more lines to 250 words. The Association of American University Presses (noncommercial publishers) is guided by its own "Resolution on Permissions" quoted below.

"1. That publications issued under our imprints may be quoted without specific prior permission in works of original scholarship for accurate citation of authority or for criticism, review, or evaluation, subject to the conditions listed below.

"2. That appropriate credit be given in the case of each quotation.

"3. That waiver of the requirement for specific permission does not extend to quotations that are complete units in themselves (as poems, letters, short stories, essays, journal articles, complete chapters or sections of books, maps, charts, graphs, tables, drawings, or other illustrative materials), in whatever form they may be reproduced; nor does the waiver extend to quotation of whatever length presented as primary material for its own sake (as) in anthologies or books of readings.

"4. The fact that specific permission for quoting of material may be waived under this agreement does not relieve the quoting author and publisher from the responsibility of determining 'fair use' of such material."

Obtaining legal permission to republish sections of copyrighted work is a relatively easy matter if republication of the section will not negatively affect the sales potential of the original. Most authors and publishing corporations will grant the permission out of professional courtesy—and often for the favorable publicity. In the case of anthologies, however, in which whole articles, poems, stories, etc. are reproduced, the original author may request a fee or a percentage of the royalties. (The late George Bernard Shaw would not permit any of his plays to be anthologized unless he received 50 percent of the royalties of the entire anthology.) Fees are usually nominal, however; the most expensive permission in this book cost forty dollars. The great majority were free.

The procedure for obtaining releases is simple. The owner of the copyright is determined from the information published in the original printing. He is sent a blank form similar to the one reproduced as Figure 24. (It is the form used in the preparation of this book.) The form is legal authorization to reprint what is described on the form and *in the manner indicated on the form*.

Quite often an atypical footnote will be indicated in the space headed "Credit line to be used." When this situation occurs, the pattern requested must be followed.

Most professional societies have standard procedures governing releases. The statement of policy and procedures given below is printed with the permission of the American Psychological Association.

POLICY AND PROCEDURES GOVERNING PERMISSIONS TO REPRODUCE MATERIALS FROM PUBLICATIONS OF THE AMERICAN PSYCHOLOGICAL ASSOCIATION

"1. *Overall policy*. Permission to reproduce material from publications of the Association is required, with appropriate citation of the reference. *All* requests should include: (a) complete names of all authors, as they

Authorization for Release and Publication

In my forthcoming college textbook, Writing for Professional and Technical Journals and Magazines, to be published by John Wiley & Sons, Inc., of New York in the spring of 1967, I should like to include the following selection from your copyrighted publication:

A release form is given below for your convenience. Your consideration of this request at your earliest convenience will be appreciated.

Sincerely,

Professor John Mitchell

- -

Professor John Mitchell
Department of English
University of Massachusetts
Amherst, Massachusetts

Sir, according to the terms stated below we grant the permission requested to reprint the following:

Conditions and fees:

Credit line to be used:

Date________________________________ By________________________________

____

____

Figure 24

appear in the publication; (b) complete and correct title of the article; (c) number and year of the volume; (d) name of the publication.

See sections below for specific procedures, and for exceptions.

"2. *Entire articles*. Permission to reproduce is required, with appropriate citation of the reference. Request should include, in addition to Section 1, numbers of the first and last pages on which the article appears.

"3. *Quotations*. Permission to reproduce is required if the quotation is in excess of 500 words (total from one article), with appropriate citation of the reference. Request should include, in addition to Section 1, (a) page number(s) on which the quotation appears; (b) identification by first and last phrases of the quotation.

"Permission is not required to reproduce quotations of less than 500 words (total from one article), but appropriate citation is required.

"4. *Tables*. Permission is required, with appropriate citation of the reference. Request should include, in addition to Section 1: (a) number of table; (b) page number on which the table appears.

"5. *Figures*. Permission is required, with appropriate citation of the reference. Requests should include, in addition to Section 1: (a) number of the figure; (b) page number on which the figure occurs.

"6. *Paraphrases*. Permission is not required, if the material to be used definitely constitutes a paraphrase: uses no actual wording—is not merely a shortened version of the actual wording.

"7. *Contingent permissions*. Permissions will not be granted contingent upon like permission of the author of the article in the following instances: (a) when an *entire* or major portion thereof, is to be used; (b) for tables and figures; (c) if alterations, deletions, or additions are to be made (apart from a definite paraphrase).

"When authorship is multiple, permission of only one author will be required. If there is no living author, an author's permission will be waived. Care should be taken, however, in making a *reasonable* effort to locate a living author.

"8. *Limitations*. Permission will not be considered for use of materials in any but the *forthcoming* edition, translation, or revision of a given work. Requests to reproduce materials in "all future editions, . . . etc.," will *not* be considered. *However,* the Association does not limit the *area of distribution* of the work in which the materials to be used.

"Permissions will not be granted to institutions, but to persons as representatives of institutions, or to individual persons.

"9. *Intended use*. All requests *must* include a statement of the intended use of the material—didactic, library, research, inclusion in publication, etc.

"Requests may be handled more quickly if submitted in duplicate, and in such manner that the permission may be indicated on the request itself."

Bibliographic procedures in professional journals are not standardized. The few generalities that pertain exist within disciplines. For that reason, the remainder of this section segregates bibliographic techniques according to discipline. The techniques shown are those followed by journals and *appropriate to articles*. A definitive discussion of the theory of bibliographical description appropriate for librarians may be found in Fredson Bower's *Principles of Bibliographic Description* (Princeton, 1949).

The term "applied sciences" is used here to mean the various departments of the professional schools of engineering and of business. Each department has a national society which supports one or more professional journals. The style guides of those journals are relatively slight. The engineering journals usually refer authors to the style guides of one of the physical sciences and concern themselves with private variations and addenda. The business journals usually follow the MLA *Style Sheet* or the University of Chicago *Manual of Style*.

Chapter Five reprints, with permission and without the list of "Recommended Abbreviations," the three-page guide "Information for IEEE Authors" which governs the more than thirty journals published by the Institute of Electrical and Electronics Engineers. Of significance here is that the recommended bibliographic practice follows the *Style Manual* of the American Institute of Physics except that in journal references the abbreviation "vol." is used instead of printing the volume number in roman (i.e., not italic) type.

The "Style Instructions" of the *American Economic Review* is reprinted in the same place. Of significance here is that the recommended bibliographic practice follows the MLA *Style Sheet* except that the initial footnotes carry an asterisk, and the date is not placed within parentheses.

The term "biological sciences" is here used to mean the full range of life sciences from agriculture to zoology. More than eighty journals in the biological sciences subscribe to the policies stated in the *Style Manual for Biological Journals*. This document is outstanding. Written by the Conference of Biological Editors, Committee on Form and Style, it presents a pattern of communication acceptable to and compatible with the journals of most biological disciplines.

Chapter 5 contains sections reprinted from the *Style Manual for Biological Journals* by permission of the American Institute of Biological Sciences. The quoted passages indicate the two acceptable systems for citations, and the one acceptable system for bibliographic arrangement.

This last is fully compatible with the machine techniques for information retrieval discussed earlier.

There are many overlapping and interdisciplinary areas within the biological sciences. The fields of medicine, chemistry, and psychology overlap in external areas. Certainly mathematics is the common language of all scientific method. As a result, the eighty journals that follow the AIBS *Style Manual* represent but a fraction of those printing articles whose content might be regarded as germane to the biological sciences. Writers submitting articles to journals in these allied fields should follow the style guides of those fields and, ideally, of the specific journal.

Even within the biological sciences, there are some variations in format for professional articles. In agriculture, for example, governmental agencies publish more articles and professional notes and bulletins than do the professional societies. Format in these articles is governed by documents ranging from the Government Printing Office *Style Manual* to Austin E. Showman's monograph "The Good Word: A Writing Manual for County Extension Agents and Other Leaders in Agriculture," Ohio State University Extension Service, Columbus 10, Ohio. Of significance here, however, is the difference in level and in tone. Articles printed in the journals of the professional societies are aimed at experts and near-experts. Articles printed in Agricultural Extension Bulletins are aimed at farm technicians. These last have no interest in elaborate citation systems.

Printed below is the "Official Style Guide: Research Papers for Institute of Food Technologists Journals." These journals, as the "Style Guide" indicates, follow the procedures recommended by the American Chemical Society *Handbook for Authors,* although food technology, for administrative purposes, is linked with agriculture and the biological sciences.

An article which illustrates the bibliographic recommendations of the AIBS *Style Manual* begins on pg. 214.

The term "physical sciences" is used here to mean the disciplines of astronomy, chemistry, geology, meteorology, and physics. These are the "basic" or "earth" or "hard" sciences, and their pedigrees are both ancient and honorable. Because they are ancient, mutations within disciplines have made their forms and techniques dissimilar. Because they are honorable, they guard their professional areas or empires.

The professional societies of each of the physical sciences attempt to correlate and to guide the literature *in their own disciplines*. The attempts are as commendable as they are expensive, and although the concept of disunified science may seem archaic or medieval, the high standards maintained by the societies, as well as the high cost of printing, have assured a professional literature of the highest caliber.

Official Style Guide:

Research Papers for IFT Journals

George F. Stewart, Executive Editor

GENERAL POLICIES

Typing and Submittal. Typing should be double-spaced throughout (including reference list, footnotes, tables, and legends). Use a good grade of bond paper. Margins should be at least one inch on all sides. Number the pages. Submit manuscripts **in duplicate.** Keep a third copy for your files.

Form and Style. Manuscript subdivisions will normally consist of:

Summary. Clear, concise, and factual abstract of the paper—no longer than 3% of text.

Introduction. Concise account of why study was undertaken.

Literature Review. Brief review, directly related to investigation. Cite by author(s) and year of publication.

Experimental Methods. Describe methods, equipment, and materials. Avoid repeating previously published details, unless modifications are extensive. Indicate specific experimental design and justify its use (unless obvious).

Results and Discussion. Clear, concise account of findings and their interpretation. Data should be presented in the form that is briefest and clearest. Many times this is best done with tables, graphs, and charts. Limit discussion to subject; avoid unsupported theories and speculations.

TECHNICAL DETAILS

1. **References.**

   a) **In the text.** Cite (in parentheses) by author(s) and year of publication. Use *et al.* for more than two authors. Items in a single year by identical author(s) should be distinguished by adding a, b, etc.

   b) **In the reference list.** Under "References" list in **alphabetical order** at the end of paper. List all authors. **The reference list should be typed double-spaced throughout.** Examples follow:

   Patent citation: Cole, G. M., and R. E. Cox. 1938. Method of treating pectin. U. S. Patent 2,109,792.

   Journal citation: Banfield, F. H. 1935. The electrical resistance of pork and bacon. *J. Soc. Chem. Ind.,* **54,** 411T.

   Book citation: Bernard, C. 1877. Lecons sur le diabete de la glycogenese animal. J. B. Billier & Fils, Paris, p. 576.

   Bulletin citation: 1) Bull, S., and H. P. Rysk. 1942. Effect of exercise on quality of beef. *Illinois Agr. Expt. Sta. Bull.* No. 488, p. 105.

   2) Haines, R. B. 1937. Microbiology in the preservation of animal tissues. *Food Invest. Board. Special Rept.* No. 45 H. M. Stationery Office, London.

   Thesis citation: Hoover, S. R. 1940. Physical and chemical study of ovomucin. Ph.D. Thesis. Georgetown University Library, Washington, D.C.

   Unpublished citation: Smith, John. 1947. Unpublished data. University of California, Davis.

   Journal titles should be abbreviated according to the style given in *List of Periodicals* abstracted by *Chemical Abstracts* (1956).

2. **Abbreviations.** The following are acceptable abbreviations:

Term	Abbreviation
ampere	amp
Angstrom units	A
atmospheres	atm
average	av
avoirdupois	avdp
boiling point	bp
Centigrade	C
centimeters	cm
cubic centimeters	cc
cubic millimeters	mm^3
dextro	*D-*
effective dose, 50%	ED_{50}
et alii	*et al.*
Fahrenheit	F
figure	Fig.
foot	ft
gallon (U.S.)	gal.
gram(s)	g
gravitational constant	G
hour	hr
inch	in.
kilocalories	kcal
kilograms	kg
least significant difference	LSD
lethal dose, 50%	LD_{50}
levo-	*L*
liters	L
logarithm (common)	log
logarithm (natural)	$\log_e$
melting point	mp
meter	m
micrograms	µg
micron	µ
milliamperes	ma
milliequivalents	meq
milligrams	mg
milliliters	ml
millimeters	mm
millimeters of mercury	mm Hg
millimicrons	mµ
millimoles	m*M*
milliseconds (time)	msec
millivolts	mv
minutes	min
molar	*M*
molecular weight	mol wt
normal	*N*
ounce	oz
parts per million	ppm
percent	%
pound	lb
refractive index	n_D
roentgen	r
seconds	sec
specific rotatory power	$(a)D^{20}$
square centimeters	sq cm
statistical significance	
@ 5% level	*
@ 1% level	**

The correlation of the professional literature of the physical sciences is achieved by *standardization of format by discipline*. Thus a journal of the American Chemical Society is dissimilar from one of the American Institute of Physics. This presents no problems to the professionals of either camp; their graduate schools have trained them in the bibliographic techniques of the discipline. These techniques are discussed in Part C, Chapter 5.

The word "humanities" is here used to mean literature and language. Journals in these fields generally follow the *Style Sheet* of the Modern Language Association. Over seventy journals in the humanities have accepted every rule of that document. Over forty university presses follow it but *only in the fields of the modern languages, literature, and allied disciplines*.

Section 39 of the "Supplement" to the *Style Sheet* contains a discussion of bibliographic form added at the request of the university presses and of graduate schools. The discussion was omitted from the original publication because many learned journals do not require bibliographies for manuscripts of article length and because there is no real agreement among the editors of the journals represented as to bibliographic form. Section 39 is quoted below in its entirety.

39 **Bibliography.** As a rule, put all items of a bibliography into a single list, labelling it precisely, e.g., "List of Works Consulted" or "List of Works Cited" or "A Selected Bibliography" or "A Brief Annotated Bibliography." If it seems desirable to *classify* these items (e.g., into primary and secondary sources, or manuscript and printed sources), consult your instructor. Whatever kind of list you decide upon, it will usually include all your first references in footnotes, but now *alphabetized* as in the examples below:

Sample Bibliography

Baker, Ernest A. *The History of the English Novel*. 10 vols. London, 1924–39.

———. *The Uses of Libraries*. London, 1927.

C[oleridge], D[erwent]. "An Essay on the Poetic Character of Percy Bysshe Shelley, and on the Probable Tendency of His Writings," *Metropolitan Quarterly Magazine*, II (1826), 191–203.

Fisher, John H. "Serial Bibliographies in the Modern Languages and Literatures," *PMLA*, LXVI (April 1951), 138–156.

Hall, R. A., Jr. "American Linguistics, 1925–1950," *Archivum Linguisticum*, III, Fasc. ii (1951), 101–125; IV, Fasc. i, 1–16.

Higher Education for American Democracy. 2 vols. Washington, 1947. (Report of the President's Commission on Higher Education.)

"The Laureate and His 'Arthuriad'." Anon. rev., *London Quarterly Review*, XXXIV (April 1870), 154–186.

[Lewes, George H.]. "Percy Bysshe Shelley," *Westminster Review*, XXXV (April 1841), 303–344.

Pope, Alexander. *The Works of Alexander Pope . . .*, ed. W. Elwin and W. J. Courthope. 10 vols. London, 1871–89.

Bernbaum, Ernest, Samuel C. Chew, Thomas M. Raysor, Clarence D. Thorpe, Bennett Weaver, and René Wellek. *The English Romantic Poets: A Review of Research.* New York, 1950.

Shryock, Richard H. "The Academic Profession in the United States," *Bulletin of the American Association of University Professors*, XXXVIII (Spring 1952), 32–70.

Sterne, Y. B. *Research Can Be Fun.* Urbana, 1958.

Wood, Paul Spencer, ed. *Masters of English Literature.* 2 vols. New York, 1946.

Scholarship in the fine arts tends to employ the bibliographic techniques common to the fields of history and the social sciences. Individual listings within the bibliography follow the same form as the first citation in a footnote in the field of history. The arrangement of sources for historical manuscripts of book length is discussed below. Bibliographical arrangement of sources for historical manuscripts of article length is determined by the footnotes.

The illustrative sample printed below is taken from a senior journal in the fine arts, *Architectural History*. The arithmetic progression is simply the arrangement of footnotes and citations in order of their appearance. The sample includes 14 of the 74 notes that were used in the 12-page article by Paul Thompson entitled "All Saints' Church, Margaret Street, Reconsidered." Notes 3 and 8 illustrate editorial notes. Notes 1, 5, and 14 illustrate multiple notes. Notes 2, 7, 9, 11, and 12 illustrate secondary references while notes 1, 5, and 6 are primary references. Of particular significance are the abbreviations, use of italics, and the internal punctuation. The editor of *Architectural History* expects all articles submitted for publication in his journal to follow the forms illustrated.

1. C. L. Eastlake, *A History of the Gothic Revival,* 1872, p. 253; Sir John Summerson, *Heavenly Mansions,* 1949, p. 174; H-R. Hitchcock, *Early Victorian Architecture,* 1954, chapter xvii; etc.
2. For the latter theory see especially Summerson, *op. cit.*
3. H. W. and I. Law, *The Book of the Beresford-Hopes,* 1925. A. J. Beresford-Hope (1820-87), son of Thomas Hope, writer and art patron, was one of the founding members of the Cambridge Camden Society in 1839 and its Chairman from 1843. He continued to be Chairman after its reconstitution in 1845 as the Ecclesiological Society. He was a Conservative M.P., a noted champion of High Church interests, in 1841-52 and 1857-87. He married Lady Mildred Cecil in 1842 and was proprietor of the *Saturday Review*. The most important of his other building schemes

was the reconstruction of St. Augustine's, Canterbury, as a missionary training college, with Butterfield as architect, in 1844-48. Benjamin Webb (1819-85), whom he met at Trinity College, Cambridge, was first secretary of the Cambridge Camden Society, and later priest at Kemerton (1843-51), Sheen (1851-62), and St. Andrew, Wells St. (1862-85).

4. I am deeply grateful to Mrs. Tritton for permission to use these letters.

5. T. F. Bumpus, *London Churches Ancient and Modern,* 1908, II, 239-241. Letter from Hope, *Morning Chronicle,* 13 June, 1854.

6. *Ecclesiologist,* November, 1841.

7. Law, *op. cit.,* p. 161.

8. William Upton Richards (1811-1873), graduate of Exeter College, Oxford; an assistant in the British Museum Manuscript Department until 1847; joined the Ecclesiological Society in 1847; priest at Margaret Street from 1847.

9. Law, *op. cit.,* pp. 161-63; Bumpus, *op. cit.,* pp.240-41.

10. Hope to Tritton, 18 December, 1852 (Tritton Letters).

11. *Ibid.,* 23 January, 1850; 18 December, 1852.

12. *Ibid.,* Richards to Tritton, 18 March, 1853.

13. e.g. *Builder,* 9 November, 1850.

14. Law, *op. cit.,* p. 166; Hope to Tritton, 26 July, 1852 (Tritton Letters); *Morning Chronicle,* 13 June, 1854.

Chapter 5 contains the "Note for Contributors" that establishes bibliographic form for *The Art Bulletin.* Following it is an article from that journal illustrating the application of the notes.

The journals of mathematics are organized. Almost all are guided by "A Manual for Authors of Mathematical Papers" published by the American Mathematical Society, P. O. Box 6248, Providence, Rhode Island 02904. This document is reprinted in its entirety in Chapter 5. Section 6 of the "Manual" illustrates the bibliographic technique recommended by the society.

An example of an article illustrating the principles recommended by the American Mathematical Society is reprinted by permission on pp. 310-313.

The problems of *printing* mathematical notations are discussed and partially resolved in the "Manual." The complexity of these problems is indicated by the statement "The cost of setting a page of mathematics in type varies from $10 to $30."

The problems of *typing* notations and equations are resolved in another copyrighted paper entitled "The Preparation and Typing of Mathematical Manuscripts." This excellent guide may be obtained without charge by writing to Mathematical Manuscripts, Bell Telephone Laboratories, Incorporated, 463 West Street, New York, New York 10014.

All medical journals collate footnotes and citations at the end of each article. However, four dissimilar techniques for arranging citations are used by the various journals. (a) In order of occurrence in the article. (b) Alphabetically by author's surname. (c) Chronologically. (d) On the line following the reference. The first three techniques require dissimilar bibliographic arrangements. The fourth permits none at all. The conflicts in editorial policy produce confusion in the retrieval of medical literature. The efforts of the American Medical Association and the limitations of machine retrieval techniques may force the standardization of usage.

Within citations, there is equal confusion. Some medical editors require titles; others do not. Some editors want both first and last pages indicated; others want only the first. Some editors italicize either the source journal or the volume number; others use roman type. Some editors use parentheses; most do not. A writer can only imitate back issues of the journal for which the article is intended.

Several attempts have been made to standardize bibliographic techniques in medical journals. In his witty and highly readable book entitled *Guide to Medical Writing* (New York: Ronald Press, 1957), Henry A. Davidson, M. D., Editor of the Journal of the Medical Society of New Jersey, recommends the order printed below.

Citation number
Last name of author
First name, middle initial
Title of article (do not use quotation marks)
Title of the source journal (do not abbreviate)
Volume number (use an arabic numeral, even if the source journal itself uses roman numerals for its volumes numbers)
Colon (some journals use a comma here, but the colon is commoner)
First page of the article, a hyphen, last page of the article
The month of issue of the source journal (if the journal is issued more than once a month, give the day as well)
The year of issue

The American Medical Association has published two documents dealing with this subject: *Stylebook and Editorial Manual,* 1966, and *Advice to Authors,* 1964. Each is available from the American Medical Association, 535 N. Dearborn Street, Chicago, Illinois 60610.

Morris Fishbein, M. D., in his book *Medical Writing: The Technic and the Art,* third edition (Blakiston, 1957), recommends essentially the same procedure.

Robert O'Leary, Assistant Editor of the prestigious *New England Journal of Medicine,* described the recommended reference pattern of that journal in an article entitled "Technic of Medical Communication" appearing in the April 28, 1966, issue. The section on references and the bibliography of that article are reprinted by permission below. The latter not only illustrates O'Leary's recommendations, it also serves as a current bibliography upon style in medical writing.

The term "social sciences" is used here to mean the disciplines of psychology, sociology, economics, history, government or political science, and law. The goals and techniques of each are different—and so are their bibliographic practices. For this reason, it is necessary to discuss the recommended practices of each separately.

The American Psychological Association publishes twelve journals. The fields of each journal are described on pp. 2–4 of the *Publications Manual* of the American Psychological Association, 1957 revision. Writers are cautioned to study those descriptions to determine *both content and organization appropriate for a specific journal.*

The footnote and bibliographic techniques required by the journals published by the APA are both unique and complex. Because they are undergoing a major revision (to be issued in 1968) a query to the editor of the specific journal is the best way to determine the present style of that journal.

Journals of sociology follow either the style guides created by their own editors or the University of Chicago *Style Manual.* (*The American Sociological Review* is published by the University of Chicago Press.)

Journals of economics are independent and unique. Style requirements vary according to the journal and its goal. Passages reprinted on pp. 366-370 illustrate the fairly representative pattern of the *American Economic Review.*

Political science or government journals are not standardized. The mission of each journal determines the content and format. The United Nations *Bibliographical Style Manual* (United Nations Publication ST/LIB/SER. B/8, Sales No. 63.1.5, $0.75) contains a definitive and excellent discussion of citations. The *Manual* is far too lengthy for reproduction here, but Chapter 5 contains a sample bibliography illustrating some of the recommendations of that outstanding reference work.

The majority of journals of political science loosely follow either the MLA *Style Sheet* (see p. 133) or the University of Chicago *Style Manual.* Citations of legal documents follow the apparatus recommended by the Harvard Law Review Association's *A Uniform System of Citation,* tenth edition (see below).

SPECIAL ARTICLE

TECHNIC OF MEDICAL COMMUNICATION

ROBERT O'LEARY*

BOSTON

THE goal of all medical writing being to transmit information or knowledge, anything that hinders such a communication of ideas defeats the purpose for which a medical article has been written. The failure in communication may be due to any one of a number of flaws: rhetoric that obscures or distorts the author's meaning, or prolixity that discourages the reader from any further effort to grasp the author's message; illustrations so clumsy or unclear or complex that they conceal the point they were intended to illuminate; tables so long and involved, or so cluttered with ambiguous symbols and abbreviations, that they defy comprehension; and bibliographies and lists of references so incomplete or inaccurate that they preclude any transmission of information from the author to the reader. Conversely, any device or resource that furthers a free flow of useful knowledge contributes to the fundamental purpose of communication.

Although the following discussion primarily concerns the mechanical problems posed by medical writing (increasingly scrutinized by a more sophisticated and critical audience) points of style and clarity of expression merit brief mention. Style is inseparable from content, or to the extent that it may be separable, is but the icing on the cake: insubstantial except in terms of the solid fare beneath. Precision and intelligibility, on the other hand, are the essentials without which no writer – especially a scientific one – can hope to achieve his goal of imparting knowledge. The problem has been aptly characterized as follows: "Authors should be encouraged to realize that their mission is to reach a larger readership by a clear and simple presentation of their investigative results, even at the risk of further painful revision."[1] For the beginner a number of authoritative guides to the finer points of writing in general, and medical writing in particular, are available – among others, *Fowler's Modern English Usage*,[2] *The Elements of Style*, by Strunk and White,[3] and *The Careful Writer*, by Bernstein.[4] From the strictly medical point of view Fishbein's[5] *Medical Writing* and the recommendations of Hartley[6] and Wilson[7] on the preparation of manuscripts are invaluable aids. Three medical dictionaries (*Dorland's*,[8] *Gould's*[9] and *Stedman's*[10]) may be consulted by any author who has doubts about medical terminology. *Bergey's Manual of Determinative Bacteriology*[11] and *Diagnostic Microbiology*, by Bailey and Scott,[12] are essential guidebooks to their special fields. In a recent issue of the *Journal* DeBakey[13] offered excellent advice on words, phrases and expressions to avoid.

The essence of the questions of form, content and diction confronting the medical writer has been succinctly defined as follows:

> The material accepted for publication is that which appears at the time to reveal something new or offers a valuable extension of something old, or at least a reasonable hypothesis concerning it. And, in respect to the literature of modern medicine, the propriety of publishing it has little relation to its palatability or the difficulty of understanding it. It represents the inevitable course of science, and no matter how rough the road to truth may be, those who would follow it must struggle on or drop behind. If, as Euclid stubbornly contended, there is no royal road to geometry, neither is it possible to make the path to an understanding of modern medical science easy traveling for the dull, the technologically inadequate or the intellectually inert.
>
> This very fact should make it all the more incumbent on medical writers to express themselves simply and clearly. Advice, however, is not easily taken, and that which has been offered on the technics of presenting scientific material seems to be surpassed only by the indifference to it of many otherwise eager authors. Basically, the principle of good writing consists of trying to make oneself understood by the greatest possible number of readers. This is infinitely more important to one who has something to communicate, if he stops to consider it, than succumbing to the temptation of trying mainly to create the impression of superior knowledge.
>
> The most effective messages are those that have been written in the fewest and the shortest words. They are those in which the smug use of superfluous specialty jargon is avoided, for it impresses no one but him who conceals his meaning and betrays his confused readers therewith. They are the ones in which the fewest baffling combinations of initial letters are used, those that are introduced being identified promptly, for there is little that impedes understanding more surely than alphabetic contractures.[14]

What follows applies to medical manuscripts that have been accepted for publication. A high percentage of the papers submitted to the *New England Journal of Medicine* have to be rejected, many for reasons that carry no stigma but others because they are so poorly constructed that no amount of expert editing or even rewriting can make a truly professional article of them. Of manuscripts accepted about 50 per cent are returned to the authors for major or minor revisions: modification of unjustified, or unproved claims; reduction in length; and elimination or reworking of illustrations and tables. This discussion concerns the major and minor flaws in text, tables, figures and references that make revision necessary.

*Assistant editor, *New England Journal of Medicine.*

The Text

Length

A manuscript should be only as long as the point to be made may require. Most medical journals are drastically limited in the space that they may allocate to each article: the *New England Journal of Medicine,* with approximately 40 pages of text available for an average of 5 original articles, 1 special article, a medical-progress report and 2 or 3 items of "Medical Intelligence," can rarely afford 10 or 12 journal pages for 1 article, even though its authors may think it is worth its weight in all other elements of the *Journal* combined. Often, excellent papers have to be rejected because of their length alone; others are made acceptable only after considerable reduction. Again, illustrations and tables may have to be eliminated, especially when they duplicate each other or concern points that can more clearly be made in the text.

The amount of space that the published article will require may be estimated on the basis of the number of typed pages of manuscript — approximately 3½ double-spaced manuscript to 1 published page. In addition, long, elaborate tables may take a full published page (or even 2 full facing pages) and roentgenograms, electrocardiograms or photomicrographs of standard size (3 inches wide by 4 inches deep) will, with their legends, fill a quarter of a page. It is thus obvious that care in controlling the length is one form of insurance that an otherwise excellent paper will stand a good chance of being accepted.

Form

Some of the points made here may seem elementary, but each one is based not on a single example but on many. First of all, a manuscript should be submitted in the original, with a duplicate copy. It must be typed clearly and legibly, with double or triple spacing, on one side of the paper only. This stipulation applies not only to text but also to case histories, references and footnotes. The margins of the typed page should be at least 1 inch on each side and top and bottom.

The title should be long enough only to indicate the scope of the paper and to include the important key words for indexing — long, involved titles discourage the reader and tempt the blue pencil of the editor. Often, the use of a subtitle makes it possible to include material that would make the main title clumsy. Below the title the authors' names are given, with keyed footnotes indicating the academic and hospital positions of each. The *Journal,* after a rear-guard action of some years' duration, sometimes bows to current fashion and acquiesces in the listing of more than 5 authors' names. Occasionally, however, authors have to be asked to reduce the number of coauthors, on the grounds that it is impossible to understand how 10 or 15 people all managed to be involved in the writing of a paper that will take 5 or 6 pages when published. It has sometimes been necessary to insist that only the senior author be listed, with coauthors named in a footnote as a committee co-operating in the research and follow-up work.

The material in an article can usually be divided into sections for emphasis and convenience. A heading such as "Materials and Methods" may be followed by "Results" or "Case Reports," as the content requires, and by a "Discussion" and "Summary" or "Summary and Conclusions." It is worth pointing out that the "Summary" should contain no data not mentioned in the text — new citations of another author's work should not be included in the summary. "Conclusions" are exactly that: judgments or inferences arising from the matter presented in the text.

Brief acknowledgments of aid given by colleagues and consultants are acceptable, but most medical journals discourage listings of the names of typists, secretaries and other personnel in the department in which the article originated. In other words, special help or advice, but not routine assistance, is acknowledged.

Regarding proprietary names of drugs, the *Journal* has adopted the policy of giving the generic name of the drug (or the chemical name, if that is better known), followed by the trade name the first time it is used — for example, "meperidine (Demerol)." If a particular manufacturer has supplied a drug for research or therapy the drug is given its generic name in the text, with a footnote referring to the proprietary name and an acknowledgment to the manufacturer.

Manuscripts published or offered for publication elsewhere and still under consideration by another periodical are not acceptable to ethical medical journals. The contents, including illustrations, are copyrighted and may be reproduced in other publications only by express permission of the holder of the copyright — the journal in which they are first published.

References

It seems seldom necessary to list more than 10 references per 1000 words of text. They should be numbered in the chronologic order in which they are cited in the text. This applies to tables and legends for figures. For example, if 10 references have appeared in the text before a table containing a citation of 2 other sources appears, those 2 others would be numbered 11 and 12; similarly, the legend of a figure that concerns some previously published work would carry the reference number 13. Bibliographies, without numbers keyed to the text, are listed in alphabetical order.

The designation of articles and books in the list of references should conform to the style of the journal to which the paper is being submitted. The *New England Journal of Medicine* includes the names and initials of the author or authors, the title of the article, the name of the journal in which it was published, the volume number, the first and last pages included and the year. For books and monographs the author's name is given first, followed by the title of the article or chapter in the book, the title of the book itself, the total number of pages in the book, the publisher and city where he is located and the year of publication. An article in a journal is cited as follows: Bigelow, H. J. Insensibility during surgical operations produced by inhalation. *Boston M. & S. J.* **35**:309-317, 1846. A reference to a book is given in the following form: Waterhouse, B. *A Prospect of Exterminating the Small-Pox.* 40 pp. Cambridge: William Hilliard, 1800.

Tables

Certain material is better presented in tabular form than in running text – because of either convenience or emphasis. Mere lists of drugs or diseases or symptoms gain nothing from such a presentation. The table should be set up so that each vertical column includes data having a common denominator, such as age, height, temperature, laboratory values, x-ray findings and outcome of case. There is no excuse for a column that contains only factors identical for each entry in the table; a single footnote will give the necessary information at a glance. Abbreviations that have to be explained in footnotes confuse the reader and defeat the purpose of the table – to present the material briefly and surely. Once more, it should be emphasized that long, elaborately devised tables cut down the amount of text that the manuscript can be allowed. In its simplest form, a table may be set up as follows:

The table should be referred to in the appropriate place in the text.

Illustrations

Of all the problems presented by the medical manuscript, none are more exacting – and sometimes more baffling – than those involved in illustrations. The difficulty may be because a roentgenogram is of poor quality, because too much material is included in one drawing or because a chart or graph takes up an inordinate amount of space.

Quality

Authors should know, first of all, that no illustration can be better than the original and that few will be as good when reproduced. Only the clearest illustrations should be submitted. Roentgenograms with barely discernible shadows or flecks of calcium may not show, when published, the abnormality that is their whole purpose. Features or areas that will reproduce only faintly should be indicated by a small but clearly distinct arrow pasted on the glossy print.

Form of Submission

Glossy prints of all illustrations are preferred; reproductions made by a multiple-copy technic are not acceptable. Photomicrographs should be cropped to the width of a column – 3 inches; thereby, reduction in magnification will be avoided, and the author will know almost exactly how his illustration will appear when it is published. If a photograph shows identifiable features of a patient's face the eyes should be blacked out, unless a written statement giving permission for publication has been obtained from the patient or his parents or guardians; a copy of the statement should accompany the photograph. Illustrations reproduced from previously published articles and books should be accompanied by a letter from the publisher granting permission (even if the figure is from previous work done by the author himself).

Figures should be submitted without Scotch-taped or pasted-on legends – it is sometimes impossible to make a suitable reproduction because of paste that shows through from the reverse of a roentgenogram, staples across the top of an electrocardiogram or paper clips that have marred a diagram. It is well to indicate, in light pencil, the author's name and the figure number on the reverse of each illustration. Identifying legends should accompany each illustration. All figures should be referred to in the appropriate place in the text.

Charts and Graphs

These figures should have all letters and numbers large enough to be legible when reduced to the width of a column (3 inches), if reduction to that extent is possible – some charts are so elaborate, or contain so much material, that they require up to the width of the full page; in such cases all lettering and numbers should bear reduction of the whole figure to $6\frac{1}{4}$ inches. For legibility, a good rule of thumb is that each item (letter, number or symbol) should be at least $\frac{1}{16}$ inch in height when reduced to the size desired.

TABLE 1. *Mean Changes during Treatment.*

Treatment	Aldolase		Creatinine Kinase	
	at 6 mo. *units/ml.*	at 12 mo. *units/ml.*	at 6 mo. *units/ml.*	at 12 mo. *units/ml.*
Androstanolone	−10±25	−22±34	− 6±30	− 2±22
Androstanolone & digitalis	−28±42	−53±36*	−30±40	−19±35
Placebo	−57±15*	–	− 1±12	–

*Significant difference.

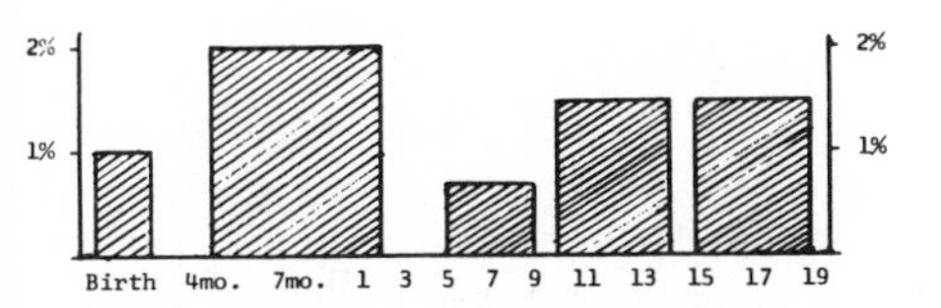

FIGURE 1. *Example of a Chart, Easily Reduced to Column Width but with Unsightly, Irregular Crosshatching.*

All lettering and signs and symbols and numbers should be distinct and unbroken, and all margins should be straight.

Occasionally, authors send in illustrations that, although clear enough even when reduced to 3-inch width, are so unsightly that it is necessary to employ an artist to copy the chart expertly. Examples of this kind of slovenly illustrative work are shown in Figures 1 and 2. It should be borne in mind that it is the responsibility of the author himself, not of the journal to which he submits his paper, to ensure that all his artwork is in the most acceptable form that he can provide. Figure 3 shows a diagram that is easy to read, easy to understand and easy to reproduce.

Color illustrations are best reproduced from the positive transparency, which may be accompanied by a glossy print to give the editor an idea of how the figure will appear when published. Most journals charge the author for reproductions in color (at present approximately $750 for each page).

Galley Proofs

Galley proofs (and proofs of figures and tables) are sent to an author for corrections of typographical errors and approval of the editorial changes made in copy — not for rewriting or rephrasing. Most scientific journals insist on what they consider their standards of expression, but an author has every right to object if an overzealous editor has, during the process of copy editing, altered the meaning of the original. It is well to bear in mind, however, before one insists that the original phraseology be restored, that at least 1 person failed to grasp the author's meaning and that, therefore, a recasting of the disputed passage may be justified. Editors try to handle all

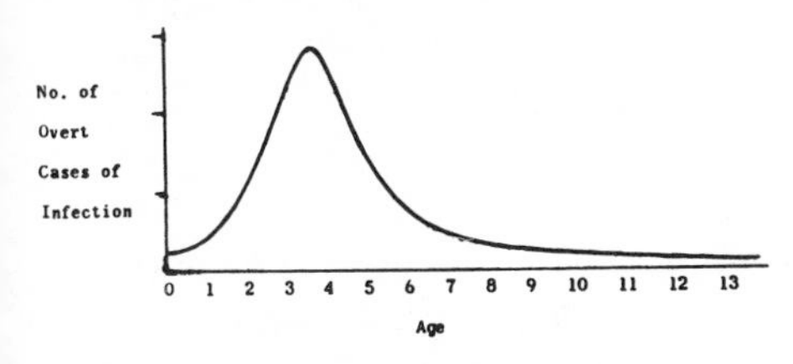

FIGURE 2. *Graph Badly Drawn, with Blurred Lines and Irregular Lettering.*

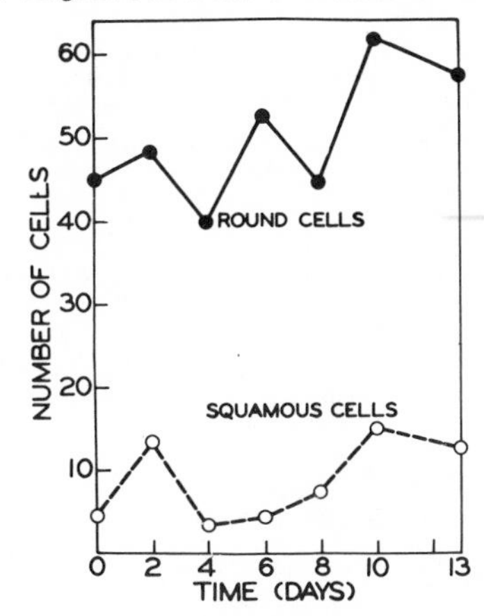

FIGURE 3. *Example of a Diagram That Is Easily Understood at a Glance.*

matter submitted in such a way that the author's own style is retained, but jargon and obscure, clumsy phraseology must be corrected.

Although it may be necessary, because of the lapse of time between submission and publication of a paper, to include an addendum citing work published after the paper was sent in, such additions should be kept to a minimum. It should never be necessary to add an illustration or a table or to alter a chart in galley proof.

If corrections are necessary they should be indicated in the margin of the proof, with instructions for deletion and addition. Each proof should be labeled "O.K. with changes or corrections" or simply "O.K." (or in some other way that will ensure that no corrections will be overlooked). All proofreader's queries should be answered.

Conclusions

This discussion has attempted to characterize the medical manuscript as an entity that is the sum of its various parts, all of importance in varying degree. Although the purely mechanical aspects have been emphasized, no component of the manuscripts should be slighted: a clear and precise presentation of text should go hand in hand with the highest possible quality of illustrative and tabular material and scrupulous attention to reference sources. The time and thought and energy involved in careful selection and revision to attain uniform excellence will help hold the reader's attention and ensure that the author's message is unmistakably imparted. The goal of communication with the reader will be achieved only if the entire manuscript represents the best effort of which the author is capable.

REFERENCES

1. Garland, J. A journal and its readers. Presented at meeting of Editors' Section, World Medical Association, London, England. September 22, 1965.
2. Fowler, H. W. *A Dictionary of Modern English Usage.* Second edition. Revised by Sir Ernest Gowers. 725 pp. London: Oxford, 1965.
3. Strunk, W., Jr., and White, E. B. *The Elements of Style.* New York: Macmillan, 1959.
4. Bernstein, T. N. *The Careful Writer: A modern guide to English usage.* 487 pp. New York: Atheneum, 1965.
5. Fishbein, M. *Medical Writing: The technic and the art.* Third edition. 262 pp. New York: Blakiston, 1957.
6. Hartley, H. L. Manuscript preparation. *Northwest Med.* **63**:35-38, 1964.
7. Wilson, G. Guidance in preparing typescript of scientific papers. *Month. Bull. Min. Health* **24**:280-307, 1965.
8. *Dorland's Illustrated Medical Dictionary.* Twenty-fourth edition. 17 pp. Philadelphia: Saunders, 1965.
9. *Blakiston's New Gould Medical Dictionary.* Edited by N. L. Hoerr a A. Osol. Second edition. 1463 pp. New York: Blakiston, 1956.
10. *Stedman's Medical Dictionary.* Twentieth edition. 1680 pp. Ba more: Williams & Wilkins, 1961.
11. Breed, R. S., Murray, E. G. D., and Smith, N. R. *Bergey's Manua Determinative Bacteriology.* Seventh edition. 1094 pp. Baltimo Williams & Wilkins, 1957.
12. Bailey, W. R., and Scott, E. G. *Diagnostic Microbiology: A textbook the isolation and identification of pathogenic microorganisms.* 327 pp. Louis: Mosby, 1962.
13. DeBakey, L. Verbal eccentricities in scientific writing. *New Eng Med.* **274**:437-439, 1966.
14. Garland, J. Shattuck Lecture: "The Proper Study of Mankin *New Eng. J. Med.* **270**:1138-1142, 1964.

Journals of history also follow the MLA *Style Sheet* or the University of Chicago *Style Manual.* Some, like the *Hispanic American Historical Review* (see p. 370) introduce subtle variations of their own. Others follow the legalistic forms of citation mentioned below. All, however, are preoccupied with citations; it is not unusual to find historical articles which contain more references than text.

Style and usage in legal articles are fixed by those graduate schools of law that publish the professional journals. Nonacademic law journals such as the *Journal of the American Bar Association* (see p. 376) usually follow the style of the *Harvard Law Review.*

Only the forms of citation and of abbreviation have been standardized in legal writing. The authoritative manual is *A Uniform System of Citation,* tenth edition, published and distributed by the Harvard Law Review Association, Gannett House, Cambridge 38, Massachusetts. This 125-page document is too lengthy to reproduce in the Appendix, but the article beginning on p. 376 illustrates some of the citation techniques recommended.

Four authorities are recognized for solving problems of style and mechanics arising in articles submitted for publication in law reviews. The journals of the Columbia, Harvard, University of Pennsylvania, and Yale law schools agree that "for spelling, syllabication, and italicization, follow Merriam-Webster's *New International Dictionary* (2nd ed., 1957); for punctuation and capitalization, follow *U.S. Government Printing Office Style Manual* (rev. ed., 1959); for grammar, follow Fowler, *Modern English Usage* (1937 ed.); for abbreviations, except for periodicals, follow Black's *Law Dictionary* (4th ed., 1951)."

The literature of education is guided by no single style manual. The most authoritative document in the discipline is the NEA *Style Manual.* This 76-page manual, numbered 381-11670, printed in 1962, and revised in 1966, is available for one dollar from the National Education Association, 1201 Sixteenth Street, N.W., Washington, D.C. 20036.

The NEA recommendations for citations are reprinted with permission on p. 380. Either these recommendations or those of the MLA *Style Sheet* are followed by the major professional journals of education.

For an annotated list of journals in this field, see L. Joseph Lins and Robert A. Rees, *Scholar's Guide to Journals of Education and Educational Psychology* (1965), Dembar Educational Research Services, Inc., Box 1605, Madison, Wisconsin 53701.

In some areas of writing, particularly in the field of history, a *bibliographic note* is replacing the alphabetized book list at the end of an article. The form is almost unique to the fine arts, humanities, and some of the social sciences. It is little used in the sciences, where the separately bound, annotated bibliography presents the same information in a more succinct form.

When the annotated bibliography is essentially an alphabetized book list, each entry of which is followed by a brief abstract, the bibliographic note is a prose discussion of the three types of literature in the field. It is an essay in three parts. The first part discusses the *books and bibliographies* that themselves discuss the literature in the field. The second part discusses *the primary sources* as to their value, significance, and validity. the third part discusses *the secondary sources, i.e.,* the articles and monographs that discuss post facto the subjects viewed at first hand by the primary sources.

The bibliographic note uses running sentences containing authors, titles, and dates to comment upon the relative merits of the sources. It is a familiar, folksy change from the clinical listing that characterizes the ordinary bibliography. As a result, the tone tends to be too personal and subjective for use with a scientific article.

An attachment differs from an appendix in two ways. First, it is a self-sufficient document usually separately bound. Second, it contains something *discussed by the article* rather than something that sheds light upon the article. For example, the digressions which make up an appendix are included in an article because they contribute in some way to the discussion. An attachment, however, is an entity such as a treaty *discussed by the article,* the record of a trial, or even an entire art form such as a sonnet series or short story.

Nonscientific journals often publish attachments in the form of *chapbooks,* which accompany an issue devoted in whole or in part to a discussion of the attachment. Scientific journals often publish them as *supplements* to an issue or announce their availability in an issue. In this last case, an additional fee may be involved.

Scientific journals, particularly those in the applied sciences, often require authors to submit *library-card abstracts* in addition to the formal abstract that appears at the beginning of the article.

A library-card abstract is not a numbered page, but an additional page bound at the end of an article. During printing, it is placed at the end of an issue of a journal with the library-card abstracts of the other articles appearing in the journal. There it may be clipped out for library reference use without mutilating the magazine.

Because library-card abstracts are primarily for the use of librarians, they are less comprehensive than formal abstracts or summaries. Because these cards are limited in size (2½″ by 4½″ card will hold 158 in. of type) important results may be omitted from the abstract material to permit inclusion of bibliographic material.

Library-card abstracts are essential sections in articles and reports submitted to the Department of Defense and other federal agencies. Because these agencies republish the cards of articles in a specific subject area *en bloc* as a service to contractors and others who have established a "need to know" in that area, the library-card abstracts of classified documents should not contain classified information.

The following sample library-card abstract is reprinted with the permission of the National Aeronautics and Space Administration.

The Defense Documentation Center requires that its own *Abstract Form* (DD Form 1473) be used in many reports. That blank form which carries its own instructions is printed in its entirety in Figure 25.

All illustrations are expensive. Their use is justified only when they present information in a highly concentrated form or when that information could not be presented so clearly in any other way. The articles reprinted in Chapter 5 contain examples of most of the techniques commonly used to illustrate technical articles.

All illustrations use abbreviations. Some of these abbreviations are listed in Chapter 5, but it is often wise to query the editor of the journal for whom the article is intended or to refer to the appropriate authority cited in the Bibliography of Abbreviation Standards at the end of this book.

In general there are three types of tables. Figure 26 illustrates a *columnar table,* usually ruled, in which similar data are grouped under boxheads. Figure 27 illustrates a *leaderwork table* in which dissimilar data are listed in rows with leaders connecting each category with the appropriate value. Figure 28 illustrates a *composite table* in which the techniques are combined. These three figures are reproduced from the NASA *Publications Manual,* and Figure 26 illustrates the headnote and footnote system recommended by that agency.

Security Classification

DOCUMENT CONTROL DATA - R & D

(Security classification of title, body of abstract and indexing annotation must be entered when the overall report is classified)

1. ORIGINATING ACTIVITY *(Corporate author)*	2a. REPORT SECURITY CLASSIFICATION
	2b. GROUP

3. REPORT TITLE

4. DESCRIPTIVE NOTES *(Type of report and inclusive dates)*

5. AUTHOR(S) *(First name, middle initial, last name)*

6. REPORT DATE	7a. TOTAL NO. OF PAGES	7b. NO. OF REFS

8a. CONTRACT OR GRANT NO.	9a. ORIGINATOR'S REPORT NUMBER(S)
b. PROJECT NO.	
c.	9b. OTHER REPORT NO(S) *(Any other numbers that may be assigned this report)*
d.	

10. DISTRIBUTION STATEMENT

11. SUPPLEMENTARY NOTES	12. SPONSORING MILITARY ACTIVITY

13. ABSTRACT

DD FORM 1 NOV 65 1473

Security Classification

Title

Footnote reference

TABLE I. — SHORT-TIME TENSILE PROPERTIES OF RECRYSTALLIZED[a] TUNGSTEN AND MOLYBDENUM

Bracketed headnote

[From ref. 2]

Boxhead

Footnote reference

Temperature, °K	Specimen type (b)	Ultimate tensile strength, N/m^2	Elongation between buttonheads, cm	Reduction of area, percent
		Tungsten		
1700	1	220 019 × 10^3	1.57	95
1900	1	131 280	1.60	75
2060	1	98 736	.69	36
2260	1	67 433	.51	25
		Molybdenum		
1650	2	93 013 × 10^3	0.95	96
1922	2	40 680	1.55	99
2255	2	14 720	1.75	99

[a]Recrystallized at 2370° K for 1/2 hour in vacuum.

[b]Specimen types are shown in figure 6.

Footnote

Figure 26

TABLE V

PERTINENT PHYSICAL CHARACTERISTICS AND DIMENSIONS OF THE TEST AIRPLANE

Total wing area, S, meters^2	130
Wing span, b, meters	35
Wing aspect ratio, A	9
Wing thickness ratio, t/c	0.12
Wing taper ratio, λ	0.42
Wing mean aerodynamic chord, $\bar{c}$, meters	4
Wing sweepback (25-percent-chord line), Λ, deg	35
Total horizontal-tail area, S_t, meters^2	24
Horizontal-tail span, b_t, meters	10
Horizontal-tail mean aerodynamic chord, c_t, meters	2.6
Horizontal-tail sweepback (25-percent-chord line), Λ_t, deg	35
Airplane weight, W, lb —	
At 1525 meters	57 260
At 10 675 meters	50 803
Center of gravity, percent $\bar{c}$	50 800

LEADERWORK TABLE

TABLE IV

ROOT-MEAN-SQUARE VALUES FOR VARIOUS MEASUREMENTS

Measurement	Root-mean-square values at —	
	1500 m	10 500 m
Pitching velocity, radians/sec	0.0078	0.0067
Rolling velocity, radians/sec	0.0205	0.0578
Yawing velocity, radians/sec	0.0090	0.0212
Normal acceleration at c.g., m/sec^2	3.43	1.94
Normal acceleration at wing tip, m/sec^2	16.58	- - - -
Lateral acceleration at c.g., m/sec^2	1.12	0.90
Gust velocity, m/sec	3.0	- - - -

COMBINATION LEADERWORK AND RULED TABLE

Figure 27

Plus signs should not be used to indicate the positive values but may be used to indicate such information as direction. If a column of values has a multiplying factor, the factor should be placed after the top value (other than zero) and not in the boxhead. If the first number in a column or under a cross rule is wholly a decimal, a cipher is added at the left. Initial zeros and multiplying factors are repeated under a cross rule. When an explanation is offered for a missing value in a column, a footnote should be added and the footnote reference mark given in parentheses at the proper place. If identical values are obtained as test values, they should be repeated or the value may be given once and an arrow drawn down the column to the next different value. If test conditions are identical for several test values, the conditions are usually given only once for the group of values to which they correspond. This grouping within columns may be shown by single horizontal rules or by spaces. If rules or spaces are impracticable, connected data may be indicated by braces.

Column headings and rows in tables should not be numbered except when necessary for reference. The numbers should then be placed above the columns or in the stub of a row in parentheses or circles and set off by a cross rule.

Leaderwork tables usually do not have rules or column headings. (See sample 10.) This type of table presents a list of different conditions pertaining to a particular subject. Frequently, a quantity, its symbol, and unit (or such information as is applicable) are given on the left with the values of the quantity given on the right. The space between the quantity and value is filled with leader dots.

Insofar as possible, each table should be self-explanatory, a unit independent of the text. Tables are numbered consecutively with Roman or Arabic numerals in order of their mention in the text. Exceptions to this are short tables which are part of the text and which follow an introductory statement, such as "Results of this test are given in the following table:". These tables are not numbered, do not have titles, and are not referred to elsewhere in the text.

The title is an integral part of the table and should be as exact and descriptive as possible. Additional information that applies to all the data in the table is given in a bracketed headnote beneath the title. (See sample 9.)

Titles should be set up in either of the two following forms and centered above the table:

TABLE IV. — EFFECT OF AGING ON CREEP PROPERTIES OF ALUMINUM

TABLE IV

EFFECT OF AGING ON CREEP PROPERTIES OF ALUMINUM

Figure 28

Corresponding forms for titles with continued parts of a table are as follows:

TABLE IV. — EFFECT OF AGING ON CREEP PROPERTIES
OF ALUMINUM - Continued

TABLE IV. — Continued

EFFECT OF AGING ON CREEP PROPERTIES OF ALUMINUM

For the last page of a continued table the word "Concluded" is used instead of "Continued."

If subtitles are necessary they should be centered beneath the title or beneath the bracketed headnote and designated by letters in parentheses.

Boxheads should be brief; if necessary they may be amplified by footnotes. Boxheads usually contain such information as quantity, symbol, and unit, separated by commas. Only the initial letter of the first word is capitalized. Material in boxheads should run crosswise if possible. Periods are not used following boxheads but a dash (or, sometimes, a colon) is used if boxhead material reads into the following matter. (See sample 11.)

Sketches may be part of a table, if the combination causes no makeup difficulties. A sketch that is part of a table is never considered a numbered figure.

Simple tables up to four columns may be left unruled. Single rules are used except for special cases, such as division in doubled-up tables.

If a table has subdivisions for different conditions, the heading for the first condition should be inserted directly after the lower cross rule of the boxheads. (See sample 9.) It should be centered and separated from the data by a complete horizontal rule. Vertical rules do not cross subdivisions. Comparable subdivision headings should be written in the same style. Boxheads and units should not be repeated if they are alike for all conditions, but the columns should be alined.

Footnotes may be used to explain or amplify a title, boxhead, or value. (See sample 9.) Footnote reference marks are usually lowercase letters and begin anew for each table. If a footnote pertains to an entire column, the reference mark is enclosed in parentheses and placed at the bottom of the boxhead, just above the column rule. Footnote reference marks are placed before numbers and after words and are introduced in a table from left to right and from top to bottom. The reference mark for a footnote referring to the entire title of a table is placed at the end of the title; the mark for a footnote referring to a part of the title is placed with that part.

Where practicable, tables should be inserted in the text as near (preferably following) their first mention as possible. A table may be placed on a page upright (preferably) or sideways, or the typed text may run above, below, or around a small table, or several tables may be grouped together on a page. Where tables are voluminous or their insertion unduly interrupts the flow of the text they may be grouped in proper sequence following the text. (See section on "Makeup," page 19.)

For any type of makeup, the final size of the table determines the size of lettering or typing used in preparing the original. Ideally, most typewritten letters and numbers should not be reduced more than one-half. Large tables planned for reduction to page size should be typed to dimensions proportional to page image area, namely 6-5/8 by 8-5/8 inches. In order to obtain the proper proportions, long narrow tables may be divided into halves, thirds, and so forth, and doubled up with two vertical rules separating the parts. In such doubled-up tables the boxheads are repeated.

An increasing number of reports contain data electronically tabulated. These data are not typed but are printed directly from the printout sheet to avoid error. Usually, a transparent overlay with rules or boxheadings is used in photographing the printout sheets taken from the computing machine. If possible, authors should have copy and setup checked by an editor before having overlays prepared to insure that symbols, spellings, and so forth are correct. No changes (such as adding initial ciphers to decimal fractions) should be made in printout data.

Because of the difficulty of preparing composite tables for printing, terminology and general arrangement have been standardized in some disciplines. Figure 29, which is reproduced by permission from the American Psychological Association *Style Manual,* identifies and illustrates the major parts of a table.

For other examples of tables within this book refer to pp. 181, 360, and 396.

Tables should be self-explanatory and independent units whenever possible. They should be numbered consecutively with arabic numerals in order of their mention in the text. Table numbers and titles should be centered above the table, and if a table is continued on additional pages, indication of this should be made in the title. For example;

Table 3. Effect of LSD on Schizophrenics—Continued

or

Table 3—Continued.

Effect of LSD on Schizophrenics

On the last page of a continued table the word "concluded" is used instead of "continued."

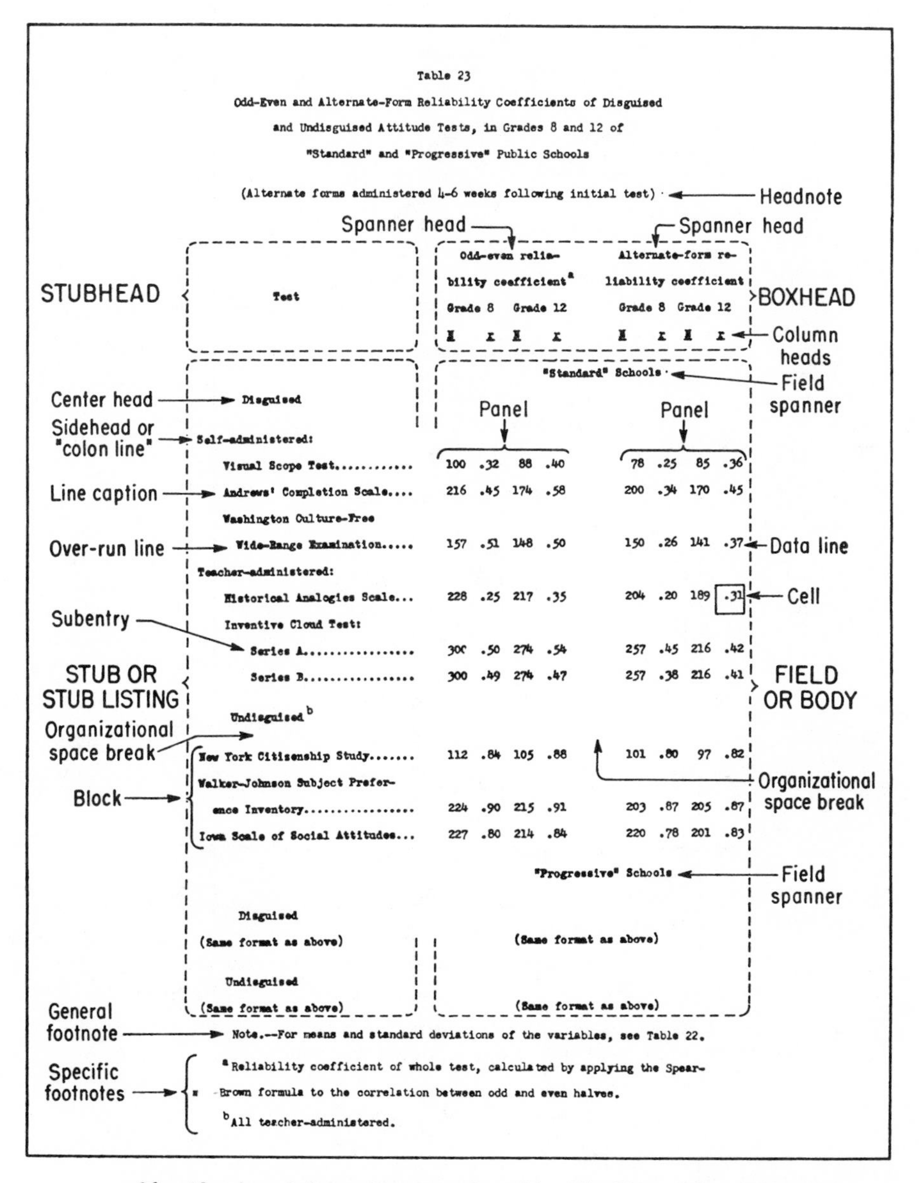

Table 23

Odd-Even and Alternate-Form Reliability Coefficients of Disguised and Undisguised Attitude Tests, in Grades 8 and 12 of "Standard" and "Progressive" Public Schools

(Alternate forms administered 4-6 weeks following initial test)

Test	Odd-even reliability coefficient[a]				Alternate-form reliability coefficient			
	Grade 8		Grade 12		Grade 8		Grade 12	
	N	r	N	r	N	r	N	r
	"Standard" Schools							
Disguised								
Self-administered:								
Visual Scope Test............	100	.32	88	.40	78	.25	85	.36
Andrews' Completion Scale....	216	.45	174	.58	200	.34	170	.45
Washington Culture-Free Wide-Range Examination.....	157	.51	148	.50	150	.26	141	.37
Teacher-administered:								
Historical Analogies Scale...	228	.25	217	.35	204	.20	189	.31
Inventive Cloud Test:								
Series A.................	300	.50	274	.54	257	.45	216	.42
Series B.................	300	.49	274	.47	257	.38	216	.41
Undisguised[b]								
New York Citizenship Study.......	112	.84	105	.88	101	.80	97	.82
Walker-Johnson Subject Preference Inventory.................	224	.90	215	.91	203	.87	205	.87
Iowa Scale of Social Attitudes...	227	.80	214	.84	220	.78	201	.83
	"Progressive" Schools							
Disguised								
(Same format as above)	(Same format as above)							
Undisguised								
(Same format as above)	(Same format as above)							

Note.--For means and standard deviations of the variables, see Table 22.

[a] Reliability coefficient of whole test, calculated by applying the Spearman-Brown formula to the correlation between odd and even halves.

[b] All teacher-administered.

Identification of the major parts of a table. The illustration also provides an example of the arrangement, spacing, and alignment of a well-prepared typewritten table. Note that overrun lines are indented two spaces; subentries are indented four spaces.

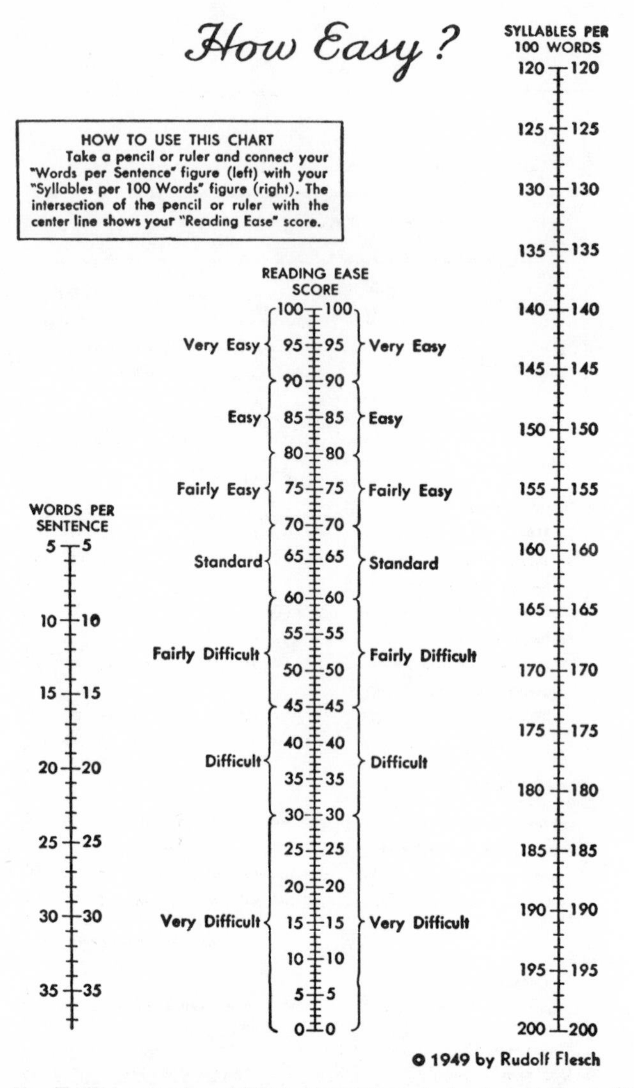

From *How To Write, Speak, and Think More Effectively*, by Rudolf Flesch (New York, Harper & Bros., 1960). Copyright 1960 by Rudolf Flesch. Used by permission.

To Find "How Easy" Your Writing Reads:

(1) Take a sample of 100 words.
(2) Divide the number of sentences in the sample into 100 to get the average sentence length.
(3) Find the corresponding number on the "words per sentence" scale.
(4) Count the number of syllables in the 100-word sample.
(5) Find that number on the "syllables per 100 words" scale.
(6) Follow the instructions in the box above entitled "How To Use This Chart" to get a rough estimate of reading ease.

Figure 30

Graphs and curves should be used to show relationships between or among varying data. *Bar graphs* should be used when the data are *discontinuous*. *Curves* should be used when the data are *continuous*. An example of an article based entirely upon curves and graphs is printed on pp. 394–397. Figure 30, reproduced by permission of Rudolph Flesch, illustrates the effect that can be achieved by juxtaposing graphs of dissimilar data.

Pie cuts and bar graphs tricked out as soldiers of varying heights, ships of varying length, or any other symbol representative of the data being compared should be avoided. Professionals will suspect such tricks as misleading.[9] For example, a graph showing two ships, one twice as long as the other, may attempt to show something twice as long as something else. Some readers, however, will see only *total area* and form the erroneous opinion that a magnitude of four is involved.

Because most readers are suspicious—and intelligent—bar graphs are almost always suspect. The three bar graphs in Figure 31 show one reason why. Graph A is misleading, for it suggests an increase of 100 percent from 1952 to 1956. Graph B is misleading, for it suggests that no striking change occurred between 1952 and 1956. Graph C is best, for it not only stresses the change but also alerts the reader to the rigging of line lengths.

[9]Darrell Huff, *How to Lie with Statistics* (New York: Norton and Company, 1954), p. 27. See also Chapter 8 "Figures Can Lie" in Henry A. Davidson, M.D., *Guide to Medical Writing* (New York: Ronald Press, 1957), pp. 142–161.

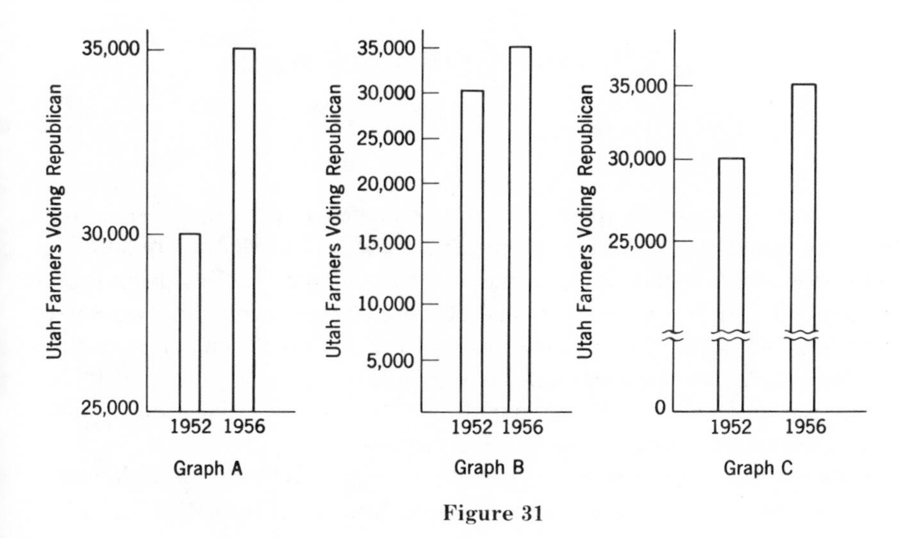

Figure 31

Curves are differentiated by solid and dashed lines. The line order for the introduction of different kinds of curves in a figure is shown in Figure 32.

Figure 32

In any one article, the same style of line should be used to represent the same condition in related figures. Nonstandard lines such as the arrowed line in Figure 32 should only be used to stress a specific situation.

Figure 33 illustrates how three charts may be linked to save printing costs. Such linkages should only be made when the meaning is obvious. Figure 33 would be too complex for some editorial tastes.

Points on curves may be indicated by symbols in the following order of introduction. If more symbols are needed, flags may be added in different positions on those symbols. Solid or partly filled-in symbols and the signs + and × should not be used.

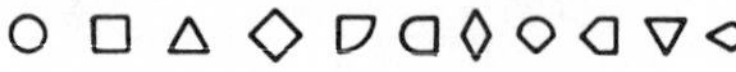

Figure 34

The charts and maps used to illustrate professional articles are often specially drawn (see examples on pp. 261–265). The selection of suitable projection and the selection of a meaningful area are vital in a brief presentation of article length. For this reason, the excellent and accurate charts and maps that the Hydrographic Office and the Department of Agriculture made available through the Government Printing Office will not always serve. However, if a map is drawn to order and distorted for emphasis, the reader must be told of the distortion.

Because charts and graphs are usually complex and difficult to do well, it is possible and often wise to imitate the successful solutions made by

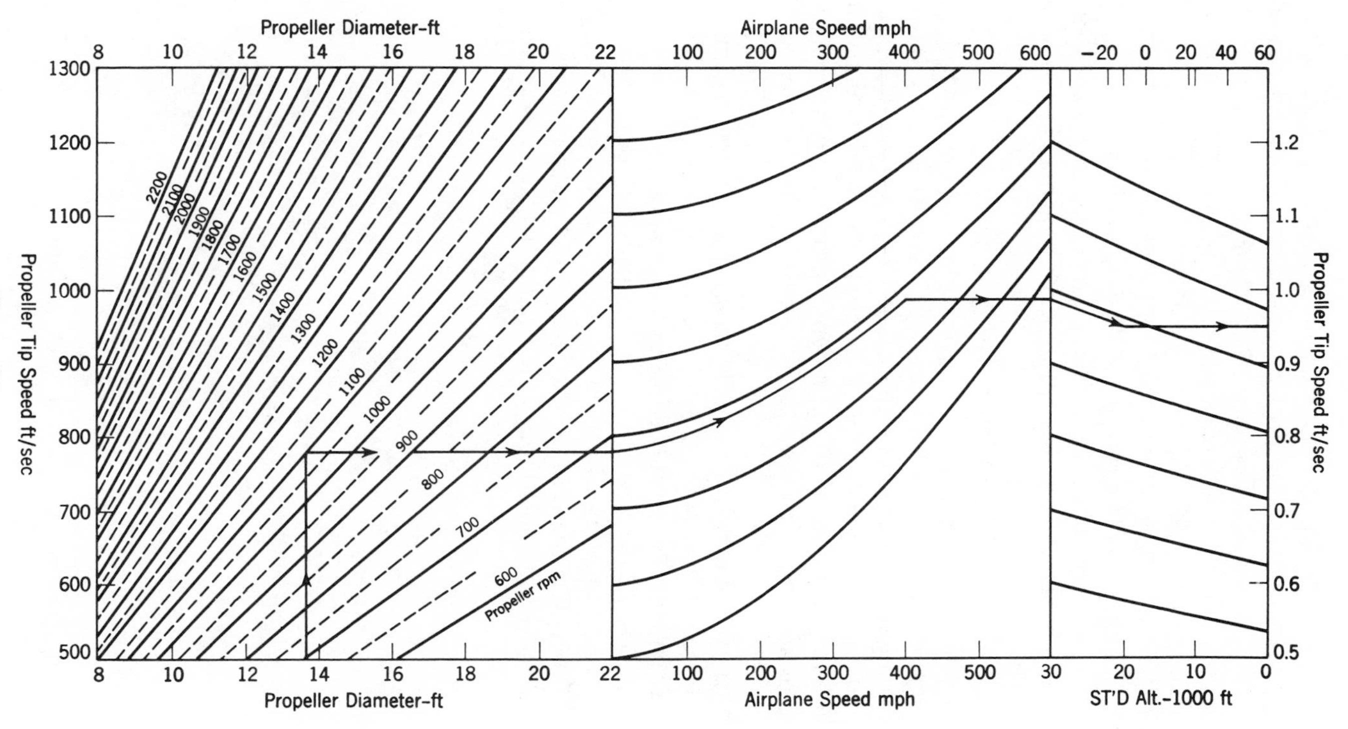

Propeller Diameter-ft
Airplane Speed mph
Propeller Tip Speed ft/sec
ST'D Alt.-1000 ft
Propeller rpm
600
700
800
900
1000
1100
1200
1300
1400
1500
1600
1700
1800
1900
2000
2100
2200

Figure 33

others to similar problems. A good source book is *Making the Most of Charts; An ABC of Graphic Presentation,* by K. W. Haemer, published by Tecnifax Corporation, Holyoke, Massachusetts.

An easy-to-understand discussion of the techniques for drawing and reproducing illustrations of all types is contained in Section III of T. O. 00-5-12 "Guide for the Preparation of Air Force Technical Orders." This Technical Order is available at most Air Force facilities.

The availability of artists, art services, photographers, and photograph and picture syndicates is shown in classified telephone directories and in the two annuals *Literary Market Place* (R. R. Bowker Company, New York) and *Writer's Market* (Writer's Digest, Cincinnati).

Few amateur photographers are capable of taking the sharp, black-and-white, glossy photos or prints essential for a printer. Even fewer can retouch photographs effectively. The availability of professionals is indicated in the preceding paragraph. This book is concerned only with the procedure for submitting glossy photographs to illustrate journal articles. The use of such photographs to clarify prose is illustrated on pp. 216–217.

Most journal editors will use only large, glossy prints which may be cropped or reduced without destroying quality. Should the writer wish to indicate cropping lines, he may do so *on a transparent overlay,* not on the print itself. Should the writer wish to add arrows or other markings, he may do so *on the overlay*. Marks on the overlay should be made lightly or with a felt-tipped marker, for a sharp pencil will make an imprint on the surface of the photograph through the overlay, and the imprint will be reproduced.

5

SAMPLE STYLE GUIDES AND ARTICLES

The illustrative style guides and articles collated in this chapter are representative; the selections are not all-inclusive. Other style guides exist and might have been included. Those selected, however, are either "standard" or definitive in the disciplines they serve.

A. APPLIED SCIENCE JOURNALS

The journals of engineering and business are more numerous than those of all the other disciplines combined. This may be partially explained by the need for immediacy and by the vast amount of information that is essential to progress in a technical society. It may also be explained by the existence of "applied" areas in every discipline. In all events, the journals of engineering and business synthesize wide subject areas and impose professional standards of their own.

For convenience the journals of the applied sciences of engineering and business are discussed separately below. Complete separation is impossible, however, for overlapping will exist in any research and development situation. Certainly, economic reality is interwoven with the function of the laboratory.

Engineering Journals

The dissimilar methods and goals of the disciplines or divisions of engineering have caused each to create its own style guide. The guide printed below, by permission of the IEEE, is loosely followed by the thirty journals of the Institute of Electrical and Electronics Engineers. As indicated in the sections quoted, the goals of these journals are themselves dissimilar.

In practice writers observe the minimum requirements of a specific journal and follow the manual of one of the physical sciences when complex communication or printing problems arise; for example, the illustrative article concerns a problem in communication. It was printed in *The Journal of Industrial Engineering,* but it follows the *Style Manual* of the American Chemical Society.

INFORMATION FOR IEEE AUTHORS

An author should familiarize himself with the following information before he prepares and submits a manuscript to any of the more than 30 journals published by the IEEE. Review and editorial processing of a manuscript are facilitated if it is in the proper form.

These instructions are for the guidance of authors preparing papers to be considered for publication in the following:

"1. IEEE SPECTRUM, a monthly technical magazine received by all wide variety of subjects of interest and importance to its diverse readership. Included are tutorial articles and articles reviewing the state of the art in specific fields.

"2. The PROCEEDINGS OF THE IEEE, a monthly research-oriented journal containing fundamental papers of broad significance and long-range importance to electrical science and technology. In addition to contributions describing original work, the PROCEEDINGS publishes papers which are high-level reviews of specific areas of current interest.

"3. The various IEEE TRANSACTIONS, published by each of the IEEE Groups at frequencies ranging from two to 12 times a year. Each of the TRANSACTIONS is devoted principally to specialized papers relevant to the field of interest of the sponsoring group.

"4. The IEEE STUDENT JOURNAL, a publication for all undergraduate IEEE Student members, which contains technical and nontechnical material intended to assist them with their career decisions.

"Anyone may submit a paper for consideration for an IEEE publication; membership in the IEEE is not a requirement. A manuscript should be submitted exclusively to a single IEEE journal, should not have been published before, and should not be under consideration for publication elsewhere.

"These instructions have been divided into two sections: 'General Information' and 'Information for Specific Publications.'

GENERAL INFORMATION

Preparation of text

"Manuscripts should be typewritten, using *double* spacing,* on one side of the sheet. Allow ample margins of at least one inch or 2.5 cm on each side. Number all pages, illustrations, footnotes, and tables. All illustrations should be referred to in the text.

Units

"*Metric* units are preferred for use in IEEE publications in light of the non-national character of the IEEE, the international readership of its journals, and the inherent convenience of these units in many fields. In particular, the use of the International System of Units (Système Internationale d'Unités, or SI Units) is advocated. This system includes as a subsystem the MKSA units, which are based on the meter, kilogram, second, and ampere.

"If an author expresses quantities in English units, he is urged to give the metric equivalents in parentheses, for example, a distance of 4.7 inches (11.9 cm); however, this practice may be impractical for certain industrial specifications, such as those giving drill sizes or horsepower ratings of motors.

Symbols and abbreviations

"The use of unit symbols and other abbreviations is optional. They should be employed only when desirable for brevity.

"The IEEE 'Standard Symbols for Units' (IEEE No. 260) should be followed; details are given in the Appendix. Avoid other abbreviations except those that are generally accepted or are listed in Table II of the Appendix. Abbreviations and symbols used on illustrations should conform to those used in the text.

Mathematical notation

"To prevent errors by the printer, subscripts, superscripts, Greek letters, and other symbols should be clearly identified (by means of a note in the margin, if necessary). In particular, a clear distinction should be made between the following:

*Manuscripts to be submitted to the TRANSACTIONS of the Power Group and the Applications and Industry Group should be typewritten using single spacing for possible photoreproduction.

1. Capital and lower-case letters, when used as symbols.
2. Zero and the letter 'o.'
3. The small letter'el,' the numeral one, and a prime sign.
4. The letters 'k' and kappa, 'u' and mu, 'v' and nu, 'n' and eta.

"Care should be observed in using the solidus (slant), radical sign, and parentheses and brackets to avoid ambiguities in equations.

"To facilitate the reading of numbers and to eliminate confusion from differing uses of the comma and period in various parts of the world, IEEE editorial practice is to separate the digits by a space into groups of three, counting from the decimal sign toward the left or right. In numbers with four digits, the space is not necessary. If the magnitude of a number is less than unity, the decimal sign should be preceded by a zero. Examples are:

12 531 7465 9.2163 0.102 834

References

"References should usually be footnotes at the bottom of the pages on which they are cited, but extended bibliographies may be placed at the end of the paper. References should be complete and in the form below.

"For a periodical: J. A. Rich and G. A. Farrall, 'Vacuum arc recovery phenomena,' *Proc.* IEEE, vol. 52, pp. 1293–1301, November 1964.

"For a book: J. D. Kraus, *Antennas*. New York: McGraw-Hill, 1950, pp. 100–108.

"References should be to commonly available publications and books; references to reports of limited circulation should be avoided.

Illustrations

"Finished drawings in black ink on white paper or tracing cloth, or photographic prints of original drawings, should be provided for all journals, with the exception of SPECTRUM. Photographs should be glossy prints not exceeding 8½ by 11 inches (21.6 by 27.9 cm). Most illustrations are reduced in size to a 3½-inch (8.9-cm) column width when printed, so it is important that all letters, numbers, and lines be so drawn as to remain legible after reduction. Letters and numbers should be at least $\frac{1}{16}$ inch (1.6 mm) high *after* reduction to be readable.

"All information to be reproduced on illustrations should be lettered in ink, not typewritten. Graphs should be drawn with only the major coordinate lines showing, since a chart containing a large number of closely spaced lines will not reproduce legibly. Short "ticks," extending a short distance from the axes, may be provided for convenience in reading intermediate values.

Captions

"All captions, with figure numbers, should be listed on a separate sheet. The captions should be self-explanatory, not merely labels. If also included on the drawing itself, a caption should not appear within the area to be reproduced.

Organization

"IEEE papers usually consist of five parts: title, abstract, introduction, body, and conclusions. Two additional divisions, a glossary of symbols and an appendix, are sometimes desirable. SPECTRUM and the STUDENT JOURNAL papers may be less formally organized than those for the PROCEEDINGS and the TRANSACTIONS.

"The *title* should clearly indicate the subject of the paper as briefly as possible. Since a paper is indexed by significant words in the title, and many readers select papers to read on the basis of the title, it should be chosen with considerable care.

"An informative *abstract* of less than 200 words is needed for all PROCEEDINGS and TRANSACTIONS papers. It should state concisely, but not telegraphically:

1. What the author has done.
2. How it was done (if that is important).
3. The principal results (numerically, when possible).
4. The significance of the results.

The abstract should be informative, *not* merely a list of general topics that the paper covers, because it will probably appear later in an abstract journal.

"The text of a paper can sometimes be simplified by following the abstract with a *glossary of symbols* if the paper contains equations in which many symbols are used.

"The *introduction* orients the reader with respect to the problem and should include the following:

1. The nature of the problem.
2. The background of previous work.
3. The purpose and significance of the paper.

Where applicable, the following points may also be included:

4. The method by which the problem will be attacked.
5. The organization of the material in the paper.

"The *body* contains the primary message of the paper in detail. The writer should bear in mind that his object is to communicate information efficiently and effectively to the reader. Even workers in the same field appreciate clear indications of the line of thought being followed, and frequent guideposts are essential for nonspecialists who want to understand the general nature and significance of the work. The use of trade names, company names, and proprietary terms should be avoided.

"The *conclusions* should be clearly stated, and should cover the following:

1. What is shown by this work and its significance.
2. Limitations and advantages.

Where applicable, the following points should also be included:

3. Applications of the results.
4. Recommendations for further work.

"Mathematical details which are ancillary to the main discussion of the paper, such as many derivations and proofs, may be included in one or more *appendixes*.

INFORMATION FOR SPECIFIC PUBLICATIONS

IEEE Spectrum

"A SPECTRUM article should be written with a very general audience in mind, as the magazine's circulation is more than 100,000. The subject should have broad significance and appeal, and should be presented so that it is understandable to the 'average' engineer. SPECTRUM articles are often revised by the IEEE editorial staff to enhance their readability; alterations appear on the galley proofs that authors receive prior to publication.

"Because all drawings are redone, final inked drawings are not necessary; legible sketches will suffice. Biographies and photographs of authors are required for publication.

"Three copies of manuscripts intended for SPECTRUM should be sent, with an identifying letter, to IEEE SPECTRUM, 345 East 47th Street, New York, N.Y. 10017. Because many articles are written for SPECTRUM in response to invitation, it is recommended that prospective authors inquire in advance concerning their proposed subject.

Proceedings of the IEEE

"Papers for this journal should be of significance and interest to a broad cross section of research-oriented engineers. A rough guide is that a PROCEEDINGS paper should be important to specialists in two or more Groups. It should be written so as to be intelligible to workers not engaged in the immediate field of the paper.

"The PROCEEDINGS also publishes letters announcing new research results, for which fast publication is desirable. Length is limited to approximately four double-spaced typewritten pages, with each illustration counted as a half page. To avoid delay, the author should indicate that his letter is intended for the PROCEEDINGS; he should provide illustrations suitable for reproduction; and he should follow the instructions accompanying the proof he will receive if his letter is accepted.

"Three copies of manuscripts intended for consideration for the PROCEEDINGS should be sent, with an identifying letter, to PROCEEDINGS OF THE IEEE, 345 East 47th Street, New York, N.Y. 10017. Biographies and photographs of authors of papers, but not of letters, are required for publication.

IEEE Transactions

"A paper or letter for one of the TRANSACTIONS should be submitted to the Editor listed on the inside front or back cover of the TRANSACTIONS, or in the April and October issues of SPECTRUM. If there is uncertainty as to whom to address the copies of a manuscript, they should be sent to the IEEE Managing Editor for forwarding to the appropriate TRANSACTIONS Editor. Most TRANSACTIONS require three copies of the manuscript and author photographs and biographies.

IEEE Student Journal

"An article for the STUDENT JOURNAL should be of interest and importance to undergraduate students of electrical engineering and technology, and it should be written so that it is comprehensible, useful, and appealing to them. It is advisable for a prospective author to write to the SJ Editor for comments and suggestions prior to preparing an article.

"Three copies of manuscripts intended for the STUDENT JOURNAL should be addressed to IEEE STUDENT JOURNAL, 345 East 47th Street, New York, N.Y. 10017. Biographies of authors are required for publication.

APPENDIX

Unit symbols

The IEEE Standard entitled 'Standard Symbols for Units' (IEEE Standard No. 260, dated January 15, 1965) lists symbols that may be used in place of the names of units. Symbols from this Standard for some important units, together with other common abbreviations, are given in

Table II. Their form is the same for both singular and plural usage, and they are not followed by a period. The distinction between upper-case and lower-case letters should be carefully observed.

"When a compound unit is formed by the multiplication of two or more units, its symbol consists of the symbols for the separate units joined by a raised dot, e.g., N·m for newton meter. When a compound unit is formed by division of one unit by another, its symbol consists of the symbols for the separate symbols either separated by a solidus (slant) or multiplied using negative powers, e.g., m/s or $m \cdot s^{-1}$ for meter per second.

Prefixes

"Prefixes indicating decimal multiples or submultiples of units and their symbols are given in Table I. Compound prefixes, such as 'micromicro' for 'pico' and 'kilomega' for 'giga,' are discouraged.

Abbreviations

"In general, most abbreviations of technical terms are capitalized, but there are notable exceptions such as ac, dc, and rms. In addition to the unit symbols, Table II lists many common technical abbreviations in their standard IEEE editorial forms. Note that periods are not used, and that the abbreviation is the same regardless of whether it is used as a noun or an adjective. A symbol which is new or not generally accepted should be defined when first used."

I. Recommended prefixes

Multiple	Prefix	Symbol
10^{12}	tera	T
10^{9}	giga	G
10^{6}	mega	M
10^{3}	kilo	k
10^{2}	hecto	h
10	deka	da
10^{-1}	deci	d
10^{-2}	centi	c
10^{-3}	milli	m
10^{-6}	micro	μ
10^{-9}	nano	n
10^{-12}	pico	p
10^{-15}	femto	f
10^{-18}	atto	a

IEEE Recommended Practice for Units in Published Scientific and Technical Work

Prepared by IEEE Standards Coordinating Committee 14 (Quantities and Units)

Chester H. Page, Chairman

Bruce B. Barrow *Wayne Mason*
J. G. Kreer, Jr. *W. T. Wintringham*

Approved by IEEE Standards Committee, October 14, 1965

"The recommendations contained in this document are based upon the following premises, which are believed to represent the broadest base of general agreement among proponents of the major unit systems:

a. That for most scientific work and technical work of an analytic nature the International System of Units (officially designated "SI")* is generally superior to other systems; that this is particularly true for the fields of electrical science and technology since the common electrical units (ampere, volt, ohm, etc.) are included among the SI units; that the International System is more widely accepted than any other as the common language in which scientific and technical data ought to be expressed.

b. That various units of the British and American systems (hereafter referred to as "British-American units"), particularly the inch and the pound, are the fundamental units used in the standards followed by a large part of the world's manufacturing industry; that this will continue to be true for some time.

c. That unit usage can and should be simplified, particularly in the fields of interest to the IEEE, that one means toward such simplification is the identification of obsolete and unneeded units, and that another is the adoption of more rational links between SI units and units of other systems.

*The International System includes as subsystems the MKS system of units, which covers mechanics, and the MKSA system, which covers mechanics, electricity, and magnetism. Appendix A describes the International System of Units.

1. RECOMMENDATIONS REGARDING PREFERRED SYSTEM OF UNITS

Technical and scientific data, except in cases such as those cited below, should be given in units of the International System, followed if desired by the equivalent data in other units given in parentheses.† *This recommendation applies only for the publication of data and does not in any way affect the choice of units for manufacturing processes or for industrial standards.*

1.1 Exceptions

1.1.1 When a nonmetric industrial standard is referred to, the nonmetric data should be given first place, with the SI equivalent data given in parentheses. Examples include standard inch sizes of nuts and bolts, American Wire Gauge sizes of electrical conductors, and the like.

1.1.2 Where, in special fields, for reasons related to the field, a unit not in the Internation System offers significant advantages, such a unit may be used, with the SI equivalent data given in parentheses. Examples include the use of the electronvolt in physics, the nautical mile in navigation, and the astronomical unit in astronomy.

1.1.3 Experimental data taken in non-SI units may be quoted without change, followed by the SI equivalents in parentheses.

1.1.4 When a non-SI unit is at present too widely used to be eliminated immediately, its use may be continued with the SI equivalent given in parentheses. Examples include the use of the horsepower to measure mechanical power and the use of the torr to measure air pressure. This exception is not intended as a permanent protection for irrational practices, however, and as future standards are drafted such anachronisms as the horsepower should be eliminated.

1.1.5 Some nonelectrical quantities are frequently expressed in units that are decimal multiples of SI units. Such units include the liter, the hectare, the bar, the tonne (metric ton), and the angstrom. These units may be used in appropriate fields. The SI units *newton* and *joule* are, however, to be used in place of the *dyne* and *erg*.

1.2 Specific Recommendations

1.2.1 The various CGS units of electrical and magnetic quantities are no longer to be used. This includes the various ab- and stat- designations (abvolt, statcoulomb, etc.) and the gilbert, oersted, gauss, and maxwell. It

†The size or description of apparatus, nominal or nonprecise dimensions, and other measurements not entering into the presentation of scientific data need not be "force-translated" into the SI; e.g., "The interferometer mirror, mounted on 1-inch rods, was advanced in 10-nanometer increments."

is of course recognized that in some cases it may be desirable to state CGS equivalents in parentheses following data given in the International System.

1.2.2 The use of prefixes‡ to express decimal multiples of SI units is permitted. Thus, although the SI unit for current density is the ampere per square meter, use of the ampere per square centimeter where convenient is entirely permissible.

1.2.3. Use of metric-system units that are not decimal multiples of SI units, such as the calorie and the kilogram-force, is especially to be avoided. Note that the 9th General Conference on Weights and Measures has adopted the joule as the unit of heat, recommending that the calorie be avoided wherever possible.

1.2.4 Where it is absolutely necessary to use British-American units, the conversion to SI units should be kept as obvious as possible. For example, if an electric generator is built to inch specifications, it may be inconvenient to express magnetic flux density in teslas (webers per square meter). In such a case webers per square inch should be used, not maxwells ("lines") per square inch.

2. RECOMMENDATION ON UNIT CONVERSIONS

Unit conversions are to be handled with careful regard to the implied correspondence between accuracy of data and the number of figures given.

2.1 Remarks

2.1.1 Some applicable rules for conversion and rounding, from American Standard B48.1, are quoted in Appendix B.

2.1.2 When a dimension or other quantity is specified with tolerances, care must be taken to ensure that the range of values permitted after unit conversion lies within the original tolerances.

2.1.3 Whenever possible, unit conversions should be made by the author of the document, for he is best able to determine how many figures are significant and should be retained after conversion. For example, a length of 125 feet converts exactly to 38.1 meters. If, however, the 125-foot length had been obtained by rounding to the nearest 5 ft, the conversion should be given as 38 m; and if it had been obtained by rounding to the nearest 25 ft, the conversion should be given as 40 m.

‡The official list of prefixes is given in Appendix A. Compound prefixes (e.g., millimicro-) are not to be used. Multiplication beyond the range covered by the prefixes should be handled by using powers of ten.

3. RECOMMENDATIONS WHERE A UNIT HAS MORE THAN ONE NAME

3.1 General Principles

3.1.1 Many of the derived units in the International System have been given special names, which may be used as alternatives to the compound forms. When the special names are formally recognized by the General Conference on Weights and Measures (CGPM), they have a standing which is equal to that of the compound forms. Both names are technically correct, and the choice between them must be made partly on the basis of taste, keeping in mind the particular application.
Example: The unit of electrical resistance is the *ohm,* which is equivalent to the *volt per ampere*. Resistance is almost always expressed in ohms, but there are occasions when the explicit expression in volts per ampere is desirable. Such usage is entirely correct.

3.1.2 Some names that have been proposed for derived SI units have not yet been recognized by the CGPM. If such a proposed name is used, care must be taken to make sure that the intended meaning is clear.
Example: The name *pascal* has been proposed for the SI unit of pressure, the newton per square meter.

3.1.3 The use of special names for decimal multiples of SI units is not recommended. Some well-established, nonelectrical units are, however, recognized as exceptions (see 1.1.5).

3.1.4 The decimal prefixes are not recommended for use with British-American units. An exception is made for the *microinch,* a unit that is frequently used in precision machine work.

3.1.5 Except for the unit of electrical conductance (see 3.2.3), the practice of giving special names to reciprocal units is to be discouraged. In particular, use of the name *daraf* for the *reciprocal farad* is not recommended.

3.2 Specific Recommendations

3.2.1 *Frequency*. The CGPM has adopted the name *hertz* for the unit of frequency, but *cycle per second* is widely used. Although *cycle per second* is technically correct, the name *hertz* is preferred because of the widespread use of *cycle* alone as a unit of frequency. Use of *cycle* in place of *cycle per second,* of *kilocycle* in place of *kilocycle per second,* etc., is incorrect.

3.2.2 *Magnetic Flux Density*. The CGPM has adopted the name *tesla* for the SI unit of magnetic flux density. The name *gamma* shall not be used for the unit *nanotesla* (see 3.1.3).

3.2.3 *Electrical Conductance*. The CGPM has not yet adopted a short name for the ampere per volt (or reciprocal ohm), the SI unit of electrical conductance. The International Electrotechnical Commission has recommended the name *siemens* for this unit, but the name *mho* has been more widely used. In IEEE publications the name *mho* is preferred.

3.2.4. *Temperature Scale.* In 1948 the CGPM abandoned *centigrade* as the name of a temperature scale. The corresponding scale is now properly named the *Celsius* scale, and further use of *centigrade* for this purpose is deprecated.

3.2.5. *Luminous Intensity.* The SI unit of luminous intensity has been given the name *candela,* and further use of the old name *candle* is deprecated. Use of the term *candlepower,* either as the name of a quantity or as the name of a unit, is deprecated.

3.2.6. *Luminous Flux Density.* The common British-American unit of luminous flux density is the *lumen per square foot.* The name *footcandle,* which has been used for this unit in the U.S., is deprecated.

3.2.7. *Micrometer and Micron.* Although the name *micron* has been widely used for the micrometer, it is not recommended. The name *nanometer* is preferred over *millimicron,* which is deprecated.

3.2.8. *Gigaelectronvolt.* Because *billion* means a thousand million in the United States but a million million in most other countries, its use should be avoided in technical writing. The term *billion electronvolts* is deprecated; use *gigaelectronvolt* instead.

4. RECOMMENDATIONS CONCERNING BRITISH-AMERICAN UNITS

4.1 In principle the number of British-American units in use should be reduced as rapidly as possible.

4.2 Quantities are not to be expressed in mixed units. For example, a mass should be expressed as 12.75 lb, rather than as 12 lb, 12 oz.

4.3 As a start toward implementing the recommendation of 4.1, above, the following should be abandoned:

British thermal unit
horsepower
Rankine temperature scale
US dry quart, US liquid quart, and UK (Imperial) quart, together with their various multiples and subdivisions*
footlambert†

*If it is absolutely necessary to express volume in British-American units, the cubic inch or cubic foot should be used.

†If it is absolutely necessary to express luminance in British-American units, the candela per square foot or lumen per steradian square foot should be used.

APPENDIX A

THE INTERNATIONAL SYSTEM OF UNITS

The following is a translation, from the original French, of the principal resolution on the International System adopted by the General Conference on Weights and Measures.

Resolution of the 11th General Conference on Weights and Measures (1960)

International System of Units (Resolution No. 12)

The Eleventh General Conference on Weights and Measures,
Bearing in mind:

Resolution No. 6 of the Tenth General Conference on Weights and Measures by which it adopted the following six units to serve as a basis for the establishment of a practical system of measures for international purposes:

length....................	meter	m
mass.....................	kilogram	kg
time.....................	second	s
electric current............	ampere	A
thermodynamic temperature	Kelvin degree	°K
luminous intensity.........	candela	cd

Resolution No. 3 adopted by the International Committee on Weights and Measures in 1956,

The recommendations adopted by the International Committee on Weights and Measures in 1958 concerning the abbreviation for the name of this system and the prefixes to be used for the formation of multiples and submultiples of units,

Decides:

1° the system based on the six basic units mentioned above is designated by the name International System of Units;

2° the international abbreviation for the name of this System is: SI;

3° the names of multiples and submultiples of units are formed by the use of the following prefixes:

Factor by Which the Unit Is Multiplied	Prefix	Symbol
1 000 000 000 000 = 10^{12}	tera	T
1 000 000 000 = 10^{9}	giga	G
1 000 000 = 10^{6}	mega	M
1 000 = 10^{3}	kilo	k
100 = 10^{2}	hecto	h
10 = 10^{1}	deka‡	da
0.1 = 10^{-1}	deci	d
0.01 = 10^{-2}	centi	c
0.001 = 10^{-3}	milli	m
0.000 001 = 10^{-6}	micro	μ
0.000 000 001 = 10^{-9}	nano	n
0.000 000 000 001 = 10^{-12}	pico	p
0.000 000 000 000 001 = 10^{-15}	femto§	f
0.000 000 000 000 000 001 = 10^{-18}	atto§	a

4° in this system the units given below are employed without prejudice to other units that could be added in the future.

Supplementary Units

Plane angle. . . .	radian	rad
Solid angle.	steradian	sr

Derived Units

Area.	square meter	m^2
Volume.	cubic meter	m^3
Frequency	hertz	Hz
Density.	kilogram per cubic meter	kg/m^3
Velocity.	meter per second	m/s
Angular velocity	radian per second	rad/s
Acceleration. . .	meter per second squared	m/s^2
Angular acceleration.	radian per second squared	rad/s^2
Force.	newton	N
Pressure (stress)	newton per square meter	N/m^2
Kinematic viscosity. . . .	square meter per second	m^2/s
Dynamic viscosity. . . .	newton second per square meter	$N \cdot s/m^2$

Work, energy, quantity of heat.	joule	J
Power.	watt	W
Electric charge .	coulomb	C
Voltage, potential difference, electromotive force.	volt	V

‡Translator's note: This prefix is spelled "déca" in French.
§The prefixes "atto" and "femto" were incorporated into the International System in 1964 by the 12th CGPM.

Electric field strength.	volt per meter	V/m
Electric resistance. . . .	ohm	Ω
Capacitance. . .	farad	F
Magnetic flux. .	weber	Wb
Inductance. . . .	henry	H
Magnetic flux density.	tesla	T
Magnetic field strength.	ampere per meter	A/m
Magnetomotive force.	ampere	A
Luminous flux. .	lumen	lm
Luminance. . . .	candela per square meter	cd/m^2
Illumination. . . .	lux	lx

[*End of Resolution*]

Definitions of the fundamental units of the International System are given below, translated from the original French.

Meter

The 11th CGPM, 1960, has adopted the following:

The meter is the length equal to 1 650 763.73 wavelengths in vacuum of the radiation corresponding to the unperturbed transition between the levels $2p_{10}$ and $5d_5$ of the atom of krypton-86.

Kilogram

The 3rd CGPM, 1901, has declared:

The kilogram is the unit of mass; it is represented by the mass of the International Prototype Kilogram [a particular cylinder of platinum-iridium alloy preserved in a vault at Sèvres, France, by the International Bureau of Weights and Measures].

Second

The 12th CGPM, 1964, considering that, in spite of the results obtained in the use of cesium as an atomic frequency standard, the time has not yet come for the General Conference to adopt a new definition of the second, a fundamental unit of the International System of Units, because of the new and important progress which may arise from current researches, and considering also that it is not possible to wait any longer to base physical measurements of time on atomic or molecular frequency standards, empowered the International Committee on Weights and Measures to designate the atomic or molecular standards of frequency to be used temporarily.

The International Committee then acquainted the CGPM with the following declaration:

The standard to be used is the transition between the hyperfine levels $F = 4$, $M = 0$ and $F = 3$, $M = 0$ of the fundamental state $^2S_{1/2}$ of the cesium-133 atom unperturbed by external fields. The value 9 192 631 770 hertz is assigned to the frequency of this transition.

Ampere

The 9th CGPM, 1948, has adopted the following:

The ampere is the constant current that, if maintained in two straight parallel conductors that are of infinite length and negligible cross section and are separated from each other by a distance of 1 meter in a vacuum, will produce between these conductors a force equal to 2×10^{7} newton per meter of length.

Degree Kelvin

The 10th CGPM, 1954, has adopted the following:

The 10th General Conference on Weights and Measures decides to define the thermodynamic scale of temperature by means of the triple-point of water as a fixed fundamental point, attributing to it the temperature 273.16 degrees Kelvin, exactly.

Candela

The 9th CGPM, 1948, has adopted the following:

The magnitude of the candela [unit of luminous intensity] is such that the luminance of a blackbody radiator at the freezing temperature of platinum is 60 candelas per square centimeter.

APPENDIX B

RULES FOR CONVERSION AND ROUNDING*

Number of Decimal Places To Be Retained

In all conversions between inches and millimeters the number of decimal places retained should be such that precision is not sacrificed in the process of conversion and at the same time such that the converted value is not carried to an implied precision that is not justified. A general rule that is satisfactory in most cases is: On converting inches to millimeters, carry the millimeter equivalent to one less decimal place than the number to which the inch value is given; and on converting from millimeters to inches, carry the inch equivalent to two more places than the number to which the millimeter value is given.

As a special case under the above general rule it should be pointed out that in converting integral values of either inches or millimeters consideration must be given to the implied or required precision of the integral value to be converted. For example, the value "4 inches" may be intended to represent 4″, 4.0″, 4.00″, 4.000″, 4.0000″, or even a still higher precision. Obviously, the converted value should be carried to a sufficient number of decimal places to maintain the precision implied or required even though the value to be converted is given only as a whole number.

Method of "Rounding Off" Decimal Values

When a decimal value is to be rounded off to a lesser number of places than the total number available, the procedure should be as follows:

(a) When the figure next beyond the last figure to be retained is less than 5, the last figure retained should not be changed. Example: 3.46325, if cut off to three places, should be 3.463; if cut off to two places, 3.46.

(b) When the figures beyond the last place to be retained amount to more than 5 in the next place beyond that to be retained, the last figure retained should be increased by 1. Example: 8.37652, if cut off to three places, should be 8.377; if cut off to two places, 8.38.

* Excerpts from "American Standard Practice for Inch-Millimeter Conversion for Industrial Use," ASA B48.1-1933 (re-affirmed 1947).

(c) When the figure next beyond the last place to be retained is exactly 5, with only zeros beyond, the last figure retained, if even, should be unchanged; if odd it should be increased by 1. Example: 4.365, when cut off to two places, becomes 4.36; 4.355 would also be cut off to the same value, to two places.

This method of rounding off even fives results, in the long run, in the same number of values being raised as are lowered, and thus the average value is correct, whereas if the even five were always retained or always discarded, the final value, in the long run, would be too large or too small.

APPENDIX C

SOME FACTORS FOR CONVERSION INTO UNITS OF THE INTERNATIONAL SYSTEM

Length

1 inch = 2.54 centimeters (exactly)
1 foot = 0.3048 meter (exactly)
1 mile = 1609.3 meters
1 nautical mile = 1852 meters (exactly)
1 micron = 1 micrometer (exactly)
1 angstrom = 0.1 nanometer (exactly)

Area

1 square inch = 6.4516 square centimeters (exactly)
1 square foot = 0.092 903 square meter
1 circular mil = 5.0671×10^{-4} square millimeter
1 acre = 4046.9 square meters
1 barn = 10^{-28} square meter (exactly)
1 hectare = 10 000 square meters (exactly)

Volume

1 cubic inch = 16.387 cubic centimeters
1 cubic foot = 0.028 317 cubic meter
1 fluid ounce (UK) = 28.413 cubic centimeters
1 fluid ounce (US) = 29.574 cubic centimeters
1 gallon (UK) = 4546.1 cubic centimeters
1 gallon (US) = 3785.4 cubic centimeters
1 barrel (US) (for petroleum, etc.) = 0.158 99 cubic meter
1 acre foot = 1233.5 cubic meters
1 liter = 1000 cubic centimeters (exactly)

*NOTE: "ESU" means "electrostatic CGS unit"; "EMU" means "electromagnetic CGS unit."

Speed
1 foot per minute = 5.08 millimeters per second (exactly)
1 mile per hour = 0.447 04 meter per second (exactly)
1 knot = 0.514 44 meter per second
1 kilometer per hour = 0.277 78 meter per second

Mass
1 ounce (avoirdupois) = 28.350 grams
1 pound = 0.453 59 kilogram
1 slug = 14.594 kilograms
1 short ton = 907.18 kilograms
1 long ton = 1016.0 kilograms
1 tonne = 1000 kilograms (exactly)

Density
1 pound per cubic foot = 16.018 kilograms per cubic meter
1 pound per cubic inch = 27 680 kilograms per cubic meter

Force
1 poundal = 0.138 25 newton
1 ounce-force = 0.278 01 newton
1 pound-force = 4.4482 newtons
1 kilogram-force = 9.806 65 newtons (exactly)
1 dyne = 10^{-5} newton (exactly)

Pressure
1 poundal per square foot = 1.4882 newtons per square meter
1 pound-force per square foot = 47.880 newtons per square meter
1 pound-force per square inch = 6894.8 newtons per square meter
1 conventional foot of water = 2989.1 newtons per square meter
1 conventional millimeter of mercury = 133.32 newtons per square meter
1 torr = 133.32 newtons per square meter
1 normal atmosphere (760 torr) = 101 325 newtons per square meter (exactly)
1 technical atmosphere (1 kgf/cm^2) = 98 066.5 newtons per square meter (exactly)
1 bar = 100 000 newtons per square meter (exactly)

Energy, Work
1 foot poundal = 0.042 140 joule
1 foot pound-force = 1.3558 joules

1 British thermal unit (thermochemical) = 1054 joules
1 British thermal unit (International Table) = 1055 joules
1 calorie (thermochemical) = 4.184 joules (exactly)
1 calorie (International Table) = 4.1868 joules (exactly)
1 electronvolt = 1.602×10^{-19} joule
1 erg = 10^{-7} joule (exactly)

Power
1 foot pound-force per second = 1.3558 watts
1 horsepower (metric) = 735.50 watts
1 horsepower (British) = 745.70 watts
1 horsepower (electrical) = 746 watts (exactly)
1 British thermal unit (I.T.) per hour = 0.2931 watt
1 erg per second = 10^{-7} watt (exactly)

Quantities of Light
1 footcandle = 10.764 lux (lumens per square meter)
1 footlambert = 3.4263 candelas per square meter

*Quantities of Electricity and Magnetism**
1 ESU of current = 3.3356×10^{-10} ampere
1 EMU of current = 10 amperes (exactly)
1 ESU of electric potential = 299.79 volts
1 EMU of electric potential = 10^{-8} volt (exactly)
1 ESU of capacitance = 1.1126×10^{-12} farad
1 EMU of capacitance = 10^{9} farads (exactly)
1 ESU of inductance = 8.9876×10^{11} henrys
1 EMU of inductance = 10^{-9} henry (exactly)
1 ESU of resistance = 8.9876×10^{11} ohms
1 EMU of resistance = 10^{-9} ohm (exactly)
1 gilbert = 0.795 77 ampere
1 oersted = 79.577 amperes per meter
1 maxwell = 10^{-8} weber (exactly)
1 gauss = 10^{-4} tesla (exactly)
1 gamma = 10^{-9} tesla (exactly)

The following illustrative article was printed in *The Journal of Industrial Engineering*, but it follows the *Style Manual* of the American Chemical Society. It is reprinted here by permission.

Manufacturing Assembly Instructions[1]

by Stephan A. Konz,
George L. Dickey,
Carl M. McCutchan
and Roger W. Daniels
Department of Industrial Engineering, Kansas State University

ABSTRACT ■ *A series of experiments was designed to give an initial evaluation of the advantages and disadvantages in the use of audio-visual equipment and programmed learning in industry. The primary objective of the experiments was to investigate the differences between forms of work instruction. The most obvious result was that pictorial presentation is the best, using either a time or an error criterion.*

■ Many jobs in industry today use a typed list of step-by-step instructions as the technique of communicating to the worker how to do the job. This technique traditionally is accepted and usually nothing else is considered. In fact, many might be inclined to say, "What more is necessary?"

In view of the increasing use of audio-visual equipment and programmed learning, the following series of experiments was designed to give an initial evaluation of the advantages and disadvantages of the various alternatives. The information gathered is interesting from both an applied and a research point of view.

Interest in programmed learning has increased since Skinner's article, "The Science of Learning and the Art of Teaching," in 1954, but most of the work has been oriented toward academic research rather than work training. A decade later, Schramm (13) compiled 160 papers of original research on programmed learning. Only 18 dealt with work related tasks and more than 80 percent of the 160 experiments used student subjects.

Communication through programmed instruction can be of two types; *work instruction*, in which the operator always has a detailed procedure before him, and *work training*, in which the operator is expected to memorize a particular procedure during a training period. Although research in both types of communication currently is in progress in the Industrial Engineering Department at Kansas State University, the remainder of the article will be concerned only with work instruction.

[1] Under sponsorship of American Society of Tool and Manufacturing Engineers Research and Educational Grant, June, 1964.

Work Done Elsewhere

Communication between the engineer and the worker typically is through the visual and/or auditory channels. Some feel that the visual channel is better than the auditory channel for transmitting information (7, 11). Cardozo and Leopold (2) had subjects transcribe letters and numbers. Visual input resulted in fewer errors than auditory input. Roshal (12) compared the efficiency of presenting information through slides and motion pictures for a knot-tying task and found motion pictures better. However, he stated, "For simple tasks, the most important consideration may be the accurate representation of the product of the perceptual motor skill." When the subject tied the knot at the same time as the motion picture was being shown, Roshal noted that this divided attention seemed to produce a conflict. McGuire (9) felt that motion picture portrayals of performance tend to present information faster than the subject's input capacity. If, however, the film speed is slowed down so that all operators can follow it, it tends to act as a pacing device for the faster operators. Slides, on the other hand, can be actuated by the individual operator at his own pace.

Use of the word "visual" is deceiving. Both a typed list of assembly instructions and step-by-step colored pictures of an assembly are visual. Cross, Noble and

Trumbo (3), in a study on controls, found that it was more effective to present a picture of what the operator should duplicate (that is, matching behavior) than to give the operator typed directions on how to move (that is, move control left 5 inches and then down 3 inches).

The auditory channel, however, also has its partisans. Goldman and Eisenberg (5) found that auditory input of information on an assembly task was more advantageous than posting a list of instructions in front of the operator. Erlick and Hunt (4) cite as an advantage of using the auditory channel that, while visual information generally requires either shifting the operator's head orientation or locating the equipment in an often already crowded work space, auditory information can be presented by equipment located in less critical work areas. Note also that, if the operator's eyes are used in the task, this means he reads the information and then performs the task (that is, works sequentially). Giving information to the operator through the auditory channel allows him to work "in parallel" as he can obtain information and work at the same time. Thus, it appears that both channels have their strong and weak points and that any general statement on which is best must be applied with caution.

The traditional work instruction device is a written or typed list of steps that have been developed by the organization's Industrial Engineering department. This list is posted in front of the operator. Recently, more sophisticated equipment (1, 15, 16, 17, 18 and 19) has been used which has tape recorded commentaries in conjunction with color slides. On at least one unit (6) the voice is hushed automatically at the end of the instruction so the operator can hear background music while carrying out the previous instruction. There is very little published evidence of controlled experiments on work method communication, thus the impetus for the following series of investigations.

Experiment One

Task

For ease of experimentation, the task should take a relatively short period of time to complete, the parts should be reusable, and the assembly should be easy to disassemble. It is desirable that both the quantity and quality of work could be evaluated. A complex task was desirable so that the subjects would make more errors and thus permit more accurate discrimination between the effects of various techniques of information presentation.

The task, depicted in Figure 1, used a pegboard with a 4×4 matrix of $\frac{1}{4}$ inch diameter 3 inch long wooden dowels on $1\frac{1}{4}$ inch centers. The columns were labeled A–D and the rows 1–4. Three types of wooden washers were made: One to fit over one dowel (singles), one to fit over two dowels (doubles) and one to fit over three (triples). Since components in many industrial assembly tasks must be placed not only in the proper position but also must have a specific orientation, the washers were colored red on one side of one end of the doubles and triples and white on the other side of the other end. The singles were red on one side and white on the other. Errors, therefore, could be classified four ways:

1. Position (on wrong pegs).
2. Orientation (on correct pegs but with ends reversed *or* part upside down).
3. Color (on correct pegs but with ends reversed *and* part upside down).
4. Omission (part omitted).

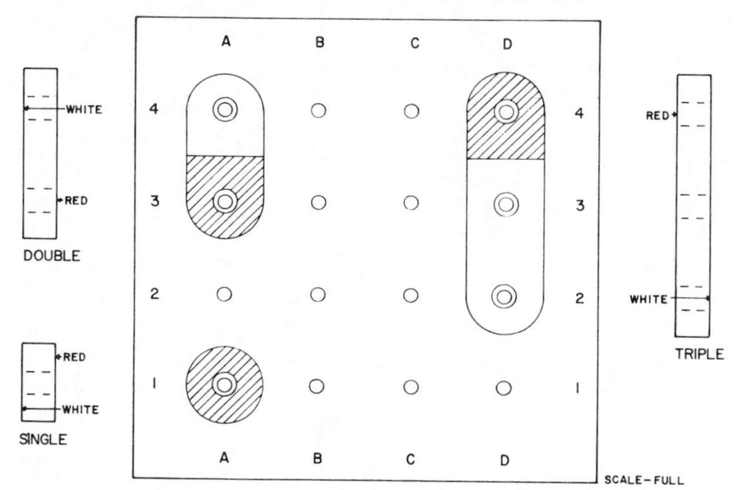

Figure 1. Pegboard Assembly Configuration Used in Experiments One, Two, Three and Four.

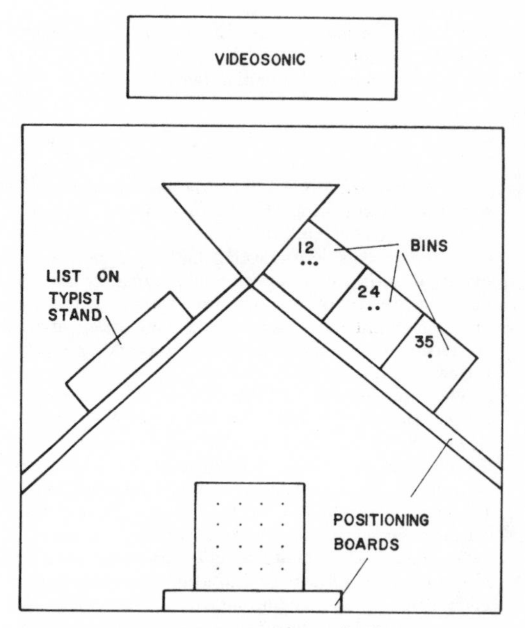

Figure 2. Workplace Layout. Scale 1″ = 6″.

Although color and orientation errors have been differentiated, some readers may wish to combine both categories into a single category of orientation error. Five singles, four doubles and one triple composed a layer of nonoverlapping washers. Twelve different patterns of layers were designed. For Experiment One, these 12 layers were grouped into four tasks of three layers each.

The workstation, shown in Figure 2, was designed for consistency of method rather than efficiency. Each operator worked only with the right hand.

Subjects

Twenty women were hired from the unskilled labor pool of the Kansas State Employment Service. Each woman worked for four hours and was paid by the hour. Years of schooling ranged from 10 to 17 with an average of 13.0.

Media

Two methods of work communication were investigated:

1. The conventional typed list shown in Figure 3.
2. Colored picture slides taken from the operator's viewing angle.

There were three slides for each layer—one for the singles, one for the doubles and one for the triple. For clarity, the "background" components were not shown on the slides. The slides were rear-projected on the screen of a Hughes Model 202 Videosonic and were changed by actuation of a footswitch. Thus, both media were operator-paced. Note that only a small portion of the total information required for the task was available on any specific picture whereas the total task information was available on the list.

Experimental Procedure

Each subject used both media and all four tasks. In order to balance practice effects 10 subjects were assigned randomly to the list-picture-picture-list (l-p-p-l) sequence and the other half to the p-l-l-p sequence. Each subject completed a total of 30 assemblies (15 per task) for the list media and 30 for the picture. The sequence of tasks was randomized with the restriction that each task appeared equally often in each of the four time periods.

While the subject worked, the experimenter timed her with a decimal minute stop watch. Upon completion of an assembly of three layers, the experimenter disassembled the unit and noted the type and frequency of errors. This information was not given to the subject. Although the inspection time varied depending on the number and type of errors, the subject was idle for approximately two to three minutes between each assembly. The subjects were instructed to work "quickly but accurately."

Task I

1. Get 5 Singles
2. White at A1, C2, D3
3. Red at B4, C3

4. Get 4 Doubles
5. White right at B1, C1
6. Red up at B2, B3
7. Red down at D1, D2
8. Red right at C4, D4

9. Get 1 Triple
10. White down at A2, A3, A4

11. Get 5 Singles
12. White at A4, D1, D4
13. Red at A3, D2

14. Get 4 Doubles
15. White down C1, C2
16. White right B4, C4
17. White left A2, B2
18. Red right A1, B1

19. Get 1 Triple
20. Red left at B3, C3, D3

21. Get 5 Singles
22. White at D3, D4
23. Red at A1, B3, B4

24. Get 4 Doubles
25. White up at C3, C4
26. White up at D1, D2
27. Red down at A3, A4
28. Red left at B1, C1

29. Get 1 Triple
30. Red right at A2, B2, C2

Figure 3. Typical Instruction List for Experiment One.

Table 1

Position Errors on Experiment One

Media	Layer			Total	Total per Layer
	Top	Middle	Bottom		
List					
1st Series	25	26	3	54	
2nd Series	24	25	1	50	
Total per 1668 Layers	49	51	4	104	.062
Pictures					
1st Series	1	0	2	3	
2nd Series	5	3	1	9	
Total per 1668 Layers	6	3	3	12	.007

Results

Five of the subjects took so long when using the list that they would not have been able to complete 15 assemblies of each of the four tasks within the four-hour experimental session. For example, Subject 17 used the list for the first task and still was taking over three minutes per assembly after 12 assemblies. At that point, the experimenter transferred her to the second task and pictures. She completed 15 assemblies of the second task and 12 assemblies of the third task using slides. She was able to complete 15 assemblies of the fourth task using the list before the four-hour limit. The data beyond the minimum number completed for any one of the four tasks were not used. Thus, for Subject 17, only the first 12 cycles of the four tasks were used.

There was an average of .060 errors per layer when using the pictures and .380 errors per layer when using the list; it required an average time of .383 minutes per layer for pictures and .815 minutes per layer for the list. More detailed information can be found in Tables 1 through 5. The statistical significance of total errors, position errors, orientation errors, color errors, omission errors and time was tested with the non-parametric Wilcoxon Matched Pairs Signed Ranks test (14). The differences between media for each of the above criteria were significant ($p<.01$).

Tables 1 through 5 indicate that there might be a layer effect. There seem to be considerably more errors for the middle and top layers which must be assembled against a background of parts than for the bottom layer which is assembled with an "empty" background. This was tested with another nonparametric test, Friedman's two-way analysis of variance (14). Both time and position errors had a chi-square value that could have occurred by chance less than one time in 100 so it was concluded that the background did affect position errors and time. The probabilities of the chi-square occurring by chance for omission errors, orientation errors and color errors were approximately 10 percent, 20 percent and 20 percent, respectively, so these three are considered to have occurred by chance.

Table 2

Orientation Errors on Experiment One

Media	Layer			Total	Total per Layer
	Top	Middle	Bottom		
List					
1st Series	69	62	57	188	
2nd Series	50	64	20	134	
Total per 1668 Layers	119	126	77	322	.193
Pictures					
1st Series	9	8	13	30	
2nd Series	8	5	6	19	
Total per 1668 Layers	17	13	19	49	.029

Next, the effect of the educational level of the subject was determined. The average errors and time per layer were calculated for each subject. The difference between the errors for each medium for each subject was plotted against the subject's educational level. A similar procedure was followed for times. The resulting Spearman rank correlation coefficient (14) was $-.494$ for errors and $-.721$ for times. Both are significant at $p<.05$. Thus, although every woman had more errors and more time for the list than the pictures, the difference was related inversely to the educational level and those with less formal schooling benefitted most from the pictures.

Table 3

Color Errors on Experiment One

Media	Layer			Total	Total per Layer
	Top	Middle	Bottom		
List					
1st Series	24	32	16	72	
2nd Series	14	27	11	52	
Total per 1668 Layers	38	59	27	124	.074
Pictures					
1st Series	3	5	2	10	
2nd Series	2	3	4	9	
Total per 1668 Layers	5	8	6	19	.011

After the experiment, the subjects were asked which medium they preferred and why. Nineteen of the 20 preferred the pictures. One preferred the list because "it is more of a challenge."

Discussion

It was felt that the requirement of "a complex operation with a long cycle time" before it was profitable to change from the conventional typed list (5) was incorrect. McCormick (8) states that in performing any type of work a person engages in essentially three kinds of functions:

1. Obtaining information.
2. Making decisions.
3. Acting upon the decisions.

It was felt that visual instruction would reduce the

Table 4

Omission Errors on Experiment One

Media	Layer			Total	Total per Layer
	Top	Middle	Bottom		
List					
1st Series	29	28	8	65	
2nd Series	8	9	3	20	
Total per 1668 Layers	37	37	11	85	.051
Pictures					
1st Series	3	3	3	9	
2nd Series	10	1	2	13	
Total per 1668 Layers	13	4	5	22	.013

time for obtaining information and making decisions but probably would not change the time to carry out the decision (for example, soldering time and the time necessary to select the parts and assemble the components). Thus, the tasks were designed with a high information and decision content and relatively low motor work content so that the task was complex but with a short time per cycle.

The list was selected as one of the media since it is the commonly accepted method while the color slides were used because some investigators have reported them as a promising method. The feedback of time and errors was withheld from the operator as part of the experimental control.

The data indicate that pictorial presentation is more effective than a typed list. Educational level also is a variable, with subjects with less education benefitting more from pictures. Adding a component against a complex background seems to affect some types of errors.

In order to verify some of these results, a second experiment was run which used college students as subjects. Since in Experiment One there may have been too much information being presented at one time, an attempt was made to equate the instructional content of the two media. Both media were presented by means of the Videosonic unit to remove any novelty effects of the machine.

Table 5

Total Assembly Time (in minutes) for Experiment One

Media	Layer			Total per 1668 Layers	Minutes per Layer
	Top	Middle	Bottom		
List					
1st Series	268.80	259.18	226.18	754.16	
2nd Series	210.22	208.27	187.07	605.56	
Total	479.02	467.47	413.25	1359.74	.813
Pictures					
1st Series	110.44	109.44	110.95	331.38	
2nd Series	101.67	100.55	104.49	306.71	
Total	212.11	210.54	215.44	638.09	.383

Experiment Two

Task

Five assemblies of the same type as in Experiment One were used, but with each assembly consisting of only two layers each instead of the three layers per assembly of Experiment One.

Subjects

Twenty male undergraduate and graduate students in Industrial Engineering served without pay.

Media

In this experiment, the typed instructions from the list were photographed and presented on a slide for each layer. The pictures also were on a one slide per layer basis instead of the three slides per layer of Experiment One. The viewing time for both media was under the control of the subject.

Experimental Procedure

Ten subjects used the list slides and 10 used the picture slides. The five assemblies were arranged in five

Table 6

Summary of Analysis of Variance for Assembly Times and Error Scores

Source	d.f.	Time Scores		Error Scores	
		M.S.	F	M.S.	F
Between Subjects	19				
Display (D)	1	8.4914	46.707**	2.000	4.348*
Blocks (B)	4	.2446	1.345	.930	2.022
D×B	4	.2341	1.287	.450	.978
Error	10	1.8175			
Within Subjects	180				
Practice (P)	4	.3555	44.438**	.718	2.124
Layer (L)	4	.0185	2.312	.080	.237
Order (LB)	1	.0103	1.287	.020	.059
P×LB	4	.0688	8.600**	.258	.763
T×LB	4	.0103	1.287	.045	.133
D×P	4	.1708	21.350**	.913	2.701*
D×L	4	.0166	2.750*	.275	.814
D×LB	1	.0134	1.675	.021	.059
B×LB	4	.0063	.788	.620	1.834
D×LB×L	4	.0012	.150	.020	.059
D×B×LB	4	.0119	1.487	.270	.799
D×LB×P	4	.0071	.887	.283	.837
Error	138	.0080		.338	

*$p < .05$
**$p < .01$

different sequences to balance sequential effects between tasks. Two subjects used each sequence with each medium. One of these two subjects had Layer One on the bottom and Layer Two on top while the other had Two on the bottom and One on top. Thus the five tasks and sequences composed a 5×5 Latin square. A 10-second rest interval was given between each layer with a 30-second interval between tasks. The subjects were told "errors and time are equally important."

Results

Table 6 gives a parametric analysis of variance of the

time and error scores. Errors were not broken into categories as in Experiment One. There was an average of .32 errors per layer with the list and .12 errors per layer with the pictures. The average time scores were .884 minutes per layer with the list and .472 minutes per layer for the pictures. Both of these effects were statistically significant; errors at $p < .05$ and time at $p < .01$.

The significant display×practice interaction for errors can be attributed to the large number of errors committed by the list group on the first task. After the first assembly the errors were distributed fairly evenly over the remaining tasks. The average error scores for the first two layers for the picture were .05 errors per layer and .75 errors per layer for the listing. For the last two layers, the pictures had .15 errors per layer while the listing had .20 errors per layer.

The subject's time scores decreased with practice to give a significant practice effect. Also, practice×top-bottom, and display×practice effects were significant for time scores. The significant display×practice reflects the fact that the picture times started low and decreased only slightly while the list times decreased more rapidly. The average time for the first two layers for the picture group was .513 minutes and for the listing was 1.154 minutes. The average time for the last two layers was .439 minutes for the picture and .775 minutes for the listing. The significant practice×top-bottom interaction reflects the failure for all bottom layers to take longer than the associated top layers. That is, initially, the bottom layer took longer than the top layer, but, after the initial learning period, the top was taking longer than the bottom. The reason for the top taking longer than the bottom probably was due to the confusing background while building the top layer which was not present while building the bottom layer.

Discussion

The listing times started out considerably higher than the pictorial and remained so through the five assemblies; however, there were indications that the differences between the media may converge. The data indicated that the error scores were converging more rapidly than time scores.

The results of this experiment show that even with the information content (amount of instruction) of the media equated and novelty effects removed the pictorial display remains superior. To make a more direct comparison, Experiment Three was designed to compare the effect of changing the information content per presentation. The effect of the use of the audio channel of presentation also was investigated.

Experiment Three

Task

The four assemblies of Experiment One were repeated.

Subjects

Sixteen women were hired from the unskilled labor pool of the Kansas State Employment Service. Each woman worked approximately four hours and was paid by the hour. Average years of schooling completed was 12.75.

Media

Four different media were used:

1. Three picture slides per layer of Experiment One.
2. One picture slide per layer similar to those in Experiment Two.
3. The list of Experiment One. However, the list was placed in a typewriter which permitted the subject to keep track of her place as she turned the roller.
4. Audio instructions. These were given through the tape recorder portion of the Videosonic with the letters given as the words "able," "baker," "charlie," and "dog." The instructions were the same as those on the list. After the instruction to get the part, there was a pause of 1.8 seconds before the position instructions were given. There was another pause of 1.2 seconds at the end of the position instruction before the new get instruction. If desired, the tape could be stopped by depressing a footswitch so this medium as well as the others was operator-paced. The running times for the tapes if they were not stopped were 2:37, 2:35, 2:35, and 2:45 minutes for Tasks 1, 2, 3, and 4.

Experimental Procedure

Each subject used each of the media for 15 consecutive assemblies. The four media and the four tasks composed a 4×4 Graeco-Latin square of subjects by sequence. Subjects 5–8 and 9–12 used different squares but Subjects 13–16 repeated the same square as Subjects 1–4. The subjects were instructed to work "quickly but accurately."

Results

Error categories are shown in Tables 7 to 10 with an average of .290 errors per layer when using the list, .592 when using audio instruction, .071 when using one picture per layer, and .103 when using three pictures per layer. Using a two-tail Wilcoxon test for comparing these errors, the difference between the one picture per layer and three pictures per layer was significant at $p < .05$ and the differences between three pictures and list and between list and audio were significant at $p < .01$.

For position errors alone (Table 7) the difference between the one picture per layer and three pictures per layer was not significant, but both the differences between three pictures per layer and the list, and between the list and audio, were significant at $p < .01$. Therefore, the difference between one picture per layer and the list also is significant.

For orientation errors alone (Table 8) the difference between the one picture per layer and three pictures per layer was not significant, and neither was the difference between three pictures per layer and the list. However, the difference between three pictures per layer and audio was significant ($p<.05$) as was the difference between the one picture per layer and the list ($p<.01$).

For color errors alone (Table 9) the differences between the list, one picture per layer and three pictures per layer were not significant, but the difference between three pictures per layer and audio was significant ($p<.05$).

The possible layer effect (that is, the effect of a confusing background) indicated in Tables 7 through 10 was tested with Friedman's two-way analysis of variance (14). For position errors (Table 7), the layer effect was significant ($p<.05$) only for the audio and list media. For orientation errors (Table 8), the layer effect was not significant for any of the media. For color errors (Table 9), the layer effect was significant only for the audio medium. For omission errors (Table 10), the layer effect was not significant for any of the media.

The time per layer for the three picture per layer group was .47 minutes, for one picture per layer was .50 minutes, for the audio was 1.01 minutes, and for the list was 1.08 minutes. The differences between the .47 and .50, and 1.01 and 1.08 were not significant but the difference between the .50 and 1.01 was significant ($p<.01$).

Table 7

Position Errors in Experiment Three

Media	Layer			Total	Total per 720 Layers
	Top	Middle	Bottom		
One Picture per Layer	4	3	3	10	.014
Three Pictures per Layer	7	4	7	18	.025
List	47	19	8	74	.103
Audio	87	82	42	211	.293

The errors made by each subject while using the one picture per layer were subtracted from the errors she made while using the list. The difference was correlated in the expected direction with education but Spearman's correlation coefficient ($r=-.240$) was not statistically significant. The difference between the list and three pictures per layer had a correlation of $-.200$, the difference between the audio and the one picture per layer had a correlation of $-.378$, and the difference between the audio and three pictures per layer had a correlation of $-.354$. These three correlations are also in the expected direction but not statistically significant. The correlation between differences in time between the list and three pictures was significant ($-.515$) but the differences in time between audio and three pictures per layer were not significant.

Each subject was asked to indicate which medium was "best" and which "worst." Best was scored as "+1" and worst as "−1." When the scores for the media were totaled for the 16 subjects, the three pictures per layer had a score of 8, one picture per layer had 4, audio was 0 and the list was −12.

Table 8

Orientation Errors in Experiment Three

Media	Layer			Total	Total per 720 Layers
	Top	Middle	Bottom		
One Picture per Layer	10	4	12	28	.039
Three Pictures per Layer	12	10	22	44	.061
List	27	30	26	83	.115
Audio	34	37	19	90	.125

Discussion

When comparing audio presentation versus pictures, audio seemed to be a very poor way of giving work instructions regardless of whether time or error scores are used as a criterion. Listing was only slightly better than the audio. A possible reason for these differences between media is referability:

1. To the same instruction.
2. To the previous instruction.

When the subject used the audio tape, it was not possible for him to repeat the instruction if he missed it much less refer back to the previous instruction. When using the pictures, he could refer back to the present instruction since it was not "destroyed" as soon as it was presented. Previous instructions were lost. When using the list, both the present instruction and previous instructions were available. Although some portion of the referability of each of the media is an artifact of the experimental instructions, it, to some extent, reflects the inherent qualities of the media. It is difficult and expensive to make the tape referable, slides can be "backed up" but this requires the operator to adjust the machine, and a list inherently is easy to refer to unless one takes special precautions to prevent it.

The reduced errors when one picture per layer was used versus three pictures per layer may be caused by the greater referability of the one slide per layer.

Since the list had not been compared "head on" with the one slide per layer technique, Experiment Four was performed. It also permitted testing a modification

Table 9

Color Errors on Experiment Three

Media	Layer			Total	Total per 720 Layers
	Top	Middle	Bottom		
One Picture per Layer	3	4	0	7	.010
Three Pictures per Layer	4	5	2	11	.015
List	20	15	15	50	.069
Audio	43	34	9	86	.119

Table 10

Omission Errors on Experiment Three

Media	Layer			Total	Total per 720 Layers
	Top	Middle	Bottom		
ist	0	2	0	2	.003
ne Picture per Layer	3	1	2	6	.008
hree Pictures per Layer	5	3	1	9	.012
udio	12	18	10	40	.055

f the list technique and the use of college students ather than non-students as subjects.

xperiment Four

Task

Two six-layer assemblies were formed by combining ask 1 and 2, and Task 3 and 4 of Experiment One.

ubjects

Ten male undergraduate students were paid by the our. Average years of education completed was 14.5. ll subjects were right-handed.

Media

Two media were used:

1. The typed list of Experiment One. The subjects, however, vere told to keep their places on the list with a finger of their left and.
2. The one picture slide per layer of Experiment Three.

Both media were operator-paced.

xperimental Procedure

Each subject used each medium for eight assemblies. he experiment was counterbalanced by having Subects 1 through 5 assemble four units for each condition sing the sequence: List-Task 1, Picture-Task 2, Picture-Task 2, List-Task 1. Subjects 6 through 10 followed the sequence P-1, L-2, L-2, P-1. They were instructed to work "quickly but accurately."

Table 11

Errors for Experiment Four

Media	Layer	Position	Orientation	Omission	Color	Total
List	Bottom	0	8	1	3	12
	2	6	10	0	5	21
	3	0	7	0	2	9
	4	5	9	0	1	15
	5	6	6	1	8	21
	Top	15	5	0	9	29
otal Errors per 480 Layers		32	45	2	28	107
verage Errors per Layer		.066	.094	.004	.058	.223
Picture	Bottom	0	2	1	2	5
	2	1	4	0	1	6
	3	0	0	1	1	2
	4	0	2	0	1	3
	5	4	4	0	2	10
	Top	1	4	1	0	6
otal Errors per 480 Layers		6	16	3	7	32
verage Errors per Layer		.013	.033	.006	.015	.067

Results

The error scores are given in Table 11. Using a Wilcoxon test, the position and omission errors were not statistically different across media, but the difference in color errors was significant ($p<.05$) while orientation errors and total errors were significantly different ($p<.01$). The average time of .72 minutes per layer for the list was significantly different ($p<.01$) from the .41 minutes for the picture. The effect of background (that is, layer effect) was not significant for either medium.

Discussion

Refer to Table 12 for a comparison of the error rates and time scores for the different media for the previous experiments. Average errors per layer ranged from .060 to .133 for pictures while the list varied from .222 to .380 per layer and the audio had .592 errors per layer. Average times per layer ranged from .38 to .50 for pictures while the list varied from .72 to 1.08 and the audio had 1.01 minutes per layer. Thus, in general, pictures would be preferred over the list which would be preferred over the audio.

Table 12

Comparison of the Average Error and Time Scores per Layer for Experiments One through Four

Experiment	Media	Types of Errors: Position	Orientation	Omission	Color	Average Errors per Layer	Average Minutes per Layer
One	List	.062	.193	.051	.074	.380	.813
	Three Pictures	.007	.029	.013	.011	.060	.380
Two	List on Slide	—	—	—	—	.320	.880
	One Picture	—	—	—	—	.120	.470
Three	List (typewriter)	.103	.115	.003	.069	.290	1.080
	Three Pictures	.025	.061	.012	.015	.133	.470
	One Picture	.014	.039	.008	.010	.071	.500
	Audio	.293	.125	.055	.119	.592	1.010
Four	List (finger)	.066	.094	.004	.058	.222	.720
	One Picture	.013	.033	.006	.015	.067	.410

A closer look at the types of errors is interesting. The advantage of a place-keeping device for the list seems to result in a decrease in omission errors and orientation errors while color errors decrease only slightly and position errors even seem to increase. Audio orientation errors are approximately three times greater than one picture orientation errors, but omission errors are seven times greater, color errors are 12 times greater and position errors are 14 times greater.

Another example of how changes in media affect errors was observed in the color errors for Experiment Four. There were five singles per layer, 48 layers per subject per medium, and 10 subjects or a total of 2400 opportunities for errors on singles. The four doubles and one triple per layer also afforded 2400 opportunities

Figure 4. Pictorial Instructions for Experiment Five.

for error. The subjects made 15 color errors on singles when using the list and six (40 percent of the list) when using the pictures. However, they reduced the color errors on doubles and triples from 13 with the list to only one (eight percent of the list) when using the pictures. Thus, even different types of parts are affected differentially by changes in media.

The reason for not using the same subjects in all experiments was due not only to availability (the experiments were run over a period of seven months) but also to study the effects on as wide a variety of subjects as possible. Education level, for example, has a very decided effect since the college student has a much better level of verbal and spatial ability than those not trained so intensively. If possible, it is desirable to use nonstudents as subjects as was done in Experiment One and Three.

Experiment Five was designed to analyze further three of the media previously studied (list, picture, and audio) but with a different type of assembly.

Experiment Five

Goldman and Eisenberg (5) showed that, for a task involving selecting, bending leads, and installing 15 resistors in a terminal board, recorded auditory work instructions were significantly better (standard time was reached in less time) than use of a printed list. Several factors were influential in the decision to replicate their experiment; the results of Experiment Three, their use of the sole criterion of number of cycles to reach standard rather than average time per cycle or errors per cycle, and the opportunity to compare the results with results obtained by another experimenter operating under different conditions.

Their experiment was modified in the following ways:

1. Another medium, colored pictures presented on slides, was added to their two media.
2. Each subject served in each condition rather than their independent groups type of experiment.
3. The number and types of errors were recorded.
4. College freshmen were used instead of industrial workers.
5. The experimental task was modified slightly.

Task

A maple board with 63, $\frac{5}{32}$ inch diameter, holes arranged in the pattern shown in Figure 4, was painted flat black. The $\frac{1}{8}$ inch high numbers were made from green pressure-sensitive plastic tape. One hundred standard IBM permanent-type plugged wires were substituted for the resistors. Each wire was $2\frac{1}{2}$ inches long and had a $\frac{7}{8}$ inch long plug on each end. These gray wires were hand-painted with three bright enamel stripes to correspond to the resistor colors. Either end of the wire could be placed in either hole so there were no orientation errors. Groups of five wires of each of the 10 color codes were placed in an unpainted pine parts board having $\frac{3}{8}$ inch holes one inch apart in a 10×10 matrix. Thus, there were 100 holes for the 50 wires. Three highly similar tasks were used, Task A, Task B, and Task C. Task B used the same sequence of color codes as Task A but the hole locations were mirror images of those in Task A. Task C was Task A in reverse order.

Subjects

Twelve male freshmen who were not color blind were paid by the hour.

Media

Three media were used. Figure 5 gives the typed instruction list for Task A. Subjects could use their nonpreferred hand (they could use only their preferred hand to assemble) to keep their places on the list but this was not suggested to them.

The instructions from the typed list were read verbatim onto a tape as quickly as possible while still remaining intelligible. The average time per instruction was three seconds with a 1.2 second pause between instructions. This speed, which was satisfactory for experienced operators, was too fast for beginners so they used their non-preferred hand to manipulate the on-off switch of the tape recorder. They were not permitted to reverse the tape to repeat a missed instruction.

Task A

Green-Brown-Red	to	20–26
Orange-White-Brown	to	43–50
Red-White-Orange	to	30–34
Brown-Black-Orange	to	7–8
Blue-Gray-Orange	to	22–24
Yellow-Violet-Orange	to	13–17
Brown-Green-Red	to	44–45
Red-White-Orange	to	55–60
Yellow-Violet-Orange	to	18–23
Green-Blue-Red	to	3–9
Orange-White-Orange	to	42–47
Orange-White-Orange	to	29–33
Red-White-Brown	to	11–12
Blue-Gray-Orange	to	51–58
Brown-Black-Orange	to	2–5

Figure 5. Typical Instruction List for Experiment Five.

An example of the colored slides taken for each in-
ruction is given in Figure 4. Each picture showed only
e wire to be inserted with no "background" com-
onents except the board itself. A red-tipped yellow
encil was used to indicate the location of the holes on
e board. An oversized replica of the wire color code
d the pertinent hole numbers appeared in one corner
the picture. The slide was projected on a screen
rectly in front of the operator by a Kodak Carousel
ojector. The subject controlled the indexing of the
ides with a remote hand switch and was permitted to
ck up to the previous slide if he wished.

xperimental Procedure

Every subject always worked through 15 assemblies
Task A, B, and C, in that order. The three media
icture, audio, and list) have six possible sequences.
wo subjects used each of these sequences in an experi-
ental design which required each medium to occur
rst, second, and third equally often for each task.

Each subject was given a demonstration of how to
omplete one assembly of Task 1. In addition, brief
emonstrations of how to use the slide projector and
pe recorder were given before starting the task re-
uiring those media. The subjects were given no rest
etween assemblies and five minutes between tasks. As
e subject worked, the experimenter quickly replaced
ires used in the previous assembly on the parts board
sed in the previous assembly and then recorded the
ypes of errors as they occurred. At the end of an
ssembly he recorded assembly time, removed wires
om the assembled board, and gave a new parts board
the subject. Meanwhile, the subject prepared the
edium for the next cycle. The subjects were given
o feedback on time or errors, although they requested
.

esults

A Dixon criterion (10) was used to test for outliers;
ubject 7 had a long average assembly time and Sub-
ct 10 had considerable errors but neither subject
ould be eliminated.

Errors are shown in Table 13. A parametric analysis
f variance established the significance of media and a
SD test was used to test means. The position errors
ere significantly less ($p<.01$) for the picture when
omparing them with the tape, but the list was not
gnificantly different from the picture or tape. Omis-
on errors were significantly less ($p<.01$) for the pic-
re than for both the tape or the list, but the difference
etween the tape and the list was non-significant. None
f the differences in color errors among the media were
gnificant. When all errors were combined the pictures
ere significantly better ($p<.01$) than the tape, and
e list was significantly better ($p<.05$) than the tape,
ut the difference between the picture and the list was
on-significant.

Table 13

Errors for Experiment Five

Media	Type of Error			Total
	Position	Omission	Color	
List	19	21	35	75
Picture	4	1	43	48
Tape	24	59	62	145

The average time for 15 assemblies when using the tape was 24.85 minutes, for pictures was 29.19 minutes and for the list was 31.16 minutes. Using Wilcoxon tests with two tails, the tape was significantly better than both the picture and the list but the picture was not significantly different than the list. The pictures took 17 percent longer than the tape and the list took 25 percent longer. It is quite likely that the 59 omitted wires when using the tape (approximately two percent of the 2500 assembled when using the tape) contributed to the lower time for the tape.

Discussion

The primary objective of this series of experiments was to investigate the differences among forms of work instruction. The most obvious result was that pictorial presentation is the best using either a time or an error criterion.

The types of errors committed with the different media also are interesting. Experiments Three, Four, and Five demonstrated that media do not affect all types of errors equally. For example, in Experiment Five there were almost five times as many position errors when using the list as when using the picture. This did not mean, however, that there would be five times as many color errors. There actually were more color errors for the picture than with the list; thus, it is extremely hard to generalize.

Visualization also seems to be an important considera-tion. If a subject either made a mistake or suspects he made a mistake when using the picture, he merely looked at the picture and matched what he saw in the picture with what he had assembled. When using the list, he had to decode the written words, form a mental picture and compare this mental picture with his actual assembly. When using audio instructions, not only was visualization necessary, as with the list, but also a good memory. Experiments Three and Four, when contrasted with Experiment One, demonstrated that keeping the place on a list (either with a finger or the typewriter carriage) decreased orientation and omission errors although it did not decrease color or position errors.

When using pictures, there seems to be an information content per message effect. That is, it may be that too much information is being presented on a picture. This was noticeable when the same information was presented in Experiment Three on either one slide or three separate slides. More slides for a given amount of information means the subject has less searching and interpreting to do. Fewer slides for a given amount of information means less indexing time between slides, possibly better referability since more information is available for reference and, of course, fewer slides.

Audio presentation of material seems to be the poorest method from both a time and error viewpoint. One problem is the lack of referability. Another problem is that the audio seems to act as a pacing device. The subject improves only up to the pace of the tape and then maintains the pace of the tape. A slow tape sometimes frustrates the subject because he feels he is being held back. On the other hand, it was noticed (especially in Experiment Five) that the tape performs a feedback function in that it gives some knowledge of results to the subject as he works. In Experiment Five, the subjects counted the number of times they had to stop the tape or counted the number of wires positioned before stopping the tape for the first time. Since not all characteristics of the tape are bad, it may be very useful when used in conjunction with other media.

The longest time any individual operator was studied was four hours in Experiment One and even in this experiment only two hours were spent in any one medium. Certainly longer term studies need to be made to see at what number of cycles the media become equivalent. Since synchronization of a tape and slides is expensive, investigation of other methods of presenting complex material as well as just tape and slides is in progress.

Thus, from the preceding experiments, it can be concluded that there are extreme differences among the media studied. The audio medium appears, in general, to be the poorest method for communicating work instructions, while the pictorial slide presentation is best. In between the two extremes lies the traditional typed list, being fairly poor in both time and errors. Therefore, there is a better method.

References

(1) ABBOTT, R., "Semiskilled Workers Take Their Cue from A-V Training Machines," *Product Engineering*, August, 1962, p. 74–75.

(2) CARDOZO, B., AND LEOPOLD, F., "Human Code Transmission," *Ergonomics*, April, 1963.

(3) CROSS, K., NOBLE, M., AND TRUMBO, D., "On Response-Response Compatibility," *Human Factors*, 1964, Volume 6, p. 31–37.

(4) ERLICK AND HUNT, Evaluating Audio Warning Displays for Weapon Systems, *WADC Report*, 1957.

(5) GOLDMAN, J., AND EISENBERG, H., "Can We Train More Effectively?" *Journal of Industrial Engineering*, 1963, Volume 14, p. 73–79.

(6) HARKER, W., "Audio-visual Learning—It's More Than Hear-Say," *Electronic Industries*, August, 1961, p. 103–105.

(7) HENNEMAN, R., AND LONG, W., A Comparison of the Visual and Audio Senses as Channels for Data Presentation, *WADC Report*, August, 1954, Tech. report 54-363 (AD 61558).

(8) MCCORMICK, E., *Human factors engineering*, McGraw-Hill, New York, 1964.

(9) MCGUIRE, W., "Effects of Serial Position and Proximity to "Reward" Within a Demonstration Film," In A. Lumsdaine (Ed.), *Student Response in Programmed Instruction*, National Academy Sciences, National Research Council, Washington, D. C., 196 Pub. 943, p. 209–216.

(10) NATRELLA, MARY, *Experimental Statistics*, National Bureau Standards, U. S. Government Printing Office, Handbook 91, Washington, D. C.

(11) REID, L., AND MORSE, W., The Influence of Complex Task Variabl on the Relative Efficiency of Auditory and Visual Message Present tion, *WADC Report*, April, 1955, Tech. report 54-288 (AD 88065

(12) ROSHAL, S., "Film Mediated Learning with Varied Presentation the Task, Viewing Angle, Portrayal of Demonstration, Motion, an Student Participation," In A. Lumsdaine (Ed.), *Student Response Programmed Instruction*, National Academy of Science, Nation Research Council, Washington, D. C., 1961, Pub. 943.

(13) SCHRAMM, W., *The Research on Programmed Instruction*, U. S. Department of Health, Education, and Welfare, Washington, D. C 1964, DE 34034.

(14) SIEGEL, S., *Non-parametric statistics*, McGraw-Hill, New York 1956.

(15) SILVERN, L., "Shaping and Controlling Human Behavior in Ma Machine Systems," *Institution of Mechanical Engineers*, Londo 1963, Volume 177, (34).

(16) "Colored-Sound Slides Speed Assembly," *American Machinis Metalworking Manufacturing*, April, 1962, p. 108–109.

(17) "It Doubled Output in Cedar Rapids," *Factory*, September, 196 p. 90.

(18) "Teaching Machines Brief Trouble Shooter," *Factory*, April, 196 p. 90–91.

(19) "Sales Training by Teaching Machine," *Sales Marketing Toda* April, 1965.

Comments on the Article

The preceding article illustrates most of the elements discussed in Chapter 4.

The *headnote* gives credit to the society that sponsored the research. The credit line is brief; it contains only a date for cross-referencing.

The *by-lines* or *author credits* are not in alphabetical order. This indicates that Konz was the principal investigator. The Department of Industrial Engineering at Kansas State University is also cited. This indicates that the work was done at the facilities of that campus; it does not necessarily show that the men were on the staff of that university.

The *abstract* gives the purpose, method, and scope of the research. It is an *informative* abstract; that is, it discusses the work and not the contents of the article.

The *introduction* first places the problem in historical perspective and then, under a subhead, reviews both the state of the art and the literature.

The *experimental section* discusses five experiments in parallel because the experiments were similar. Subheadings "Task," "Subjects," "Media," "Procedure," "Results," and "Discussion" are repeated under each experiment. This repetition not only aids communication, it also permits the rapid comparison and contrast of variable data. Any other nontabular arrangment would require complicated cross-referencing. The discussion subsection of Experiment 5 is a synthesis of preceding discussions and a summary of the work.

The *references* are collated in alphabetical sequence. This procedure permits use of the nonsequential footnoting system of the American Chemical Society.

There are additional merits to this article. The *tables* are similar whenever possible. Figures 3 and 5 might have been titled tables, but they are dissimilar in structure and content. Figures 1, 2, and 4 are photographs, which are relatively inexpensive to produce and to print by offset. It is apparent from the number of tables and figures that the authors are being didactic; that is, they are illustrating the result stated in the abstract: ". . . pictorial presentation is the best . . . ."

Business Journals

As is the case with the engineering disciplines, the recognizable areas of business administration have dissimilar methods and goals. Consequently, journals in these areas have dissimilar style guides. None of these guides, however, is narrowly prescriptive. All are as general and as flexible as the example printed below by permission of *Management Science*.

Because this particular journal is concerned with "science," its "Instructions" are preoccupied with mathematics.

The illustrative article meets not only the requirements of *Management Science,* the journal that printed it originally and through whose kind permission it is reprinted here, it also meets the requirements of the American Mathematical Society's *Manual for Authors of Mathematical Papers* which is reprinted on pp. 293–309. However, the requirements of the journal have priority over the requirements of the society. This is shown by the footnotes; they follow the procedure suggested in the journal's "Instructions to Authors" and are in direct contrast to the society's recommendation against the use of footnotes in a professional article.

INSTRUCTIONS TO AUTHORS

Manuscripts intended for publication may be sent to the Editor-in-Chief or to any of the Associate Editors. Manuscripts may be designated for Series A or for Series B. Undesignated manuscripts will be allocated by the Editors-in-Chief.

All manuscripts submitted for consideration should be typed on white $8\frac{1}{2}$ x 11″ paper and should be double-spaced throughout. Review of papers will be expedited if triplicate copies are submitted. All manuscripts should be preceded by a non-mathematical abstract of approximately 100 words. Papers may be submitted in any language, but English and French are the official languages of the Institute.

Mathematical notation should be selected so as to simplify the typesetting process. Authors should attempt to make mathematical expressions in the body of the text as simple as possible. Bars, tildes, and carets must be handset; hence, their use should be avoided as much as possible. Some useful alternatives are:

Expensive notations	*Alternatives*
$\bar{A}, \tilde{b}, \check{\gamma}, \hat{g}, \mathring{\Lambda}$	$A', b'', \gamma^*, g^+, \Lambda^\#$
$e^{-\frac{x^2+y^2}{a^2}}$	$\exp(-(x^2+y^2)/a^2)$
$\frac{7}{8}, \frac{a+b}{c}$	$7/8, (a+b)/c$
$\sum\limits_{i=0}^{n}, \prod\limits_{i=1}^{\infty}$	$\Sigma_{i=0}^{n}, \Pi_{i=1}^{\infty}$

Expensive notations	*Alternatives*
$\dfrac{\cos\frac{1}{x}}{\sqrt{a+\frac{b}{x}}}$	$\dfrac{\cos(1/x)}{(a+b/x)}$

Displayed material should clearly indicate the alignment that is desired. If equations are numbered, the numbers should be given in parentheses, flush with the left margin of the page.

Footnotes should be numbered consecutively, and typed double-spaced in a separate list at the end of the paper. In text, footnotes are indicated by superscript numbers. Footnote numbers should never be attached to mathematical symbols. All reference material is to be listed alphabetically by author at the end of the paper and may be referred to in text by bracketed numbers, i.e., [4].

Figures should be drafted in india ink on white paper in a form suitable for photo engraving and, if necessary, reduction. Special attention should be given to line weights and lettering. Typewritten lettering on figures is not suitable.

AN EXTENSION OF THE GOMORY MIXED-INTEGER ALGORITHM TO MIXED-DISCRETE VARIABLES*[1]

ROBERT E. DALTON[2] AND ROBERT W. LLEWELLYN[3]

The methods of R. E. Gomory for the iterative solution of the mixed-integer linear programming problem are extended directly to the case where some or all of the variables are nonuniformly discrete, *i.e.*, they are restricted to assume values from certain specified sets of unequally-spaced constants. The algorithm presented is shown to converge in a finite number of steps for a discrete-valued objective function.

I. Introduction

Any linear programming problem in which variables are constrained to take on only equally-spaced values can, by suitable changes in scale, be reduced to a mixed-integer programming problem, amenable to the method of R. E. Gomory [1]. There occur in practical applications several instances in which the variables are not equally spaced. These include:

a. the smoothing problem, with a finite number of possible rates of production (see Llewellyn [3]);
b. machine shop set-up time problems;
c. carload and airplane-load shipments; and
d. packaging problems.

This paper is an attempt to extend Gomory's methods and ideas to the linear programming problem with a subset of variables, each of which is constrained to assume a certain finite set of values which have no common increment between them.

II. Theory

We set up the linear programming problem in the form

$$(2.1) \qquad \begin{cases} \text{max} & Z = a_{00} + (\mathbf{a}^0)^T(-\mathbf{x}) \\ \text{subject to} & \mathbf{x}' = \mathbf{a}_0 + (a_{ij})(-\mathbf{x}) \\ & \mathbf{x}' \geqq \mathbf{0}, \quad \mathbf{x} \geqq \mathbf{0}, \end{cases}$$

where $\mathbf{x}'$ is an $m \times 1$ vector of slack variables, and the partial ordering $\mathbf{x} \geqq \mathbf{y}$ means $x_i \geqq y_i (i = 1, 2, \cdots, n)$.

Definition. Let x_i assume a finite number of predetermined values, which are

* Received January 1965.

[1] Presented under the title "Linear Programming with Variables Which Are Nonuniformly Discrete" at the New York meeting of the Society for Industrial and Applied Mathematics, June 7–9, 1965.

[2] Technical Staff, TRW Systems Incorporated, Cape Canaveral. This paper is a portion of the author's Ph.D. dissertation in the Department of Applied Mathematics, North Carolina State of the University of North Carolina at Raleigh.

[3] Department of Industrial Engineering, North Carolina State of the University of North Carolina at Raleigh.

rational numbers, but not necessarily integers, and which are not separated by a constant increment, say

$$x_i = X_{i0}, X_{i1}, X_{i2}, \cdots, X_{in_i} \qquad (X_{i0} \equiv 0); \tag{2.2}$$

where the $(n_i + 1)$ possible values of x_i in (2.2) include zero, are arranged in ascending order and have the property that, for some j and k such that $j \neq k$,

$$\Delta_j x_i = X_{i,j+1} - X_{ij} \neq X_{i,k+1} - X_{ik} = \Delta_k x_i .$$

The remaining variables $x_p (p \neq i)$ in a basic feasible solution to the problem (2.1) may be constrained to be discrete with other unequal or equal increments, or may not be additionally constrained at all. We call the problem (2.1) with additional constraint(s) (2.2) a *finite, mixed-discrete* problem.

In our algorithm for the finite, mixed-discrete problem, the optimum solution $\mathbf{x}'$ to the problem (2.1) is obtained by some simplex procedure. We then examine $\mathbf{x}'$ to see if all additional constraints, including those of the type (2.2) are satisfied. If not, a new constraint, or cut, is added, which admits all lattice points with components equal to admissible values of the various discrete and rational variables—but does not admit the current optimum $\mathbf{x}'$.

With this objective in mind, let

$$x'_i = a_{i0} + \sum a_{ij}(-x_j), \tag{2.3}$$

where

1. x'_i is in the linear programming basis,
2. x'_i is a discrete variable with a finite number of possible values given by (2.2),
3. $a_{i0} (> 0)$ is not one of these possible values—but is the current value of x'_i in the optimum linear programming solution $\mathbf{x}'$, and
4. $\{x_j\}$ is the current set of nonbasic variables. Write

$$a_{i0} = X_{ik} + b_{ik}$$

where X_{ik} is the next lowest admissible value for x'_i, i.e.,

$$X_{ik} = \max_p \{X_{ip} \mid X_{ip} < a_{i0}\};$$

and hence

$$0 < b_{ik} < \Delta_k x'_i \qquad \text{for} \quad 0 \leqq k \leqq n_i - 1;$$

$$0 < b_{ik}, \qquad \text{if} \quad k = n_i .$$

Theorem 1. The equality

$$s = -b_{ik} - \sum_{j \varepsilon S^+} a_{ij}(-x_j) - \sum_{j \varepsilon S^-} a^*_{ij}(-x_j),$$

where

$$\begin{aligned} S^+ &= \{j \mid a_{ij} \geqq 0\}, \qquad S^- = \{j \mid a_{ij} < 0\}, && s \geqq 0; \\ a^*_{ij} &= b_{ik}(-a_{ij})/(\Delta_k x_i - b_{ik}) > 0 && \text{if} \quad a_{io} < X_{in_i}, \\ a^*_{ij} &= a_{ij} && \text{if}^4 \quad a_{i0} > X_{in_i}; \end{aligned}$$

is satisfied by any nonnegative finite, mixed-discrete solution to (2.1) but is not satisfied by the present optimum solution $\mathbf{x}'$.

Proof. Suppose there exists an admissible solution to the finite, mixed-discrete problem and use $\hat{x}'_i$ and $\{\hat{x}_j\}$ to denote the values given to the variables in (2.3) by this solution. Hence

$$\hat{x}'_i = a_{i0} + \sum a_{ij}(-\hat{x}_j),$$

or

$$(2.4) \qquad \begin{aligned} \sum a_{ij}\hat{x}_j &= a_{i0} - \hat{x}'_i = (b_{ik} + X_{ik}) - \hat{x}'_i \\ &= b_{ik} - (\hat{x}'_i - X_{ik}). \end{aligned}$$

There are two cases.

Case I. $0 < a_{i0} < X_{in_i}$. We observe the possible values for $\hat{x}'_i$. If $\hat{x}'_i \geqq X_{i,k+1}$, then $(\hat{x}'_i - X_{ik}) > b_{ik}$. By (2.4), this implies

$$\sum a_{ij}\hat{x}_j < 0.$$

If $\hat{x}'_i = X_{ik}$, then $(\hat{x}'_i - X_{ik}) = 0$. By (2.4), this implies

$$\sum a_{ij}\hat{x}_j = b_{ik} > 0.$$

If $\hat{x}'_i \leqq X_{i,k-1}$, then $(\hat{x}'_i - X_{ik}) < 0$. By (2.4), this implies

$$\sum a_{ij}\hat{x}_j = b_{ik} + \sum_{j=k-s,s\leq k}^{k-1} \Delta_j x'_i > b_{ik} > 0.$$

Therefore this case has two well-defined subcases.

Ia. $b_{ik} - (\hat{x}'_i - X_{ik}) = \sum a_{ij}\hat{x}_j \geqq b_{ik}$, or

$$(2.5) \qquad \sum_{j \epsilon S^+} a_{ij}\hat{x}_j \geqq \sum_{j \epsilon S^+} a_{ij}\hat{x}_j + \sum_{j \epsilon S^-} a_{ij}\hat{x}_j \geqq b_{ik}.$$

Ib. Similarly, if $\sum a_{ij}\hat{x}_j < b_{ik}$, then this implies

$$\sum a_{ij}\hat{x}_j < 0, \qquad \text{or}$$

$$\begin{aligned} b_{ik} - (X_{i,k+1} - X_{ik}) &\geqq \sum_{j \epsilon S^+} a_{ij}\hat{x}_j + \sum_{j \epsilon S^-} a_{ij}\hat{x}_j \\ &\geqq \sum_{j \epsilon S^-} a_{ij}\hat{x}_j \end{aligned}$$

Multiplying by $-b_{ik}/((X_{i,k+1} - X_{ik}) - b_{ik})$, we have

$$(2.6) \qquad b_{ik} \leqq \sum_{j \epsilon S^-} b_{ik}(-a_{ij})/((X_{i,k+1} - X_{ik}) - b_{ik})\hat{x}_j.$$

[4] One might include constraints which specify upper bounds explicitly in the tableau (2.1) and thus avoid the case $a_{io} > X_{in_i}$, but the given method will be computationally more efficient by avoiding a row in the original tableau for each variable constrained to a set of nonuniformly discrete values.

Since either (2.5) or (2.6) holds and both sums involved are nonnegative, we have the general inequality

$$(2.7)\quad b_{ik} \leqq \textstyle\sum_{j\varepsilon S^+} a_{ij}\hat{x}_j + \sum_{j\varepsilon S^-} b_{ik}(-a_{ij})/((X_{i,k+1} - X_{ik}) - b_{ik})\hat{x}_j .$$

Inequality (2.7) is thus satisfied by any finite, mixed-discrete solution, but not by the present optimum $\mathbf{x}'$, since setting $x_j = 0$ for all j makes the right-hand side zero, contrary to the fact that $b_{ik} > 0$.

Case II. $0 < X_{in_i} < a_{i0}$, *i.e.*, $a_{i0} = X_{in_i} + b_{in_i}$, where X_{in_i} is the largest admissible discrete value for x'_i. In this case, it is known for sure that

$$\textstyle\sum a_{ij}\hat{x}_j = b_{in_i} - (\hat{x}'_i - X_{in_i}) \geqq b_{in_i} > 0,$$

since $(\hat{x}'_i - X_{in_i})$ must be nonpositive.

Therefore, the resulting inequality must be

$$(2.8)\qquad \textstyle\sum_{j\varepsilon S^+} a_{ij}\hat{x}_j + \sum_{j\varepsilon S^-} a_{ij}\hat{x}_j \geqq b_{in_i} .$$

Now, rewritten as an equation by introducing a nonnegative slack variable s' (2.7) becomes

$$(2.9)\qquad \begin{aligned} s = -b_{ik} &- \textstyle\sum_{j\varepsilon S^+} a_{ij}(-\hat{x}_j) \\ &- \textstyle\sum_{j\varepsilon S^-} b_{ik}(-a_{ij})/((X_{i,k+1} - X_{ik}) - b_{ik})(-\hat{x}_j). \end{aligned}$$

Similarly, (2.8) becomes

$$(2.10)\qquad s = -b_{in_i} - \textstyle\sum_{j\varepsilon S^+} a_{ij}(-\hat{x}_j) - \sum_{j\varepsilon S^-} a_{ij}(-\hat{x}_j).$$

This completes the proof of Theorem 1.

Note that in (2.10), there exists at least one $a_{ij} > 0$ $(j \varepsilon S^+)$ if a feasible finite, mixed discrete solution exists, since in this solution, all variables $\hat{x}_j$ are nonnegative, and $\sum a_{ij}\hat{x}_j > 0$ in Case II. Thus, there must be at least one negative element in the new row (2.10) to act as pivot element in order that the dual linear programming method may transform the augmented system back to primal feasible form.

Theorem 2. Let (2.3) denote a constraint from the current optimal linear programming tableau such that the assumptions 1, 2, 3, 4 following (2.3) hold. Suppose, moreover, that the equality (2.9) or (2.10) is adjoined to the indicated tableau as the new bottom row. Then this new row will be selected as the pivotal constraint for the next step with the dual linear programming algorithm. Then $\mathbf{a}_0$ is decreased lexicographically and also

1. If $a'_{i0} < X_{in_i}$,

$$(2.11)\quad (a)\qquad a''_{i0} = X_{ik} \qquad \text{if } a_{it} > 0,$$

$$(2.12)\quad (b)\qquad a''_{i0} = X_{i,k+1} \qquad \text{if } a_{it} < 0;$$

2. If $a'_{i0} > X_{in_i}$,

$$(2.13)\quad (a)\qquad a''_{i0} = X_{in_i} \quad \text{regardless of the sign of } a_{it} .$$

Here double primes denote elements of $\mathbf{a}_0$ after the dual linear programming iteration; t is the index of the column in which the pivot element occurs; $a_{i0} = X_{ik} + b_{ik}$; and the index i refers to the specific row (2.3) corresponding to the discrete variable in the basis, as in (2.2) to (2.10).

Proof. Clearly

$$a''_{i0} = a'_{i0} - b_{ik}a_{it}/a^*_{it}, \qquad \text{or} \tag{2.14}$$

$$a''_{i0} = a'_{i0} - b_{ik}a_{it}/a_{it}. \tag{2.15}$$

It can be shown that the use of the dual linear programming method guarantees that each column in successive matrices $A', A'', \cdots, A^{(k)}, \cdots$ remains lexicographically positive at all times. Therefore it follows, by (2.14) or (2.15), that $\mathbf{a}_0'' < \mathbf{a}'_0$ in the lexicographic sense, since $(-b_{ik}) < 0$ and the fact that the use of the dual linear programming method insures a pivot element $(-a^*_{it})$ or $(-a_{it})$ which is negative. For $a'_{i0} < X_{in_i}$ there are two cases.

1. If $a_{it} > 0$, then from (2.15),

$$a''_{i0} = a'_{i0} - b_{ik} = X_{ik}.$$

2. If $a_{it} < 0$, then from (2.14),

$$\begin{aligned} a''_{i0} &= a'_{i0} + \Delta_k x'_i - b_{ik} \\ &= X_{ik} + \Delta_k x'_i \\ &= X_{i,k+1}. \end{aligned}$$

If $a'_{i0} > X_{in_i}$, then from (2.10) and (2.15),

$$a''_{i0} = a'_{i0} - b_{in_i} = X_{in_i},$$

regardless of the sign of a_{it}.

The possibility $a_{it} = 0$ cannot occur because $a_{it} = 0$ also implies $a^*_{it} = 0$, so that neither a_{it} or a^*_{it} can be a pivot element.

Theorem 3. If the objective function is constrained to be one of the discrete variables, and an optimal finite, mixed-discrete solution to (2.1) exists, the algorithm described above will attain the optimum in a finite number of steps, with a proper rule of row selection. If no feasible solution to the finite, mixed-discrete problem exists, the algorithm will indicate this fact in a finite number of steps.

Proof. Assume that a dual feasible solution has been obtained and that in all succeeding transformations pivot elements a_{it} are selected (lexicographically) by the dual linear programming method, *i.e.*,

$$\mathbf{a}_t = a_{it} \max_j \{(1/a_{ij})\mathbf{a}_j \mid j \neq 0 \text{ and } a_{ij} < 0\}, \tag{2.16}$$

where $\mathbf{a}_t$ is the pivotal column of the matrix $(a_{ij})^{(k)}$ resulting after the kth elementary transformation of (a_{ij}) in (2.1).

Arrange the rows of $(a_{ij})^{(k)}$ so that the first row corresponds to the objective function and the rows immediately following the first row correspond to any other nonuniformly or uniformly discrete variables.

We use the following rule of choice: the $\mathbf{a}^0$ row is used to form the additional equality (2.9) or (2.10) when a_{00} is not an admissible value for Z. When a_{00} is an admissible value, we use the next row down which corresponds to a discrete variable, and which has an associated value a_{i0} which is not admissible value for x_i.

Use of a lexicographical dual linear programming method insures that we always have lexicographically positive columns in the current matrix. Theorem 2 states that successive values of the objective function are monotonically decreasing. Gomory's argument in [2] can therefore be used here to show that the objective function reaches its greatest lower bound in a finite number of steps. Thus, if the infimum is reached without driving the other discrete variables to admissible values, then it necessarily follows that $a_{oj} = 0, j \neq 0$, for all steps after some n^{th} step. The remaining iterations proceed on the basis of the next discrete variable x_i which has not yet reached its greatest lower bound greater than or equal to zero. Theorem 2 insures that the successive values of a_{i0} must be monotonically decreasing from the nth step until the infimum of x_i is reached, by the same logic.

If there are s discrete variables, there can be no more than $(n + s + 1)$ rows in the current A matrix, since slack variables re-entering the solution at positive values are redundant and the rows headed by these values can be dropped from the computation.

The remaining variables without nonuniform or uniform discrete constraints may assume arbitrary values, but they must be feasible, since repeated use of the dual linear programming pivot elements a_{it} in (2.16) insures this.

The use of the dual linear programming algorithm will also insure that the non-existence of a feasible finite, mixed-discrete solution will be indicated in a finite number of steps. In the discrete case, this will be indicated when none of the coefficients a_{ij} or a_{ij}^* of the cut defined in Theorem 1 are positive. Then the set of feasible solutions for (2.1) in conjunction with (2.2) is empty, since clearly otherwise $s \leqq -b_{ik}$ in (2.9) or (2.10).

III. Computational Considerations

The algorithm described above has been programmed in FORTRAN II for the IBM 1410 digital computer. It was incorporated into an earlier program employing Gomory's mixed-integer algorithm and the primal-dual linear programming algorithm.

Selection of the row in the current matrix of coefficients from which to form the augmented row may be carried out by any of the rules-of-choice given in Gomory [2] for the all-integer and mixed-integer problems. In the nonuniformly discrete case, the criterion analogous to Gomory's fractional part f_{i0} is the ratio or

$$f_{i0}^* = b_{in_i}/(a_{io} - X_{in_i}) \equiv 1, \qquad \text{if } a_{i0} > X_{in_i} .$$

These ratios $\{f_{i0}^*\}$ may be compared among the rows corresponding to the various variables which are constrained to be nonuniformly discrete in the problem and also are directly comparable with the fractions $\{f_{i0}\}$ from the rows corresponding to uniformly discrete variables. Note the use of a "maximum fractional part" criterion, in which the row index is

$$i = k \mid f_{k0} = \max \{f_{s0}, f_{s0}^* \mid x_s \text{ discrete}\}$$

will, if possible, always lead to the selection of a row with $f_{i0}^* = 1$, thereby, by (2.13), bringing the value of the nonuniformly discrete variable x_i into the desired range of admissible values immediately with the completion of the subsequent dual linear programming iteration.

References

1. Gomory, R. E., "An Algorithm for the Mixed Integer Problem", Research Memorandum RM-2597, Rand Corporation, Santa Monica, California, 1960.
2. ——, "Integer Solutions to Linear Programs", *Recent Advances in Mathematical Programming*, R. L. Graves and P. Wolfe (eds.), McGraw-Hill, New York, 1963.
3. Llewellyn, R. W., *Linear Programming*. Holt, Rinehart and Winston, New York, 1964, p. 283.

Comments on the Article

The preceding article illustrates both standard and nonstandard usage. For the journal *Management Science*, however, the treatment is entirely appropriate—because it is consistent throughout the article.

The *title* is made up of keywords. There are six words that might be used for cross-referencing and for recall or collating purposes. The meaning of the title is clear, for the authors have succeeded in arranging the six keywords in a meaningful pattern.

The dual *headnote* is nonstandard in that an asterisk is used to indicate the date the editors of the journal received the article. This usage, however, is consistent throughout the journal. The numbered headnote is standard usage. Here it indicates that the article is a reworking of a professional paper delivered earlier and perhaps available elsewhere.

The *by-lines* or *author credits* not only show the present professional positions of the authors, they also show that the article is but a part of one author's PhD dissertation, which by definition is available at the University of North Carolina at Raleigh.

The *abstract* is essentially an informative abstract which discusses the procedures, results, and findings.

The *introduction* is remarkably succinct. First, it defines and illustrates the problem. Second, it cites solutions in the literature. Finally, it indicates the methodology to be applied to the problem.

The body of the article is divided into two parts with subheadings "II. Theory" and "III. Computational Considerations." The organization, however, is based upon mathematical logic. A theorem is introduced; it is applied to a Case; and a result is attained by use of successive, numbered equations. The numbering of the equations is consistent with practice recommended by the American Mathematical Society (see p. 000). The progression is obvious and inexorable. Line headings and succinct or blunt statements such as "This completes the proof of Theorem 1" lead the reader to findings promised in the abstract.

The *references* are arranged alphabetically in a modification of MLA *Style Sheet* usage. Consecutive, bracketed footnotes are used because the journal's "Instructions to Authors" printed above require them. It is interesting to note that the American Mathematical Society advises against the use of footnotes in professional articles.

B. BIOLOGICAL SCIENCE JOURNALS

The *Style Manual for Biological Journals* fixes usages for articles upon biological science. A section of that outstanding manual is reprinted below by permission of the Conference of Biological Editors. The entire document is too lengthy for reproduction here. It is available at a cost of $3.00 from the American Institute of Biological Sciences, 2000 'P' Street, N.W., Washington, D. C. 20036.

Acknowledgments

A section headed ACKNOWLEDGMENTS may be placed between the text and LITERATURE CITED. Avoid acknowledgments as footnotes to the title or to words in the text. Do not use a credit line (with the technical assistance of . . .) immediately after the name of the author(s) because the person credited may be inadvertently cited as an author.

In this section you may acknowledge grants-in-aid, and statements, tables, or figures borrowed from published material. It is courteous to ask the author and editor for permission to reproduce even material that is not copyrighted. Copyright owners sometimes specify the phrasing for credit lines.

Literature Cited

Basic considerations in making bibliographic references are accuracy, readers' convenience, and librarians' time.

In the text, citations should be made consistent (according to the practice of the journal) by use of one of the following systems:

1) *Name-and-year system.* Depending upon the construction of the sentence, the citation will appear as Smith and Jones (1960), or (Smith and Jones, 1960). When there are three authors, name all in the first citation, e.g., Doe, Miller, and Wilson (1960), but subsequently use Doe et al. (1960). When there are four or more authors cite their paper in the form Doe et al. (1960).
2) *Number system.* Depending upon the construction of the sentence, the citation will appear as Smith and Jones (1), or (Smith and Jones, 1) or simply (1). If citations are to be numbered, number them after all additions and deletions have been made.

List literature citations or references at the end of your paper in alphabetical order by authors. Include only those references cited in the text. Do not cite unpublished work unless the paper has been accepted for publication. Unpublished results may be mentioned as such in the text with the word (*unpublished*) in parentheses after the author's name. References should contain all the data necessary to locate the source easily in a library. Check all parts of each reference against the original. An inaccurate or incomplete reference wastes time of readers and librarians and reflects on the scholarship of the author.

Most journals have their own style for capitals, italics, boldface, and so on, in literature citations. Be sure to consult a recent issue. Literature citations must be typewritten and double-spaced.

The critical items for literature citations follow:

1) *Authorship.* The family name of the first or sole author precedes the initials or given name. Cite names of all co-authors as given in the byline.

It is usually not difficult to invert the family and given names of the first or sole author in preparing a reference list. Personal names in many countries usually correspond in form to *John C. Smith,* and can be easily inverted (*Smith, John C.*). But designations of rank within a family and compound and hyphenated family names of foreign origin may present problems.

Junior (Jr.) and designations of rank within a family, such as *II* and *III,* are indicated after the initials (F. W. Day, Jr., inverts to Day, F. W., Jr.; C. G. Child II to Child, C. G. II.). The Spanish word *hijo* (h.) means *son* and is equivalent to *junior,* and should be so translated: Gonzalo Ley (hijo) becomes Ley, G., Jr. Also maintain the maternal name in Spanish; *Casimir Gómez Ortega* inverts to *Gómez-Ortega, C.* not *Ortega, C. G.,* and *Juan Pérez y Fernández* inverts to *Pérez y Fernández, J.*

Compound and hyphenated *American* family names, irrespective of origin, are treated in the same manner as other American names:

Examples	*Invert to*
Henri Vander-Brink	Vander-Brink, Henri
C. B. van Niel	Van Niel, C. B.
R. P. De Smet	De Smet, R. P.
S. Bayne-Jones	Bayne-Jones, S.
J. de Bueno	De Bueno, J.
T. l'Eltore	L'Eltore, T.

Compound family names in publications from other countries (Canada, Czechoslovakia, England, Finland, Germany, Italy, Poland, Scandinavia, Spain, USSR, etc.) are similarly inverted and the particles are capitalized.

For *Brazilian* and *Portuguese* names the particles (*do, da, dos, das*) follow the initials or given name:

Silvio do Amaral	Amaral, Silvio do
A. C. dos Santos	Santos, A. C. dos

In *Chinese* publications the family name precedes the given name (usually hyphenated):

Chen Tai-chien	Chen, Tai-Chien
Lin Ke-sheng	Lin, Ke-Sheng

But in American and British journals *Chinese* names are usually anglicized and inverted:

C. Ying Chang	Chang, C. Ying
Hsi Fan Fu	Fu, Hsi Fan

With *Dutch* names, particles and particle phrases follow initials when inverted:

L. A. de Vries	Vries, L. A. de
Willem van Eyck	Eyck, Willem van
J. van der Hoeve	Hoeve, J. van der
L. W. van Horts van Bing	Horts van Bing, L. W. van

Egyptian and other *Arabic* proper family names appear last:

Hassan Fahmy Khalil	Khalil, Hassan Fahmy
Mohamed Metwali Naguib	Naguib, Mohamed Metwali

When either prefixes and their variants (*el, ibn, abdel, abd-el, abdoul, abu, abou, aboul*) or the particle *el* alone precedes a name, either should be hyphenated to the name it precedes:

Aly Abdel Aziz	Abdel-Aziz, Aly
Youssef Abou-el-Ezz	Abou-el-Ezz, Youssef
Aziz Ibn Saud	Ibn-Saud, Aziz
Kamel el Metwali	el-Metwali, Kamel
Hedieh Khalil el Agouz	el-Agouz, Hedieh Kahalil

In compound *French* names, the definite article (*le, la, les*) or combination with the preposition *de* (*du, de la, des*) precedes the family name. *De* (or *d'*) alone follows the initials:

J. Le Beau	Le Beau, J.
R. L'Epée	L'Epée, R.
V. du Bary	Du Bary, V.
A. de Bary	Bary, A. de
B. d'Aubiac	Aubiac, B. d'

The particles (*im, von, zu, zum, zur*) and their abbreviations in *German* names follow initials when inverted, and should be spelled out:

C. von Holt	Holt, C. von
H. zur Horst-Meyer	Horst-Meyer, H. zur
Ludwig v. Obersteg	Obersteg, Ludwig von

In *Hungarian* the family name regularly precedes the given name and inversion is unnecessary.

Farkas Karoly	Farkas, Karoly
Szent-Györgyi Albert	Szent-Györgyi, Albert

If *Sen* or *Das* precedes an *Indian* name, include it with the family name:

B. C. Sen Gupta	Sen Gupta, B. C.
K. P. Das Gupta	Das Gupta, K. P.

All elements of *Vietnamese* or *Thai* names are taken in the order in which they appear in the journal, joined by hyphens, and lower case is used for the second element:

Nguyen Lam Tiep	Nguyen-lam-Tiep

2) The *year* of publication follows the authorship:
Jones, T. C., and R. Doe. 1959.

When more than one paper or book by the same author(s) has appeared in a given year, the letters *a, b,* etc., should be used after the year (e.g., Smith, 1959*a, b*). In the references each entry should be typed separately, with the same letters after the date as appeared in the text (e.g., Smith, R. P. 1959*a*, Smith, R. P. 1959*b*).

For the name of one author or of identical authors listed in the same order, the editor may substitute a 3-em dash (———) in repeated entries.

3) The *title* must appear exactly as it does on the first page of the article or on the title page of the book.
4) *Abbreviations* are commonly used for the names of serial publications, except for one-word names. Follow the abbreviations listed by The American Standards Association (*see* examples, p. 82). If no abbreviation is found, use the rules of the American Standards Association, which are summarized as follows:
   a) Never abbreviate the title of a journal consisting of a single word. Example: Phytopathology

b) Do not abbreviate the title of a journal consisting of several words so much that the journal cannot be recognized.
c) Never abbreviate personal names when they begin a journal name. Example: Hoppe-Seyler's Z. Physiol. Chem.
d) Form the abbreviation by omitting a continuous group of the final letters of the word; terminate it, if possible, after a consonant. Example: Biol., not bio. for biology.
e) The order of abbreviations in a title should be the same as the word order in the complete title. But in a long title some of the final words may be omitted. Never include abbreviations of subtitles.
f) Omit articles, conjunctions, and prepositions.
g) Capitalize the initial letter of the first element of the abbreviation. For the remainder, capitalize the first letter of each element, all letters, or none. Examples:

Amer. J. Physiol.
AMER. J. PHYSIOL.
Amer. j. physiol.

h) For compound words, abbreviate only the final element. Example: Bodenforsch. *for* Bodenforschung.
i) Use either a period or a space between abbreviations of title words. If a space is used, each element must begin with a capital letter.
j) Diacritical marks may be used in an abbreviation but are not required. Be consistent.

5) *Volume and pages of serials* appear in arabic numbers after the abbreviated name of the periodical: 2: 120–136. An issue, number, supplement, or other part within a volume is shown in parentheses only when paged independently: 2(4): 1–56; 34 (Suppl. 2): 1–26. Any special series (Ser. 3, III, or C) precedes the volume number: Ser. 3, 2: 120–136; III, 2(4): 1–56; C, 2: 120–136.
6) In *book citations* the year of publication and title follow the authorship. The following appear in sequence after the title: the edition if other than the first, the publisher's name or shortened name (according to the *Cumulative Book Index*), the place of publication, and the number of pages if one volume, but the number of volumes if more. If particular pages are cited mention them in the text.
7) *Illustrations* are not mentioned unless they are separately paged from the text or are of particular importance.

8) For *transliteration* of words from Greek and Russian, *see* p. 42 and 43
9) *Missing bibliographic details* added for clarity (names, dates, publishers, etc.) should appear in brackets. For exceptions consult Bryant (1951).
10) Unpublished *documents* and other *source material* should be cited within parentheses in the text rather than in the **Literature Cited.** For example: (R. W. Smith, *personal communication*) (J. K. Jones, *unpublished data*).

EXAMPLES

Various types or categories of problems encountered in citing references are listed below in boldface type, and examples are given.

Author: Name as in by-line; Abstract:

Hildebrandt, Albert C. 1948. Influence of some carbon compounds on growth of plant tissue cultures in vitro. Anat. Rec. 100:674. (Abstr.)

Author: Prefix in name anglicized; Miscellaneous publication:

Van Dersal, W. R. 1938. Native woody plants of the United States, their erosion-control and wildlife values. U.S. Dep. Agr. Misc. Publ. 303. 362 p.

Author: Prefix in French name not anglicized; Pages separated:

Bary, A. de. 1886. Ueber einige Sclerotinien und Sclerotienkrankheiten. Bot. Zeit. 44:377–387, 393, 404, 409–426, 433–441, 449–461, 465–474.

Author: Hyphenated name (Compound name, without hyphen); Subtitle:

Gwynne-Vaughan, Helen. 1922. Fungi; Ascomycetes, Ustilaginales, Uredinales. Cambridge Univ. Press, London. 232 p.

Author: Transliterated names; Volume omitted; Each issue numbered independently:

Gavrilov, K. A., and T. S. Perel. 1958. Earthworms and other invertebrates in the soil under forests in Vologda region [in Russian]. Pochvovedenie 1958(8):133–140.

Author: Transliterated name; English title on original; Annals of Society; Number paged separately; Summary in English:

Nishikado, Y. 1921. On a disease of the grape cluster caused by *Physalospora baccae* Cavara [in Japanese, English summary]. Phytopathol. Soc. Japan, Annu. 1(4):20–42.

Author: Committee chairman; Preposition omitted from name of publication:

Riker, A. J. [*Chairman*]. 1952. Literature citations; how biologists like them. AIBS (Amer. Inst. Biol. Sci.) Bull. 2(1):18–19.

Author: Society committee; One-word serial name not abbreviated.

American Phytopathological Society, Committee on Standardization of Fungicidal Tests. 1943. Defiinitions of fungicide terms. Phytopathology 33:624–626.

Author: Service agency, omitted as publisher:

Chemical Abstracts Service. 1961. Chemical Abstracts list of periodicals with key to library files. American Chemical Society. Washington, D.C. 397 p.

Author: State institution; Fiscal year; Special part; Bulletin:

Wisconsin Agricultural Experiment Station. 1950. What's new in farm science; 66th annual report 1948/49. Part I. Wisconsin Agr. Exp. Sta. Bull. 491. 88 p.

Author: Federal agency; Two or more volumes:

U. S. Bureau of the Census. 1927. United States census of agriculture. 1925. U. S. Government Printing Office, Washington. 3 vol.

Author: Federal agency; Pages not numbered:

U. S. Department of Agriculture. Plant Pest Control Division, Pesticide Regulation Section. 1957. A summary of certain pesticide chemical uses. Loose leaf. n. p.

Author: Federal agency (omitted as publisher); Revised edition:

U. S. Government Printing Office. 1959. Style manual. Revised ed. Washington, D.C. 492 p.

Book:

Schwarts, R. J. 1955. The complete dictionary of abbreviations. T. Y. Crowell Co., New York. 211 p.

Book, Part of:

Overstreet, H. A. 1925. The psychology of effective writing, p. 87–109. *In* H. A. Overstreet, Influencing human behavior. Norton, New York.

Bulletin:

Bryant, Margaret S. 1951. Bibliographic style. U. S. Dep. Agr. Bibliogr. Bull. 16. 30 p.

Documents not published in conventional manner:

Annual report:

McClellan, R. O., J. R. McKenney, and L. K. Bustad. 1961. Metabolism and dosimetry of cesium-137 in rems, p. 55–59. *In* Hanford biology research annual report for 1960. HW-69500 (Hanford Laboratories, Richland, Wash.)

Nuclear Science Series reference; distributor cited:

Finston, H. L., and M. T. Kinsley. 1961. The radiochemistry of cesium. Nat. Acad. Sci., Nuclear Sci. Ser. NAS-NS-3035. Office of Technical Services, Dep. of Commerce, Washington, D.C.

Progress report:

Auerbach, S. I., and R. M. Anderson. 1959. Ecological research, p. 18-54. *In* Physics Division annual progress report for period ending 31 July 1959. ORNL-2806 (Oak Ridge National Laboratory. Tenn.)

Technical Report; double reference:

Gloyna, E. F., E. R. Hermann, and W. R. Dryman. 1955. Oxidation ponds—waste treatment studies, radioisotope uptake, and algae concentration. Univ. Texas, Dep. Civil Eng. Tech. Rep. No. 2. (*Also* AECU-3113).

Translation:

Nakaya, T. 1960. Biological, geological and chemical studies on Sr^{90}, Cs^{137} in fresh water regions. [Transl. from Japanese] p. 5 to 8. USAEC-tr-4245.

Illustrations not included in pagination, and important:

Smith, E. F. 1917. Mechanism of tumor growth in crowngall. J. Agr. Res. 8:165–183; Fig. 4-65.

Newspaper (pages separated):

Maverick, M. 1944. The case against "gobbledygook." New York Times Magazine. 21 May: 11, 35–36.

Paper in a collection or book by various authors:

Link, G. K. K. 1928. Bacteria in relation to plant diseases, p. 590 to 606. *In* E. O. Jordan and I. S. Falk [ed.] The newer knowledge of bacteriology and immunology. Univ. Chicago Press, Chicago.

Patent: original not seen:

Penn, F. H. 1942. Hydrogenated butter method. U. S. Pat. 2,272,578 Feb. 10 Abstr. in Offic. Gaz. U. S. Patent Office 535:322.

Proceedings of Society; Series:

Salaman, R. N., and F. C. Bawden. 1932. Analysis of some necrotic virus diseases of the potato. Roy. Soc. (London), Proc., B. 111:53–73.

Thesis on microfilm:

Rafferty, Nancy S. 1958. A study of the relationship between the pronephros and the haploid syndrome in frog larvae. Ph.D. Thesis. Univ. Illinois (Libr. Congr. Card No. Mic. 58-5479) 41 p. Univ. Microfilms. Ann Arbor, Mich. (Diss. Abstr. 19:1146)

Transactions of society:

Vose, G. P. 1963. Thermal destruction of bone as seen with the electron microscope. Amer. Microscop. Soc., Trans. 82:48–54.

Two papers in same year, lettered when citations not numbered; Name repeated:

Magoon, M. L., R. W. Hougas, and D. C. Cooper. 1958*a*. Cytogenetic studies of tetraploid hybrids in *Solanum* from hexaploid-diploid matings. J. Hered. 49:171–178.

Magoon, M. L., R. W. Hougas, and D. C. Cooper. 1958*b*. Cytogenetic studies of complex hybrids in *Solanum*. J. Hered. 49:285–293.

In the typewritten copy of the literature citations repeat all names of multiple authors. Editors or printers may later insert a 3-em dash (———) for the names in consecutive citations after the first.

ABBREVIATIONS OF WORDS USED IN CITATIONS

Titles of journals can be abbreviated by combining the abbreviations of the words or word stems listed below (single word titles are not abbreviated). The abbreviation appears in boldface type. The list is adapted from one prepared by The American Standards Association, Sectional Committee Z39 on Library Work and Documentation.

Abhandlung-
Abstract
Abteilung
Academ-
Accadem-
Administr-
Advance-
Aerologicheskii
Aeromedica, Aeromedic-
Aeronaut-
Aerzteblatt
Africa
Agraire, Agralia, Agrar-, Agrarnyi, Agricol-, Agricult-, Agrikult-
Agrobotanica
Agrogeological
Agronom-
Akadem-
Algologi-
Allgemein
Amendment
America-, Amerika-
Anaesthes-, Anaesthetist
Anais, Anale
Anal-
Anatom-
Angewandt-
Animal-
Annaes, Annal
Anniversary
Annotation-
Announcement
Annual, Annuale, Annuario
Anorganisch
Anthropolog-
Antibiotic
Antimicrobial
Anual-, Anuar-
Apicole
Apicolt-
Apicult-
Apothecary, Apotheker
Appendix
Applicada, Applicat-, Applied, Applique
Arbeit-, Arbete-
Arboriculture
Archaeolog-

Archeolog-
Archiv-, Archiwum
Arhiv
Arkhiv
Arquiv
Asociacion
Associa-
Astronom-
Astrophys-
Atmosfaer-, Atmosfar-, Atmosfer-, Atmosphar-, Atmospher-
Atomic
Auditory
Automatic
Avance-
Avhandling-
Bacolog-
Bacteriolog-
Bakteriolog-
Batteriolog-
Behavior
Beiheft
Beilage
Beitrag
Belg-
Bericht
Bibliograf-, Bibliograph-
Bibliotec-, Bibliotek-, Bibliothec-, Bibliothek, Bibliotheque
Biennial
Biochem-
Biochim-
Biodynamica
Biofizika
Biogeochimique
Biogeograph-
Biograf-, Biograph-
Biokhim-
Bioklimatologie
Biolog-, Bioloskih
Biomedical
Biophysic-
Bioquimica
Biotheoretic-
Biuletyn, Biulleten
Bjuletin
Bodenforschung
Bodenkunde
Bohemosloven-
Boletim
Bolgarskii
Bollettino
Botan-
Bratislav-
Britain, Britanni-, British
Bryology-
Buleten
Bulgarian
Bulletin-, Bullettino
Bureau
Canad-
Cardiolog-
Cartografica, Cartographie
Catalog-
Cechoslov-
Centennial
Centraal, Central-
Ceskoslovensk-
Chemi-
Chinese
Chirurg-
Chromatography
Chroni-
Ciencia-
Cientifica
Circular
Cirkulaer
Cirugia
Class-
Climatolog-
Clini-
Colegio
Collaboration, Collaborazione
College
Comerci-, Commerce
Commission, Committee
Communic-
Company
Compar-
Compte, Comptes
Comunic-
Confederation
Conference
Congres-
Conserv-
Contribut-
Cooperat-
Corporation
Cryptogam-
Cultur-, Cultuur
Cytochem-

Cytolog-
Czechoslovak

Decennial
Demographie
Dendrolog-
Dent-
Departament-, Departement-, Department-
Dermatolog-
Deutsch-
Digest-
Direc-, Direcc-, Direct-, Direkt-
Disease
Disserta-
Divis-
Document-
Doklad-
Dokument

Ecolog-
Econom-
Edition, Editor
Educa-
Egyet-
Egyptian
Ekolog-
Electrochem-
Electrochim-
Electrolog-
Electrotechnical
Embriolog-
Embryol-
Encyclopedia
Endocrinolog-
Engineer-
Enolog-
Entomolog-
Enzymolog-
Epidemiolog-
Escola-
Espan-
Essential
Ethnograf-, Ethnograph-
Ethnolog-
Etudes
Eugenics
Europe-
Evolution
Examination
Exchange
Exhibit-
Experiment-
Extension
Extract

Facolt-, Faculd-, Facult-
Fakult-
Farmaceut-, Farmacevt-, Farmaci-, Farmaco
Farmacolog-
Federac-, Federal-
Finland-
Finn-
Fitolog-
Floricoltura
Floristica
Flugblatt
Forest-
Forsch-
Foundation
Fysiograf-
Fysiolog-

Gazet-, Gazett-
Gemolog
Genel, General-
Genet-
Genitourinary
Geochem-
Geochim-
Geodaes-, Geodaet-, Geodas-, Geodat-, Geodes-, Geodet-, Geodez-
Geograf-, Geograph-
Geolog-
Geomagnetism
Geophys-
Geriatri-
German-
Gerontolog-
Gesellschaft
Gesundheit
Gibridizatsiia
Gidrobiol-
Gidrolog-
Gigiena
Giornale
Glaciology
Graduate
Gynecolog-

Haematolog-
Helveti-
Hematolog-
Herbari-
Heredit-
Histochem-

Histolog-
Histor-
Horticol-, Horticult-, Hortikult-, Hortique
Hospit-
Hungar-
Husbandry
Hydrograf-
Hydrolog-
Hygien-

Ichthyolog-
Illustr-
Immigration
Immunitatsforschung
Immunolog-
Imperial-
Importacao, Importacion, Importation, Importazione
Imunolog-
Incorporated
Industr-
Infect-
Infekt-
Inorganic
Institucao, Institucio-, Institut-, Instytut
Interamerica
Internal
International
Investiga-
Iranicus
Itali-

Jaarboek
Jahresbericht
Japan-, Japon-
Jardim, Jardin-
Jewish
Jornal, Journal
Jugoslav-

Katalog
Kem-
Klass-
Klini-
Kommission, Kommitte
Kommun-
Konfer-
Kongres, Kongress

Laboratoire, Laborator-
Landwirtschaft-
Language
Latin, Latinus
Latinoamericana
Leaflet
Lebanese
Lebensmittel
Lectur-
Leningrad-
Librair-, Library
Lichenolog-
Limnolog-
Linguistic
Literar-, Literatur-
Lithuanian

Magazin
Malacolog-
Malariolog-
Mammalog-
Management
Mathemat-
Mechanic-
Medecin-, Medic-, Meditsin-, Medizin-, Medycyna, Medyczny
Memento, Memoir-, Memorand-, Memoryal, Memuary
Mental-
Method-
Metrolog-
Mexic-
Micologia
Microbiolog-
Microscop-
Mikologi-
Mineral-, Mineralog-
Minerolog-
Minister-, Ministr-
Miscelan-, Miscellan-
Modern-
Molecul-
Monograf-, Monograph
Morpholog-
Moskovskii
Municip-
Muse-
Mycolog-

Nation-, Natirali, Natirelles
Natur-

Naturforschung
Nederland-
Netherlands
Neurobiolog-
Neurolog-
Neurosurgery
New England
New Series
New Zealand
Nippon-
Nord-
Nuclear-

Observ-
Occupation-, Occupazione
Oceanograf-, Oceanograph-
Ocular-
Offici-
Ophthalmolog-
Optic-, Opticheskii Optik-, Optique, Optisch
Optometry
Organic-, Organicheskii, Organique
Organisat-, Organizac-, Organizat-, Organize-, Organizing, Organizzazione
Orient-
Original-, Origineel
Otolaryngolog-
Otolog-

Paleontolog-
Pamflet, Pamietnik-, Pamphlet-
Parasitenkunde
Parasitolog-
Patent
Pathogen
Patholog-
Pediatr-
Pharmaceut-, Pharmaci-, Pharmacy, Pharmazeut-, Pharmazie
Philosoph-, Philoszophia
Photograaf, Photograf-
Physica-, Physicist, Physics, Physicu-, Physik-, Physique-
Physiolog-
Phytolog-
Phytopatholog-
Polish, Polnisch, Polon-, Polski
Pomolog-
Populae, Populair, Popular-
Postgraduate
Prehistori-
Prelimin-
Proceeding
Professional, Profession-
Project-, Projekt
Psychiatr-
Psycholog-
Psychopharmacology
Publisher, Publication

Quantitativ-
Quarterly

Radiation
Radioactive
Radiobiolog-
Radiolog-
Reclamation
Record, Recueil
Registr-
Religious
Rendu, Rendus
Report
Reproduction
Repubblica, Republ-
Research
Review, Revista, Revue
Rhumatologie
Rivista
Romanian
Royal
Rumanian
Russ-

Scandinavi-
Schrift-
Schweizer-
Scien-
Scotland, Scottish
Sectio-
Seismolog-
Serie, Series
Serolog-
Silvicult-
Simposio
Social-, Sociedad-, Societ-
Sovet-
Special-
Station, Stazione

Statist-
Street
Stud-
Sumar-, Summar-
Supplement-
Surg-
Survey
Swed-
Switzerland
Sympos-
System-

Taxonom-
Techni-
Technolog-
Tijdschrift
Topograf-, Topograph
Toxicolog-
Transaction, Transazione
Translation
Travail, Travaux
Treasurer, Treasury
Tropic-, Tropik-, Tropique, Tropisch
Trud-
Turkish, Turkiye
Topograf-, Typograph

Ukrain-
Union of Soviet Socialist Republics
United Kingdom
United Nations
United States
United States of America
Universidad-, Universit-, Universytet
Urolog-

Virolog-
Virusforschung
Vitaminolog-
Viticult-
Volume

Weekblad
Wetenschapp-
Wissenschaft
Wochenschrift

Zeitschrift
Zeitung
Zentralblatt
Zhirovoi
Zhurnal
Zoolog-

The range of the AIBS *Style Manual* is from Agronomy through Zoology. All the "life sciences" are involved, which partially explains the 45, 000 abstracts published annually by *Biological Abstracts*.

Clearly there are overlapping and interdisciplinary areas within the biological sciences. Separate style manuals for some of these areas (for example, chemistry, medicine, and psychology) are discussed below. For this reason, writers submitting articles to journals in overlapping and allied fields should follow the style manuals of those fields and, ideally, of the specific journal. The AIBS *Style Manual,* however, is so outstanding that the variations introduced by individual journals are merely addenda to the *Manual.*

The illustrative article printed below by permission of the author and the editor of the *Biological Bulletin* indicates the international role of the *Manual.* The article, written at the University of Washington by a scientist currently at the University of Newcastle-upon-Tyne, would have the same content and organization had it been written in any western country. The language might be different, but the format would have been that of the AIBS *Style Manual.* The subject field happens to be zoology, but usage would have been the same had the subject been bacteriology, entomology, or oceanography.

BROODING BEHAVIOR OF A SIX-RAYED STARFISH, LEPTASTERIAS HEXACTIS[1]

FU-SHIANG CHIA

Department of Zoology and the Friday Harbor Laboratories, University of Washington, Seattle, Washington

It has been known for a long time that certain starfishes, found in all three orders of the class, exhibit the habit of caring for their young. For example, Sars described the brooding habits of *Leptasterias mulleri* (Order Forcipulata) and *Henricia sanguinolenta* (Order Spinulosa) in 1846. Thomson observed the brooding habit of *Archaster excavatus* (Order Phanerozonia) in 1878.

Since then a number of brooding species have been added to the list. Ludwig (1903), Fisher (1940), and Hyman (1955) have named brooding forms and their distributions. It was pointed out by these authors that most of the brooding species are inhabitants of deep waters of the Antarctic and sub-Antarctic areas. However, some are commonly found on both the Atlantic and Pacific coasts of North America (Verrill, 1914; Fisher, 1911, 1928, 1930). Fell (1959) recorded a New Zealand shallow-water species, *Clavasterias suteri,* which is a brooding form also.

Brooding species usually produce a small number of large, yolky eggs, which undergo the direct type of development. Brooding species of the Orders Forcipulata and Spinulosa usually brood their young at the oral region by arching their arms to form a brooding chamber. However, in *Leptasterias groenlandica* (Lieberkind, 1920; Fisher, 1930), the young actually develop in the cardiac stomach. In the family *Pterasteridae* (Order Spinulosa), it was reported that the young develop in the nidamental chamber, the space between the aboral body wall and the supradorsal membrane (Koren and Danielssen, 1856; Fisher, 1940). Yet the evidence which I have gathered in *Pteraster tesselatus* of the San Juan Archipelago, Washington, indicates that this species does not in fact brood. The development is the direct type in this species, but is planktonic. Brooding species in the Order Phanerozonia, such as *Leptychaster almus* (Fisher, 1917) and *Ctenodiscus australis* (Lieberkind, 1926), usually brood their young among the paxillae on the aboral surface of the body wall. *Leptychaster* of Greenland (Order Phanerozonia), according to Fell (1959), hatches its young in the stomach, as does *Leptasterias groenlandica.*

All the publications of which I am aware that deal with brooding behavior in asteroids are limited to brief descriptions in connection with the studies of classification. No serious study on this subject has been reported. *Leptasterias hexactis* provides ideal material for such a study; it is an abundant intertidal species which

[1] This work represents a part of a dissertation submitted to the University of Washington in partial fulfillment of the requirements for the Ph.D. degree granted in August, 1964. The author is greatly indebted to Dr. R. L. Fernald for his advice and encouragement. Present address: Department of Life Science, Sacramento State College, Sacramento, California.

makes it easier to observe in its natural environment, and when collected, its small size makes it easy to handle and observe in the laboratory. A number of observations on brooding specimens in their natural habitat, as well as in the laboratory, have been made, and a number of simple experiments have also been performed in the Friday Harbor Laboratories.

General Observations

The animals in the Friday Harbor area enter the reproductive season at the end of November and terminate in April. The eggs are yolky and measure about 0.9 mm. in diameter. The number of embryos in each brood varies from 52 to 1491 and is correlated with the size of the female animal.

In order to examine the brooding habits more closely, four to five animals selected at random were kept in a small glass aquarium equipped with running sea water. When the females were ready to spawn, they assumed a typical brooding position (Figs. 1–3). If one was located on the wall of the aquarium, she arched her arms to form a brooding chamber and began to release eggs. During the first few minutes, the eggs were not sticky and did not adhere to each other. They were heavier than the sea water and tended to fall to the bottom of the aquarium. The female had to catch them by means of her tube feet and place them in the brooding chamber. Some eggs, especially those shed from the gonopores of the lower interradii, fell away. The males were spawning prior to the females or at the same time. At some times the spawning of male animals was unobservable and at others the spawning caused the water to become milky for a couple of hours. After the eggs were fertilized, the fertilization membrane became so sticky that the eggs adhered to one another and formed an almost solid mass (Figs. 4, 5). The process of shedding eggs lasted for several hours.

In the natural environment, the animals brood on the undersurfaces or sides of rocks. Attachment of the female to the undersurface has an obvious advantage during spawning, for in this position the eggs can be collected easily in the brooding chamber and the eggs always are provided with some moisture at low tide. In addition, direct exposure to sunlight is avoided. After spawning, the female starfish usually remains in this position until the young have completed metamorphosis. If the female is disturbed, she may move to another place, but does not give up her brooding activities.

In making collections, the animals were often placed in plastic bags containing sea water. These brooding animals usually brought their arms together to form an enclosed chamber, or lay free in the water or attached themselves to the plastic bag with the tube feet at the distal ends of the arms. They often remained in this position overnight without forsaking the embryos when the plastic bag was floated in the aquarium. After being released from the bag, they moved around in the aquarium and finally found a place to attach and resumed the brooding position again.

During the first 40 days of brooding, the position remained unchanged. The female attached herself only by the use of the tube feet at the distal end of the arms. The tube feet at the other parts of the arms were used to hold the embryos and orient them by pulling and twisting. The pressure exerted on the embryos by

FIGURE 1. Lateral view of a brooding female on the wall of a Plexiglas tank. Natural size.

FIGURE 2. Oral view of a brooding female on the wall of a Plexiglas tank, showing the completely closed brooding chamber. Natural size.

FIGURE 3. Lateral view of two brooding females on the wall of a Plexiglas tank. Natural size.

FIGURE 4. Cleaving embryos; most are in four-cell stage. Living specimen. 14 ×.

FIGURE 5. Three brooding animals turned oral side up, showing the application of the tube feet on the embryo masses. Living specimens. Natural size.

FIGURE 6. Oral view of three animals shown brooding artificial embryo masses made of gelatin capsules filled with wax. Living specimens. Natural size.

the tube feet may be fairly high and even temporarily distort the embryos. In most cases, from early development through metamorphosis, the embryos did not attach themselves to the substratum; instead, they adhered to each other and to the mother, and remained as a mass within the brooding chamber (Fig. 5). After about 40 days of development, when the tube feet of the young stars started to function, they began to attach themselves to the substratum but did not move away. It was at this stage that the adult resumed the flattened position; that is, she kept her arms straight, but remained with the brood. The whole picture reminds one of a mother hen protecting her chickens under her feathers and wings. After two months of development, when the preoral lobe of the young starfish had been completely absorbed, and the mouth opened and independent existence was possible, the mother animal left.

One female starfish was observed to spawn twice in the tank. This animal just moved a few inches away from her first brooding position and spawned again. The second spawning was about one month after the first and all of the first brood of embryos were metamorphosing and attached to the substratum. However, in most cases the female animals spawn only once per breeding season.

One can remove the embryos from the brooding chamber by a water current from a pipette. There are always small sand granules and debris which accumulate at the peristomial area, but the embryos in all cases observed were completely clean. This observation suggests that one of the functions of brooding is to keep the embryos clean.

Culture in vitro of the Embryos

When the embryo masses were removed from the brooding chambers at the early cleavage stage, two methods of rearing them in the laboratory were repeatedly attempted:

(1) Ten embryo masses were placed in a small aquarium with a continuous flow of sea water. This method would simulate a situation in which the animals would not brood, but only spawn their gametes in the sea water without further care. None of the embryos treated in this manner survived more than five days: due to the sticky nature of the fertilization membrane, they were soon covered by debris and became infected by bacteria and protozoans, and the whole embryo mass decomposed.

(2) Ten embryo masses were placed in separate fingerbowls containing filtered sea water maintained at a temperature of 10° to 15° C. on sea water tables. The sea water in the fingerbowls was changed daily. The embryos seemed to develop satisfactorily until the postgastrula or even the early brachiolarian stage. However, the development of embryos in a mass was not synchronous. In addition, most of the embryos cultured in this way could not break through the fertilization membrane and hatching did not occur. These embryos soon were infected by bacteria or protozoans, and the cytoplasm became liquified and turned tea-brown in color. The embryos swelled to about twice normal size and finally burst. Occasionally, an embryo was able to get one brachiolar arm (usually the dorsal arm) of the preoral lobe out of the fertilization membrane, but only one or two per cent were capable of hatching, and then the process was much delayed.

By way of contrast, the development of the brooded embryos is synchronous and they have no apparent difficulty in hatching. They hatch simultaneously and the post-brachiolaria lasts four or five days until the first sign of metamorphosis. This suggests the possibility that the female may secrete some substances, possibly enzymatic in nature, which break down or weaken the fertilization membrane. To make a preliminary test of this possibility, some tissues of the peristome and cardiac stomach of a brooding animal whose larvae were at prehatching stage or just hatched were homogenized and this preparation was added to the dishes containing embryos at various stages of development. The embryos were observed under a microscope and allowed to develop. In no case was there any positive evidence that this juice weakens the fertilization membrane. It is thought that the hatching processes may be facilitated by the mechanical action on the fertilization membrane of the tube feet of the mother animal.

Selection of Substratum

The brooding animals not only favor the sides and undersurfaces of rocks, as stated above, but are also very sensitive to other characteristics of the rocks. Observations in 1960–1961 demonstrated that the brooding animals show preference for the dark and rough-surfaced rocks. Similar experiments were repeated in 1962–1963 with the same results.

Thirteen brooding animals were put on the floor of a small aquarium supplied with running sea water, in which there were two pieces of rock of comparable size, but one dark and the other pale. The results were recorded as follows:

Time	Dark rock	Pale rock	Wall of aquarium
0 hrs.	0	0	0
4 hrs.	6	0	7
24 hrs.	9	0	4

In another experiment all 13 animals were put on the pale rock and the results were as follows:

Time	Dark rock	Pale rock	Wall of aquarium
0 hrs.	0	13	0
1 hrs.	0	0	13
2 hrs.	2	0	11
5 hrs.	7	1	5
24 hrs.	9	1	3

The first experiment shows that 6 of the animals moved to the dark rock during the first four hours and 7 to the aquarium wall, but none of them moved to the pale rock. Within another 20 hours, about half of the animals originally settled on the aquarium wall had moved to the dark rock, but still none had moved to the pale rock.

The second experiment shows that all the 13 animals migrated to the aquarium wall from the pale rock in the first hour. Twenty-four hours later nine of them moved to the dark rock, one to the white rock and three remained on the aquarium wall.

In another aquarium two pieces of dark rock of similar size, one having a rough surface and the other smooth, were placed in with eight brooding animals. Twenty-four hours later seven animals were on the rough-surfaced rock and one on the smooth-surfaced one.

The non-brooding animals have been checked by the same method. They may show a preference for the dark and rough-surfaced rocks. However, their reaction is not as definite as that of the brooding animals.

Feeding Activities During Brooding

Since the embryo mass occupies the peristomial region of the brooding animal, whose tube feet are used in orientation of the embryos (Fig. 5), any feeding activity by the brooder would be handicapped. In fact, no feeding activities of brooding animals have been observed, either in the laboratory or in the natural environment. Is this non-feeding situation simply due to the blockage of the food passage or is it controlled by some physiological factors?

To investigate this problem a simple experiment was set up as follows: three small rocks, each covered with about a dozen barnacles (*Balanus glandula*), six limpets (*Acmaea digitalis*), six mussels (*Mytilus edulis*) and one chiton (*Tonicella lineata*), were placed in three separate aquaria, A, B, and C, and supplied with running sea water. In aquarium A five brooding animals were placed, the brood being about 10 days old. There were five brooding animals in aquarium B, but the embryo masses were removed from the brooding chamber. Five non-brooding animals were placed in aquarium C. Throughout a month of observation, no animals in aquarium A fed on the available food items; the animals in aquarium B did not feed in the first two days, but began to feed on barnacles on the third day; the animals in aquarium C started to feed within the first six hours.

This observation clearly indicates that these starfishes stop feeding only when they are actually brooding (aquarium A). This may imply that the cessation of feeding is either a simple matter of interference by the embryo masses or is regulated by physiological factors but a constant stimulation by the embryo is needed. The latent period of two days as shown by the animals in aquarium B may be the time required for readjustment after the embryos were removed.

It is of interest to note that for an average-size star (radius 2.5 cm.), it takes 24 hours to complete the feeding on an *Acmaea digitalis* (1.4 cm. diameter), 30 hours on a small *Tonicella lineata,* and 60 hours on a *Mytilus edulis.*

Brooding on Artificial Embryos

If an embryo mass is removed from a brooding female for five hours and then returned, she will pick up the embryos and continue to brood. A brooding animal also will pick up an extra embryo mass if it is available. These observations lead one to believe that the animals actually recognize embryos of their own species.

To investigate this possibility an experiment was set up by placing 10 embryo masses together with 10 non-brooding animals in an aquarium. Within 8 hours, 9 of the 10 animals had picked up the embryo masses and settled on the aquarium wall to brood. Four of the 9 animals continued to brood until all the embryos completed metamorphosis; the other five ceased brooding activity during the first week. It was proved later by dissection that these five animals included three males (all spawned) and two females (non-spawned). The four which continued to brood were apparently unspawned females, for they also spawned their own eggs, and therefore possessed embryos of two different stages in the same brood.

The question thus arises as to how the animals recognize the embryo masses. In an attempt to answer this question the following four experiments were performed by using artificial embryo masses.

(1) Artificial embryo masses were made with wax and dyed with orange G. A piece of copper wire was inserted in the center of each of the wax blocks to keep it anchored at the bottom of the aquarium. The size, color, and shape of the artificial embryo masses were somewhat similar to the natural embryo mass. Ten of these artificial embryo masses were placed with 10 non-brooding animals in a small aquarium of running sea water; no response was observed in two days.

(2) The 10 artificial embryo masses prepared for this experiment were similar to those in experiment 1 except that they were punched with numerous holes by using a small needle and then soaked overnight in an homogenate of embryos of *Leptasterias*. It was hoped that some kind of substances from the embryos might be a factor in attracting the animals to brood. But after these artificial embryo masses were given to 10 non-brooding animals again, no reaction was observed in a two-day period.

(3) In a third experiment, 10 artificial embryo masses were made by filling gelatin capsules with colored wax and pieces of copper wire. After being placed in sea water, the gelatin swelled a little and became sticky; this condition is similar to that of the fertilization membrane. Two hours later, three of the ten non-brooding animals each picked up an artificial embryo mass to brood (Fig. 6) Twelve hours later, 8 of the artificial embryo masses had been picked up by 7 animals; one animal picked up two. The brooding of the artificial embryos, however, did not last more than two days. After 24 hours the gelatin capsules had swollen so much as to become loose from the wax blocks, and were discarded by the starfishes.

(4) For this experiment 10 non-brooding animals were given 10 artificial embryo masses which included five with gelatin capsules and five without gelatin capsules. Twelve hours later, three of those artificial embryo masses with gelatin capsules were picked up; none of the artificial embryo masses without gelatin capsules were picked up. The possibility that the animals were actually feeding on the gelatin capsules instead of brooding was ruled out by the fact that no extrusion of the stomach was ever observed in any of the cases.

These four experiments suggest that the animal's acceptance of a "mass" for brooding has little to do with its size, color, or shape. It also failed to provide evidence that chemotaxis plays a significant role in stimulating the recognition. Rather, the physical character of the surface layer of the embryo is possibly the main factor in its acceptance.

Conclusion and Discussion

In a recent study Arnold (1962) demonstrated that in the cephalopod, *Loligo pealii*, an artificial egg mass can stimulate sexual behavior, which is followed by the establishment of a social hierarchy. The stimulus, according to the author, is a visual response. On the other hand, Fell (1940), reported that in *Amphipholis squamata,* a brooding brittle star, there is actually a trophic relationship between the mother and the young. This is to say that the female animal secretes some nourishing substance(s) necessary for development of the embryos. In the present species, as it has been demonstrated, the animal will pick up an artificial embryo mass to brood, but this is apparently a response to the physical contact between the mother and the embryos. Meanwhile, no trophic relationship between the mother and young was indicated, because the embryos, after being removed from the mother, develop normally up to pre-hatching larval stage, and, if the fertilization membrane is removed manually, they will develop normally up to the young adult stage without any sign of morphological abnormality. In other words, the stored yolk in the ooplasm is sufficient to provide all necessary nutrients for development to a young adult. Thus, the major functions provided by brooding are protection, cleaning, maintenance of a uniform environment, and initiation of the hatching process.

Compared with all the other intertidal sea-stars at the San Juan Island area, *Leptasterias hexactis* has been very successful in spite of the fact that it produces only a small number of eggs and lacks the pelagic larvae to disperse the progeny. Embryos protected in the brooding chamber not only avoid exposure to severe effects of some environmental factors such as desiccation but also are safe from predators. The only predators for adult *Leptasterias* I have observed are other starfishes such as *Solaster, Crossaster,* and *Pycnopodia.* These were observed to prey on *L. hexactis* in the laboratory; however, species of these genera rarely exist together in the natural habitat.

The movement of cilia and tube feet is responsible for removing dirt and debris which become attached to the embryos. It is also responsible for maintaining a water current which insures a uniform environment and facilitates synchronous development. As for breaking through the fertilization membrane, there are two possible ways by which this may be achieved: (1) the embryo itself may secrete a digestive enzyme to dissolve the membrane; (2) the brooding animal may provide other chemical or mechanical means to rupture the membrane; possibly, a combination of these is involved. The failure of isolated embryos to hatch, as demonstrated in this study, may imply that the brooding mother is fully responsible for this process, but the possibility of the failure of embryos to secrete such a digestive enzyme because of the isolation from the mother is not excluded.

One of the most interesting problems, of course, is the control mechanism of brooding behavior. When an animal assumes a brooding position (Fig. 8) from a pre-brooding position (Fig. 9), one can easily see that at least three obvious changes occur in functional morphology. (1) Muscular system: the contraction of some muscles (longitudinal adambulacral muscles, for example) enables the animal to form a brooding chamber to house the embryos. This position lasts for at least 30 days. (2) Ovary: it is filled at the pre-brooding stage and emptied at the brooding stage, during which all the eggs are fertilized in the brooding chamber; they then form the embryo mass. (3) Pyloric caeca: materials such as lipids, polysaccharides, acidophilic granules, and zymogen granules, which are abundant in the epithelial cells of the pre-brooding animal, disappear after 30 days of brooding. Details of this phenomenon will be reported in a separate paper.

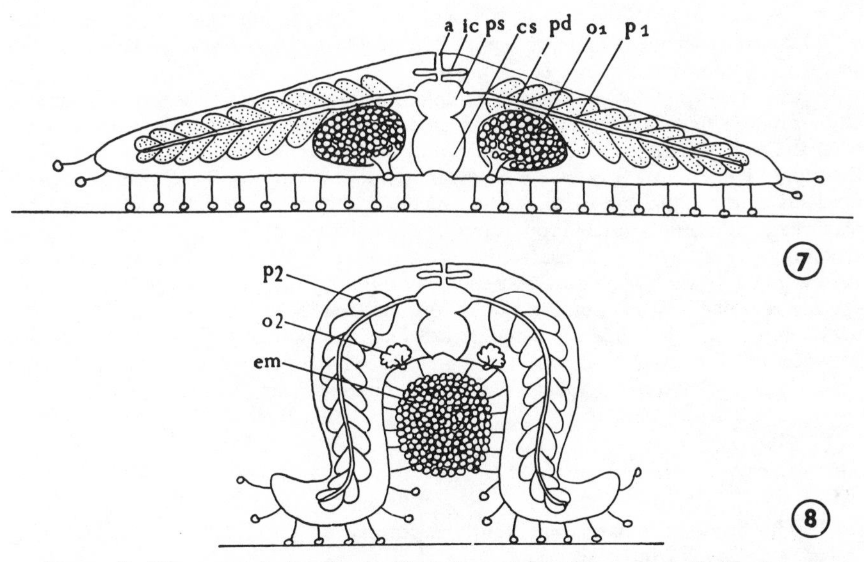

FIGURE 7. Diagrammatic drawing to illustrate a prebrooding animal (feeding). a. anus; ic, intestinal caecum; ps, pyloric stomach; cs, cardiac stomach; pd, pyloric duct; O_1, ovary before spawning; P_1, pyloric caecum of a feeding animal.

FIGURE 8. Diagrammatic drawing to illustrate a brooding animal. em, embryos; O_2, ovary after spawning; P_2, pyloric caecum of a brooding animal.

A number of other behavioral changes also occur during brooding, such as increased sensitivity to the substratum and cessation of feeding.

To maintain the brooding position, as shown in this study, a direct contact between the mother and the embryos is necessary, since the animal straightens her arms to a normal position and resumes feeding if the embryos are removed. Furthermore, the stimulus resulting from the contact between the embryos and the mother is thought to be physical rather than chemical in nature. But how does the stimulus pass from the contact area to the target organs? In other words, what is the coordination system? Taking the pyloric caeca and the muscular system as target organs, a hypothetic coordination process can be presented as follows:

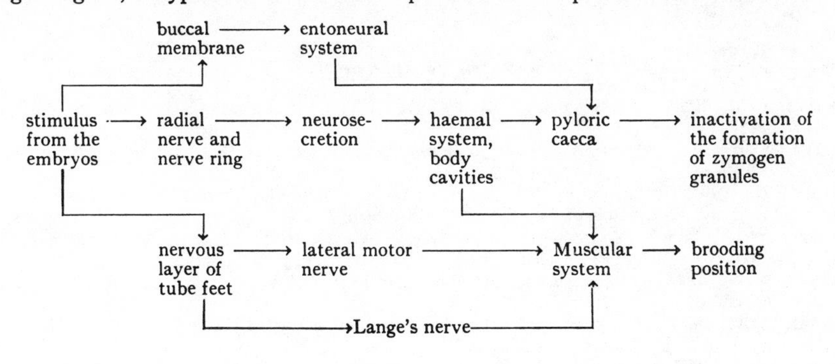

In support of this hypothesis, a number of facts, such as the nature of the stimulus and the changes in neurosecretion during brooding, have to be substantiated and studies along this line are being pursued.

Recent studies of Chaet (1964), Kanatani and Noumura (1962) and Noumura and Kanatani (1962) have shown that an intracoelomic injection of water extract from the radial nerve cord induces spawning in several species of asteroids. Furthermore, Unger (1960, 1962) has demonstrated the presence of neurosecretory products in the radial nerve of *Asterias glacialis* and their effect on motor activity, osmoregulation and pigmentation; therefore, it is reasonable to assume that neurosecretion may be involved in controlling the brooding behavior in the present species. A preliminary study on neurosecretion by histological methods in the present species revealed a scant number of neurons with secretory granules in the radial nerve cords, but it failed to demonstrate any substantial differences in quality between the brooding and non-brooding animals. However, the possibility that the brooding behavior is influenced by neurosecretion is far from exhausted.

Summary

1. The brooding habit is apparently necessary for these animals; none of the embryos survived without brooding.

2. The main functions provided by brooding are protection, cleaning, maintenance of a uniform environment, and initiating the hatching process.

3. During brooding, the animals are particularly sensitive to the substratum. They not only favor the under and vertical surface of rocks, but also prefer the darker-colored and rougher-surfaced rocks.

4. At the time of brooding, *Leptasterias hexactis* stops feeding completely, but the animals will take food after a latent period of two days if the embryos are removed from the brooding chamber. Thus, a constant stimulation by the embryos is apparently necessary to prevent the animals from feeding.

5. During the breeding season, all the adult animals (female, male, spawned and non-spawned) are able to recognize their embryos and will pick them up to brood. This recognition is not due to the size, color, or shape of the embryo masses but rather to the physical nature of the fertilization membrane.

6. The significance and possible control mechanism of brooding were discussed.

LITERATURE CITED

Arnold, J. M., 1962. Mating behavior and social structure in *Loligo pealii*. *Biol. Bull.*, **123:** 53–57.

Chaet, A. B., 1964. A mechanism for obtaining mature gametes from starfish. *Biol. Bull.*, **126:** 8–13.

Fell, H. B., 1940. Culture *in vitro* of the excised embryo of an Ophiuroid. *Nature*, **144:** 73.

Fell, H. B., 1959. Starfishes of New Zealand. *Tuatara*, **7:** 127–142.

Fisher, W. K., 1911. Asteroidea of the North Pacific and Adjacent Waters. Part 1. Phanerozonia and Spinulosa. *Bull. U. S. Nat. Mus.*, **76:** 1-419.

Fisher, W. K., 1917. *Trophodiscus*, a new sea star from Kamckata. *Proc. U. S. Nat. Mus.*, **52:** 367–372.

Fisher, W. K., 1928. Asteroidea of the North Pacific and Adjacent Waters. Part 2. Forcipulata. *Bull. U. S. Nat. Mus.*, **76:** 1–245.

Fisher, W. K., 1930. Asteroidea of the North Pacific and Adjacent Waters. Part 3. Forcipulata. *Bull. U. S. Nat. Mus.*, **76:** 1–356.

FISHER, W. K., 1940. Asteroidea. *Discovery Rep.*, **20**: 69–306.
HYMAN, L. H., 1955. The Invertebrates. Vol. IV, Echinodermata. New York: McGraw-Hill Book Co. Inc.
KANATANI, H., AND T. NOUMURA, 1962. On the nature of active principles responsible for gamete-shedding in the radial nerves of starfishes. *J. Fac. Sci. Univ. Tokyo*, **9**: 403–416.
KOREN, J., AND D. C. DANIELSSEN, 1856. Observations sur le développement des astéries. *Fauna Littoralis Norvegiae*, **2**: 55–59.
LIEBERKIND, I., 1920. On a starfish (*Asterias groenlandica*) which hatches its young in its stomach. *Vid. Medd. Dansk. Nat. Hist. Foren.*, **72**: 121–126.
LIEBERKIND, I., 1926. *Ctenodiscus australis*, a brood-protecting asteroid. *Vid. Medd. Dansk. Nat. Hist. Foren.*, **82**: 184–196.
LUDWIG, H., 1903. Seesterne. Expedition Antarctique *Belge*, Résultats Voyage *Belgica* 1897–1899. 72 pp.
NOUMURA, T., AND H. KANATANI, 1962. Induction of spawning by radial nerve extracts in some starfishes. *J. Fac. Sci. Univ. Tokyo*, **9**: 397–402.
SARS, M., 1846. Beobachtungen über dei Entwicklung der Seesterne. *Fauna Littoralis Norvegiae*, **1**: 47–62.
THOMSON, W., 1878. Peculiarities in the mode of propagation of certain echinoderms of the south sea. *J. Linn. Soc. London, Zool.*, **13**: 55–79.
UNGER, H., 1960. Neurohormone bei Seesternen (*Marthasterias glacialis*). *Symposia Biologica Hungarica* (1958), **1**: 203–207.
UNGER, H., 1962. Experimentelle und histologische Untersuchungen über Wirkfaktoren ans dem Nervensystem von *Asterias glacialis*. *Zool. Jb. Physiol.*, **69**: 481–536.
VERRILL, A. E., 1914. Monograph of the shallow-water starfishes of the North Pacific Coast from the Arctic Ocean to California. Smithsonian Inst. Harriman Alaska Ser. Vol. 14. 408 pp.

Comments on the Article

The *title* of this article contains the keywords necessary for effective cross-referencing and recovery. The meaning is clear, and the limitation of the article to "Brooding Behavior" is given.

The *headnote*, required by the AIBS *Manual* fixes the article as a part of a PhD dissertation and indicates where the complete dissertation may be found. The headnote also gives credit to the thesis advisor and gives the then-current address of the author.

The *author credit* indicates in italics the laboratories and facilities where the work was conducted.

The first three paragraphs review the literature. Arrangement is essentially chronological. Exceptions are made to correct the error of another, but the temporal sequence is retained. A less effective writer might have luxuriated in pointing out the error and written an obvious digression.

The fourth paragraph summarizes the present state of knowledge and indicates how the work under discussion extends that knowledge.

The first section of the text proper is headed "General Observations" and discusses the selection of specimens, the reactions of specimens, and the phenomena observed prior to experimentation. This background information is essential to an understanding of the research methodology to be described.

The second section of the text is headed "Culture *in vitro* of the Embryos" and is divided into two numbered laboratory phases. Each phase is discussed in a single paragraph, and a third paragraph is devoted to a statement of what normally occurs in nature. Because the natural hatching process suggests that "the female may secrete some substances which . . . weaken the fertilization membrane . . . ," the author discounts this obvious suggestion by disproving it in three sentences devoted to an experimental test. Discounting the suggestions leaves only the possibility of "mechanical action" which is stated as a linking device to the third section.

The third section of the article is headed "Selection of Substratum," and it proves that brooding animals prefer dark and rough-surfaced rocks.

The fourth section, "Feeding Activities During Brooding," is essential because the animal broods in a position that blocks the food passage. An experiment is described which proved that "starfishes stop feeding only when they are actually brooding."

The fifth section, "Brooding on Artificial Embryos," describes four numbered experiments which showed "that the animal's acceptance of a 'mass' for brooding has little to do with its size, color, or shape."

The first paragraph of the sixth section, "Conclusion and Discussion," concludes upon the basis of the experiment described in the preceding section that "the major functions provided by brooding are protection, cleaning, maintenance of a uniform environment, and initiation of the hatching process." Subsequent paragraphs in this section discuss the conclusion.

The seventh section, "Summary," is a factual, numbered list of the findings that might be drawn from the tests. There is no equivocation and no justification; this section is simply a professional's statement of what the work showed.

The concluding section, "Literature Cited," illustrates the usage required by AIBS journals. The changes in type face, the unique abbreviations, and the sequence and positioning of units such as dates are essential to publication in the journals of the biological sciences.

C. CHEMICAL AND PHYSICAL SCIENCE JOURNALS

The professional societies of chemistry and physics attempt to correlate and to guide the literature in their own disciplines. Writers submitting articles to the journals published by the American Chemical Society and the American Institute of Physics must follow the ACS *Handbook for Authors* or the AIP *Style Manual*. These documents are sold by the American Chemical Society Publications, 1155 Sixteenth Street, N. W., Washington, D. C. 20036 and the American Institute of Physics, 335

East 45th Street, New York, New York 10017. Each is excellent. The recommendations do not equivocate. Almost all problems that might arise in written communication within the disciplines have been anticipated and resolved.

The literature of astronomy and meteorology usually follows the stylistic recommendations of either specific journals or of the AIP *Style Manual.* Additional problems are introduced, however, by the government agencies that support much of the work in these two areas. The GPO *Style Manual* or the NASA *Publications Manual* must be followed by writers submitting articles and reports for publication by federal agencies or facilities.

The literature of geology tends to follow the ACS *Handbook for Authors.* Subsidization of research by either governmental or private industrial agencies leads to problems similar to those arising in astronomy and meteorology. An article on lunar geology published by NASA would follow the NASA *Publications Manual.* An article released by a private industry might follow the University of Chicago *Style Manual.* An article printed in a journal published by the American Geological Society would follow the unique "Style Guide" of that journal. The following documents offer some clarification.

J. W. Low and J. Braunstein, "Preparation of a Technical Article;" *Bull. A. A. P. G.*, Vol. 48, 1965, pp. 1837–1846.

H. D. Miser, et al., "Suggestions to Authors of the Reports of the U.S. Geological Survey," fifth edition, Washington: Government Printing Office, 1958.

Chemical Journals

Although the requirements for notes, communications, and special features vary among the journals of chemistry, a standard pattern exists for the articles that are published by all. That pattern has been discussed in detail in Chapter 4. The discussion is based upon *Handbook for Authors of Papers in the Research Journals of the American Chemical Society,* copyright 1965 by American Chemical Society Publications.

The article below is reprinted from the *Journal of Chemical and Engineering Data,* Vol. 11, No. 3, July 1966, pp. 343–346. Copyright 1966 by the American Chemical Society and reprinted by permission of the copyright owner. It illustrates procedures recommended by the ACS for chemical articles. The references and footnotes illustrate practice acceptable to the majority of research journals of chemistry. Practice varies, however, from journal to journal, and that of *Biochemistry* is quite dissimilar.

Solubility of Niobic Oxide and Niobium Dioxyfluoride in Nitric Acid–Hydrofluoric Acid Solutions at 25° C.

LESLIE M. FERRIS
Chemical Technology Division, Oak Ridge National Laboratory, Oak Ridge, Tenn.

The solubilities of niobic oxide and niobium dioxyfluoride were determined at 25° C. in hydrofluoric acid solutions that initially were up to 25*M* in HF and in nitric acid–hydrofluoric acid solutions that initially were up to 20*M* in HNO_3 and 5*M* in HF. In all cases, the niobium concentration in the saturated solution was about one fifth the total fluorine concentration, indicating that niobium was present in these solutions primarily as the $NbOF_5^{-2}$ ion. The solid phases at equilibrium always contained NbO_2F, indicating that the equilibria involved were $2NbO_2F_{(s)} + 8HF_{(aq)} \leftrightarrows 2H_2NbOF_{5(aq)} + 2H_2O$ and $Nb_2O_{5(s)} + 6HF_{(aq)} \leftrightarrows NbO_2F_{(s)} + H_2NbOF_{5(aq)} + 2H_2O$, depending on the initial compound used.

A PRIOR STUDY of the solubility of niobic oxide in hydrofluoric acid solutions was made by Nikolaev and Buslaev (*9*), who concluded that the oxide reacted with 0 to 25*M* HF according to the equation $Nb_2O_{5(s)} + 10HF_{(aq)} \leftrightarrows 2H_2NbOF_{5(aq)} + 3H_2O$, and that $Nb_2O_5 \cdot 2H_2O$ was the solid phase at equilibrium. During studies at this laboratory on the dissolution of niobium oxide in nitric acid–hydrofluoric acid solutions (*2*), it was noted that a solid, identified as NbO_2F by x-ray diffraction analysis, precipitated under certain conditions. This observation prompted the present investigation of the solubility of niobic oxide and the dioxyfluoride in HF-HNO_3 solutions.

EXPERIMENTAL

Materials Used. Hydrous niobic oxide was obtained by dissolving sintered Nb_2O_5 (Kawecki Chemical Co.; total metal impurities, less than 300 p.p.m.) in hydrofluoric acid, then precipitating the hydrous oxide with ammonium hydroxide. The precipitate was washed alternately with water and 0.1*M* HNO_3 and then air-dried at room temperature before use. The hydrous oxide contained about 50% niobium, 25% water, 0.1% fluorine, and less than 0.5% nitrogen.

Hydrated dioxyfluoride was obtained by allowing high purity niobium metal to react with boiling 16*M* HNO_3–1*M* HF (F/Nb atom ratio of less than 5). The hydrated dioxyfluoride flaked off the surface of the metal and settled to the bottom of the Teflon reaction vessel; it was then collected by filtration, washed with acetone, air-dried, and finally dried over Drierite in a desiccator. Analyses: Nb, 61.6%; F, 12.4%; H_2O, 5.12%. Calculated for $NbO_2F \cdot \frac{1}{2}H_2O$: Nb, 60.8%; F, 12.4%; H_2O, 5.89%. The existence of a hydrate was confirmed by infrared analysis. The water deformation band appeared at 1630 $cm.^{-1}$ and the –OH stretching bands at 3240 and 3360 $cm.^{-1}$. The x-ray powder pattern of $NbO_2F \cdot \frac{1}{2}H_2O$ was the same as that reported for NbO_2F by Frevel and Tinn (*3*). This is not surprising, since Frevel and Rinn found no change in lattice parameter with NbO_2F having water contents between 1.9 and 5.7%. The formula NbO_2F is used throughout this paper since, in general, the presence of the dioxyfluoride in equilibrium solid phases was confirmed by x-ray diffraction and not by chemical analysis. The hydrated dioxyfluoride was fairly stable on heating in dry helium at about 3° per minute in a thermobalance. Noticeable weight loss occurred only above about 200° C. On heating to 900° C., complete decomposition of the dioxyfluoride to Nb_2O_5 occurred.

All solutions were prepared from reagent grade acids and distilled water.

Procedure. Series of samples were prepared by adding excess oxide or dioxyfluoride to hydrofluoric acid or HF-HNO_3 solutions. The samples were equilibrated at 25° ± 1° C. Periodic analyses showed that samples originally containing the hydrous oxide attained equilibrium in a few days, but equilibration in samples initially containing the dioxyfluoride required nearly 2 years. After equilibration, samples of the saturated solutions were removed, clarified by centrifugation at 25° C., and analyzed. In some cases, wet residues were removed to allow determination of the equilibrium solid phase by the method of Schreinemakers (*10*). In all cases, the residual solids were recovered by filtration, washed with water, and subjected to x-ray diffraction and chemical analyses. No detectable hydrolysis of the dioxyfluoride occurred during washing.

Analytical. Niobium was determined both by an x-ray absorption method (*1*) and by ignition of samples to Nb_2O_5 at 900° C. The results from the two methods generally agreed within 3%. Total fluorine in solution and in wet residues was determined by differential potentiometric titration using standard NaOH as the titrant. This method is similar to the one used by Nikolaev and Buslaev (*9*). The over-all reaction (with the end point at pH 7 to 8) is: $2H_2NbOF_5 + 10NaOH \rightarrow Nb_2O_5 + 10NaF + 7H_2O$. Pyrohydrolysis (*4, 8*), after drying of the sample in the presence of an equimolar mixture of WO_3 and Na_2WO_4, was used for the determination of fluorine in solid compounds and to confirm fluorine values obtained by the titration method. The free fluoride concentration in solution was determined by a spectrophotometric titration method (*6, 7*) using a thorium-bearing solution as the titrant and SPADNS [the trisodium salt of 4,5-dihydroxy-3-(*p*-sulfophenylazo)-2,7-naphthalene disulfonic acid] as the indicator. The absorbency of the sample solution is recorded during the titration. A sharp increase in absorbency occurs at the end point. Nitrogen was determined by a modified Kjeldahl method.

X-ray powder patterns were obtained with a Debye-Scherrer 114.59-mm.-diameter camera using filtered Cu$K\alpha$ radiation. Infrared spectra were obtained with a Beckman IR7 spectrometer, using both the Nujol mull and KBr pellet methods.

RESULTS

The solubilities of hydrous niobic oxide and the dioxyfluoride in hydrofluoric acid solutions increased linearly with increasing total fluorine concentration. The niobium concentration in the saturated solutions was about one fifth the total fluorine concentration (Tables I and II, Figure 1). The plot of niobium concentration in the saturated solutions *vs.* total fluorine concentration (Figure 1) was a straight line of slope 0.220, corresponding to a F/Nb atom ratio of 4.54 in solution. This result suggested that the niobium was present in solution primarily as the $NbOF_5^{-2}$ ion. Raman spectra of selected saturated solutions were identical to that of the $NbOF_5^{-2}$ ion reported by Keller (*5*). Thus, the saturated solutions can be considered as solutions of H_2NbOF_5. The densities of these solutions at 25° C. increased linearly with increasing niobium concentration: ρ(g./ml.) = 1.000 + 0.1355 (Nb concn., *M*).

An attempt was made to determine the solid phase present at equilibrium by applying Schreinemakers' wet residue method (*10*) to several of the samples that originally contained hydrous oxide. No single compound was indicated as the equilibrium solid phase (Figure 2). Instead, the solid phases appeared to be mixtures of the oxide and the dioxyfluoride. The results of chemical and x-ray analyses of solids recovered from the wet residues after thorough water-washing were consistent with this hypothesis. Solids that were indicated by Schreinemakers' method to contain mostly niobic oxide gave x-ray patterns similar to that of the original hydrous oxide, but also contained appreciable amounts of fluorine. On the other hand, samples that were indicated to contain mostly the dioxyfluoride yielded the x-ray pattern for this compound; the F/Nb atom ratios in these solids were generally about 0.9, indicating the presence of some niobic oxide.

Residues from the equilibration of $NbO_2F \cdot 1/2H_2O$ with hydrofluoric acid solutions were recovered by filtration and washed with water before x-ray diffraction and chemical

Table I. Solubility of Niobic Oxide in Hydrofluoric Acid Solutions at 25° C.

(Samples equilibrated about 4 months)

	Saturated Solution						Wet Residue, Wt. %	
Sample	Density, g./ml.	Nb concn., *M*	Nb_2O_5 concn., wt. %	F concn., *M*	HF concn., wt. %	F/Nb atom ratio	Nb_2O_5	HF
1	1.0079	0.053	...	0.237	...	4.47	...	...
2	1.0334	0.237	...	1.06	...	4.47	...	...
3[a]	1.0773	0.589	7.26	2.63	4.87	4.46	27.8	5.34
4[a]	1.1326	0.956	11.2	4.44	7.83	4.64	35.2	6.85
5	...	1.23	...	6.26	...	5.09	...	...
6[a]	1.2594	1.88	19.9	8.69	13.8	4.62	37.0	11.5
7	1.3143	2.34	...	10.6	...	4.55	...	...
8[a]	1.3805	2.74	26.4	12.4	18.0	4.53	37.1	15.9
9	1.4394	3.23	...	14.3	...	4.41	...	...
10[a]	1.4644	3.40	31.0	15.6	21.2	4.57	41.5	19.4
11	1.5003	3.69	...	17.4	...	4.70	...	...
12[a]	1.5857	4.34	36.4	18.9	23.8	4.35	43.8	22.0
13	...	5.29	...	25.2	...	4.76	...	...

[a] Free fluoride concentration in saturated solution less than 0.003 *M*.

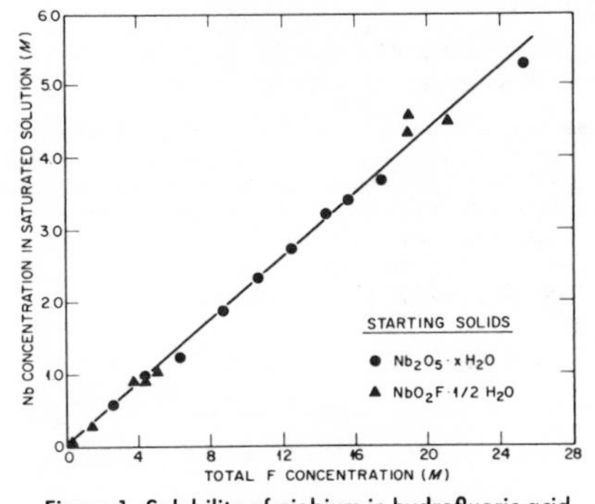

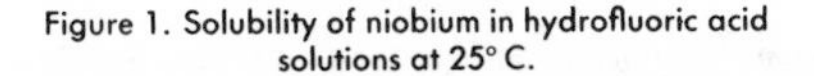

Figure 1. Solubility of niobium in hydrofluoric acid solutions at 25° C.

Slope of line, 0.220, corresponding to F/Nb ratio of 4.54 in saturated solutions

Table II. Solubility of NbO_2F in Hydrofluoric Acid Solutions at 25° C.

(Samples equilibrated about 2 years)

	Saturated Solution				Composition of Solid Phase[a]		
Sample	Nb concn., *M*	F concn., *M*	F/Nb atom ratio	Density, g./ml.	Nb, wt. %	F, wt. %	F/Nb atom ratio
14	0.082	0.41	5.00	1.0144	60.4	12.7	1.03
15	0.145	0.67	4.62	1.0366	60.3	10.6	0.86
16	0.293	1.47	5.02	1.0448	59.5	11.9	0.98
17	0.932	3.89	4.17	...	52.2	13.6	1.28
18	0.921	4.42	4.80	...	...	...	...
19	1.06	4.58	4.32	...	57.4	12.6	1.07
20	1.06	5.01	4.74	...	...	...	...
21	4.60	19.0	4.12	...	60.9	13.8	1.11
22	4.50	21.2	4.71	...	...	...	...

[a] Solids were recovered by filtration, washed with water, and air-dried at room temperature. NbO_2F detected in each solid by x-ray diffraction analysis.

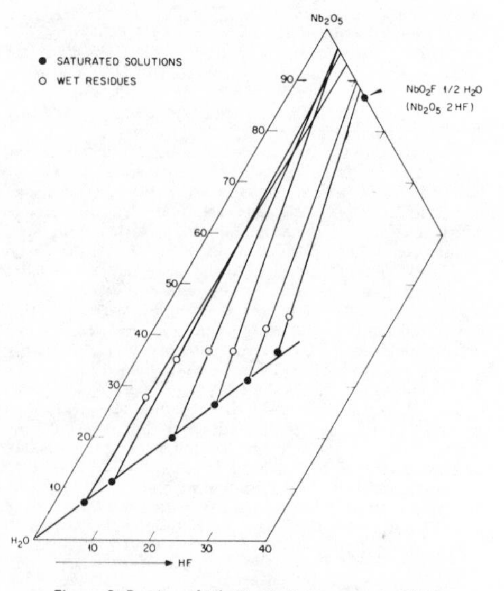

Figure 2. Portion of Nb_2O_5-HF-H_2O system at 25° C.
Composition in weight %

Table III. Solubility of Niobic Oxide in Nitric Acid–Hydrofluoric Acid Solutions at 25° C.

(Samples equilibrated about 1 year)

Sample[a]	Concentration in Saturated Solution, M HNO$_3$	Nb	F	F/Nb Atom Ratio
23	5.35	0.087	0.353	4.07
24	4.93	0.138	0.758	5.49
25	4.94	0.449	1.58	3.51
26	4.68	0.572	3.37	5.89
27	4.42	1.02	5.06	4.96
Av.	4.86			
28	9.66	0.080	0.421	5.27
29	9.62	0.132	0.742	5.62
30	8.89	0.361	1.82	5.04
31	9.30	0.493	2.68	5.45
32	8.98	0.932	4.74	5.08
Av.	9.29			
33	12.4	0.031	0.200	6.45
34	12.5	0.204	1.03	5.05
Av.	12.4			
35	21.3	0.030	0.198	6.69
36	20.6	0.173	0.916	5.29
37	19.7	0.243	1.56	6.41
Av.	20.5			

[a] NbO_2F identified in each solid phase by x-ray diffraction analysis.

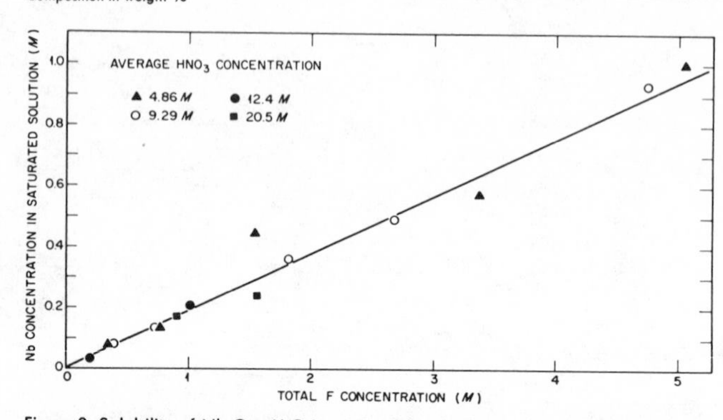

Figure 3. Solubility of $Nb_2O_5 \cdot xH_2O$ in nitric acid-hydrofluoric acid solutions at 25° C.
Slope of line 0.189, corresponding to F/Nb atom ratio of 5.28 in saturated solutions

analyses. Not only was NbO_2F detected by x-ray analysis in each residue but also the chemical analyses, within experimental error, were those expected for $NbO_2F \cdot \frac{1}{2}H_2O$ (Table II).

Only limited studies were made of the solubility of niobium compounds in HF-HNO_3 solutions. Data were obtained by equilibrating the hydrous oxide with solutions that initially were 0 to 20M in HNO_3 and up to 5M in HF. Over this concentration range, the niobium concentrations in the saturated solutions were again about one fifth the total fluorine concentrations (Table III, Figure 3). The dioxyfluoride was detected in each solid phase by x-ray diffraction analysis.

DISCUSSION

The results of this study indicate that equilibration of either niobic oxide or niobium dioxyfluoride with hydrofluoric acid and hydrofluoric acid–nitric acid solutions at 25° C. results in solubilization of the niobium as the $NbOF_5^{-2}$ ion, since the F/Nb atom ratio in the saturated solution was always about 5 and direct analysis showed the free fluoride ion concentrations in the solutions to be negligibly low. Furthermore, the solid dioxyfluoride apparently is the thermodynamically stable solid phase. Thus, the equilibria involved are

$$2NbO_2F_{(s)} + 8HF_{(aq)} \rightleftarrows 2H_2NbOF_{5\,(aq)} + 2H_2O \quad (1)$$

and

$$Nb_2O_{5\,(s)} + 6HF_{(aq)} \rightleftarrows NbO_2F_{(s)} + H_2NbOF_{5\,(aq)} + 2H_2O \quad (2)$$

depending on the initial compound used. Hence, only when the dioxyfluoride is used for the equilibration does the composition of the solid phase remain constant. Reaction of hydrofluoric acid solutions with the oxide is analogous to an acid-base reaction. When an excess of oxide is present, hydrofluoric acid is completely consumed according to the stoichiometry of Equation 2 and the solid phase becomes a mixture of the oxide and the dioxyfluoride.

The conclusion reached in this study, suggesting the $NbOF_5^{-2}$ ion as the primary niobium-containing species in 0 to 5*M* H_2NbOF_5 solutions, is the same as that arrived at by Nikolaev and Buslaev (*9*). Although the results of the present study are in good agreement with those of Nikolaev and Buslaev regarding the composition and densities of the saturated solutions, there is marked disagreement as to the composition of the equilibrium solid phase. Nikolaev and Buslaev, who equilibrated the hydrous oxide with hydrofluoric acid solutions and used the wet residue method to determine the solid phase, concluded that $Nb_2O_5 \cdot 2H_2O$ was the equilibrium solid phase. The dioxyfluoride was always present in the solid phases obtained in this study. This disparity in results is inexplicable unless it is assumed that Nikolaev and Buslaev used very large amounts of oxide relative to the amounts of hydrofluoric acid present in their samples. If this were the case, all the hydrofluoric acid could have been consumed without markedly depleting the amount of niobic oxide present.

ACKNOWLEDGMENT

The author is indebted to J.F. Land for his technical assistance in this work. Analyses were provided by the ORNL Analytical Chemistry Division: x-ray absorption analyses by H.W. Dunn, other chemical analyses under the supervision of W.R. Laing, x-ray diffraction analyses by R.L. Sherman, and infrared analyses by L.E. Scroggie. The author also thanks O.L. Keller, Jr., for many helpful discussions and for determining the Raman spectra of several samples.

LITERATURE CITED

(1) Dunn, H.W., *Anal. Chem.* **34**, 116 (1962).
(2) Ferris, L.M., Oak Ridge National Laboratory, unpublished data, 1962.
(3) Frevel, L.K., Rinn, H.W., *Acta Cryst.* **9**, 626 (1956).
(4) Horton, A.D., U. S. Atomic Energy Comm. Rept. **TID-7015**, Sec. 1 (1958).
(5) Keller, O.L., Jr., *Inorg Chem.* **2**, 783 (1963).
(6) Laing, W.R., U. S. Atomic Energy Comm. Rept. **ORNL-3750**, 70 (1965).
(7) Layton, F.L., Laing, W.R., Division of Analytical Chemistry, Southeast–Southwest Regional Meeting, ACS, Memphis, Tenn., December 1965.
(8) Menis, O., U. S. Atomic Energy Comm. Rept. **TID-7015**, Suppl. 3 (1961).
(9) Nikolaev, N.S., Buslaev, Yu. A., *Russ. J. Inorg. Chem.* **4**, 84 (1959).
(10) Schreinemakers, F.A.H., *Z. Phys. Chem.* **23**, 417 (1897).

RECEIVED for review December 10, 1966. Accepted April 4, 1966. Research sponsored by the U.S. Atomic Energy Commission under contract with the Union Carbide Corp.

Comments on the Article

The article is a first-rate example of the style and format followed by the journals of the American Chemical Society. That format, discussed in detail in Chapter 4, predetermines the organization of chemical articles. It also permits the rapid recovery and collation of chemical data.

The *title* is outstanding. The first work, "Solubility," fixes the general topic to be discussed. (For the same reason, the first work of the title of this book is "Writing.") Then the names of the specific compounds to be discussed are given. They are followed by the names of the solvents and, finally, the temperature in degrees Centigrade at which the solubility tests were conducted. Six keywords are thus furnished reference librarians and searchers of the literature.

The author's by-line and credits immediately follow the title. Ferris's degrees are not given, but the specific division of the national laboratory at which the work was conducted is cited.

An informative *abstract,* written by the author and suitable in form and length for separate publication in anthologies such as *Chemical Abstracts,* is the first formal section of the article. It is printed in a type script dissimilar from that of the rest of the article, and it is both complete and specific; that is, it is composed of positive statements that assert the facts precisely and lead to a specific set of equations. The use of equations in a chemical abstract is desirable; it saves space, and it clarifies meaning.

The first paragraph of the article is both an introduction to the problem and a review of the literature.

The *Experimental* section is subdivided into three subsections: "Materials Used," "Procedure," and "Analytical." The first subsection states not only the origin and purity of the materials, it also discusses techniques used to maintain or to increase this purity.

The second subsection, "Procedure," discusses not only what was done but also why it was done. This last reassures the reader, particularly when careful and time-consuming procedures are described. The clause "but equilibration in samples initially containing the dioxyfluoride required nearly 2 years" is as impressive as are the references to "the method of Schrienemakers." This last, together with its footnotes, places the procedure within the literature on the problem.

The third subsection, "Analytical," is also specifically documented. Footnotes and references to standard methodologies are used. Because parts of this subsection can best be discussed with equations, equations are used when possible. Only standard abbreviations and terms, however, are used. The term SPADNS is defined because it is nonstandard. (The abbreviations and symbols used in *Chemical Abstracts* and acceptable to the ACS are listed after these comments.)

The *Results* section summarizes the data presented in the figure and the tables. Whenever possible, reference or comparison is made to other work in the literature.

The *Discussion* section synthesizes the data contained in the Results section. The section closes with a conclusion which is partially dissimilar from a conclusion reached by others and reported elsewhere in the literature. The dissimilarity is explained with data taken from the Results section.

The Acknowledgment section indicates the contributions—and the responsibilities—of others.

The *Literature Cited* section illustrates the bibliographic practice recommended by the American Chemical Society. Volume numbers are printed in roman type, and page numbers are not given.

The *Endnote* indicates the date the article was submitted, the sponsor of the work, and the laboratory that performed the work.

ABBREVIATIONS AND SYMBOLS USED IN *CHEMICAL ABSTRACTS*

A. angstrom unit(s)
abs. absolute

abstr. abstract
Ac acetyl (CH_3CO, not CH_3COO)
a.c. alternating current
addn. addition
addnl. additional, additionally
alc. alcohol, alcoholic
alk. alkaline (not alkali)
alky. alkalinity (alkys, for alkalinities is not approved)
amp. ampere(s)
amt. amount (as a noun)
anal analytical (not analysis), analytically
anhyd. anhydrous
app. apparatus
approx. approximate (as an adjective), approximately
approxn. approximation
aq. aqueous
assoc. associate(s)
assocd. associated
assocg. associating
assocn. association
at. atomic (not atom)
atm. atmosphere(s), atmospheric
av. average (except as a verb)
b. (followed by a figure denoting temperature) boils at, boiling at (similarly b_{13}, at 13 mm. pressure)
bacteriol. bacteriological, bacteriologically
Bev. or Gev. billion electron volt(s)
biol. biological, biologically
B.O.D. biochemical oxygen demand
b.p. boiling point
Btu. British thermal unit(s)
Bu butyl (normal)
Bz benzoyl (C_6H_5CO, not $C_6H_5CH_2$)
c- centi- (as a prefix, e.g., cm.)
c. curie(s)
cal. calorie(s)
calc. calculate(s)
calcd. calculated
calcg. calculating
calcn. calculation
cc. cubic centimeter(s)
c.d. current density

chem. chemical (as an adjective), chemically (not chemistry)
clin. clinical, clinically
coeff. coefficient
com. commercial, commercially
compd. compound (as a noun)
compn. composition
conc. concentrate(s) (as a verb)
concd. concentrated
concg. concentrating
concn. concentration
cond. conductivity
conds. conductivities
const. constant
contg. containing
cor. corrected
cp. centipoise(s)
crit. critical, critically
cryst. crystalline (not crystallize)
crystd. crystallized
crystg. crystallizing
crystn. crystallization
d- deci- (as a prefix, e.g., dl.)
d. density (d^{13}, density at 13° referred to water at 4°; d_{20}^{20}, at 20° referred to water at the same temperature)
D. debye unit
d.c. direct current
decomp. decompose(s)
decompd. decomposed
decompg. decomposing
decompn. decomposition
deriv. derivative
det. determine(s)
detd. determined
detg. determining
detn. determination
diam. diameter
dil. dilute(s)
dild. diluted
dilg. diluting
diln. dilution
dissoc. dissociate(s)

dissocd. dissociated
dissocg. dissociating
dissocn. dissociation
distd. distilled
distg. distilling
distn. distillation
d.p. degree of polymerization
elec. electric, electrical, electrically
emf. electromotive force
en ethylenediamine (used in Werner complexes only)
E.P.R. electron paramagnetic resonance
equil. equilibrium(s)
equiv. equivalent
esp. especially
E.S.R. electron spin resonance
est. estimate(s) (as a verb)
estd. estimated
estg. estimating
estn. estimation
esu. electrostatic unit(s)
Et ethyl
ev. electron volt(s)
evap. evaporate(s)
evapd. evaporated
evapg. evaporating
evapn. evaporation
examd. examined
examg. examining
examn. examination
expt. experiment (as a noun)
exptl. experimental, experimentally
ext. extract
extd. extracted
extg. extracting
extn. extraction
f. faraday, farad
f.p. freezing point
g. gram(s)
γ microgram(s)
geol. geological, geologically
ha. hectare(s)

histol. histological, histologically
inorg. inorganic
insol. insoluble
ir infrared
iso-Bu, iso-Pr isobutyl, isopropyl
I.U. International Unit
j. joule
k- kilo- (as a prefix, e.g., kg.)
kc. kilocycle(s)
kev. kiloelectron volts
l. liter(s)
lab. laboratory
L.C.A.O. linear combination of atomic orbitals
L.D. lethal dose
m- milli- (as a prefix, e.g., mm.)
m. meter(s); also (followed by a figure denoting temperature) melts at, melting at
m molal
M molar
ma. milliampere(s)
manuf. manufacture(s)
manufd. manufactured
manufg. manufacturing
math. mathematical, mathematically
max. maximum(s)
Mc. megacycle(s)
Me methyl (not metal)
mech. mechanical, mechanically
meq. milliequivalent(s)
Mev. million electron volts
min. minimum(s); also minute(s) (time unit only)
misc. miscellaneous
mixt. mixture
mol. molecule, molecular (not mole)
m.p. melting point
μ micron(s); also micro- (as a prefix, e.g., μl.) (μg. is not approved)
n refractive index (n_D^{20} for 20° and sodium D light)
N normal (as applied to concn.)
neg. negative (as an adjective), negatively
n.m. nuclear magneton(s)
N.M.R. nuclear magnetic resonance

no. number
oe. oersted(s)
org. organic
oxidn. oxidation
pathol. pathological, pathologically
p.d. potential difference
Ph phenyl
pharmacol. pharmacological, pharmacologically
phys. physical, physically
physiol. physiological, physiologically
pos. positive (as an adjective), positively
powd. powdered (as an adjective)
ppb. parts per billion
ppm. parts per million
ppt. precipitate(s)
pptd. precipitated
pptg. precipitating
pptn. precipitation
Pr propyl (normal)
prep. prepare(s)
prepd. prepared
prepg. preparing
prepn. preparation
psi. pounds per square inch
psig. pounds per square inch gage
psia. pounds per square inch absolute
py pyridine (used in Werner complexes only)
qual. qualitative, qualitatively
quant. quantitative, quantitatively
r. roentgen
redn. reduction
rem. roentgen equivalent man
rep. roentgen equivalent physical
resp. respective, respectively
rpm. revolutions per minute
R.Q. respiratory quotient
sapon. saponification
sapond. saponified
sapong. saponifying
sat. saturate(s)
satd. saturated

satg. saturating
satn. saturation
S.C.E. saturated calomel electrode
sec. second(s) (time unit only)
sec secondary (with alkyl groups only)
sep. separate(s), separately
sepd. separated
sepg. separating
sepn. separation
sol. soluble
soln. solution (as in a solvent)
soly. solubility (solys. for solubilities is not approved)
sp. specific (used only to qualify a physical constant)
sp. gr. specific gravity
spp. species (with genus names only)
sq. square
sym. symmetrical, symmetrically
tech. technical, technically
temp. temperature
tert tertiary (with alkyl groups only)
titrn. titration
U.S.P. United States Pharmacopeia
uv ultraviolet
v. volt(s)
vol. volume (not volatile)
w. watt(s)
wt. weight

BIOCHEMICAL ABBREVIATIONS

ACTH adrenocorticotropin
ADP adenosine 5′-diphosphate
AMP adenosine 5′-monophosphate
ATP adenosine 5′-triphosphate
ATPase adenosinetriphosphatase
CDP cytidine 5′-diphosphate
CM-cellulose carboxymethyl cellulose
CMP cytidine 5′-monophosphate
CoA coenzyme A
CTP cytidine 5′-triphosphate
DEAE-cellulose diethylaminoethyl cellulose

DNA deoxyribonucleic acid
DNase deoxyribonuclease
DPN diphosphopyridine nucleotide (NAD)
DPNH reduced DPN
FAD flavine adenine dinucleotide
FMN flavine mononucleotide
FSH follicle-stimulating hormone
GDP guanosine 5′-diphosphate
GMP guanosine 5′-monophosphate
GTP guanosine 5′-triphosphate
IDP inosine 5′-diphosphate
IMP inosine 5′-monophosphate
ICSH interstitial cell-stimulating hormone
ITP inosine 5′-triphosphate
LH luteinizing hormone
MSH melanocyte-stimulating hormone
NAD nicotinamide adenine dinucleotide (DPN)
NADH reduced NAD
NADP nicotinamide adenine dinucleotide phosphate (TPN)
NADPH reduced NADP
NMN nicotinamide mononucleotide
RNA ribonucleic acid
RNase ribonuclease
TEAE-cellulose triethylaminoethyl cellulose
TPN triphosphopyridine nucleotide (NADP)
TPNH reduced triphosphopyridine nucleotide
Tris tris(hydroxymethyl)aminomethane
TSH thyroid-stimulating hormone
UDP uridine 5′-diphosphate
UMP uridine 5′-monophosphate
UTP uridine 5′-triphosphate

LANGUAGE OF ARTICLES

Following the journal reference, the language of the original article is indicated in parentheses. The following abbreviations are used:

Bulg – Bulgarian
Ch – Chinese
Croat – Croatian
Dan – Danish
Dut – Dutch

Eng – English
Fle – Flemish
Fr – French
Ger – German
Hung – Hungarian
Ital – Italian
Japan – Japanese
Norweg – Norwegian
Pol – Polish
Port – Portuguese
Rom – Romanian
Russ – Russian
Slo – Slovak
Span – Spanish
Swed – Swedish
Ukrain – Ukrainian

For all other languages, no abbreviations are used.

Physics Journals

The *Style Manual* of the American Institute of Physics governs the literature of physics. Because the document is available from the AIP as indicated above, only the section containing rules for presentation of mathematical expressions is reprinted by permission below. These rules are dissimilar from those established by the American Mathematical Society as indicated by their *Manual for Authors of Mathematical Papers* which is reprinted on pp. 293-309.

The illustrative article, reprinted below with the kind permission of the *Journal of Mathematical Physics,* shows footnoting practice appropriate for the discipline of physics. Comparison with the footnotes to the chemical article reprinted above illustrates the stylistic variations between the disciplines.

A. Presentation of Mathematical Expressions

"In preparing a manuscript involving considerable mathematics, the author should bear in mind that correct typographical presentation is an important factor in understanding equations. He should also recognize certain standard methods of presentation that have been worked out to make possible machine setting and to avoid handwork, which under all circumstances is expensive. Authors of mathematical papers should therefore read carefully the instructions given in this section of the manual.

1. MARKING OF MANUSCRIPT

"The importance of legible, precise, and carefully aligned mathematical copy cannot be overemphasized. The author should always keep in mind that those who have the responsibility of converting his manuscript to a printed version will reproduce what they *see*, not what the author *knows*. The compositor is neither a mathematician nor a physicist. He should be regarded virtually as an automaton who sets type directly from copy and who cannot be expected to apply editorial judgment.

"Both typed and handwritten mathematical equations have shortcomings which the author must surmount. Both have problems of alignment, which will be discussed later in this section.

"Typed material has the advantages of legibility and consistent formation of letters, but some confusion is possible. On most typewriters the letter '*l*' and the number '1' are the same, and the author must distinguish between them. To do this, the letter *l* should always be looped when it is used as a mathematical symbol. In rare cases when a script letter is called for, the word 'script' should be written and circled lightly above it. As an additional safeguard, when the typewritten character means one, the word 'one' should be written and circled lightly above it. Similar precautions should be taken to distinguish between typewritten zeros and capital O's and between X and a multiplication sign. Writing and circling 'zero' and 'multi' close to the symbol will prevent misinterpretation.

1 unity   *l* italic letter ℓ script letter

"In handwritten material, alignment is more easily achieved than on the typewriter, but special care is necessary to make all letters clear, without possibility of misinterpretation. For example, carelessly written c's and e's may be confused; u's and n's may look alike; a k may be confused with an h or a Greek kappa. There is the additional danger that consistency of size may not be maintained; e's and l's and letters with similar lower-case and capital formation may be confused. Such letters as c, k, o, p, s, u, v, w, and z are among the most difficult to distinguish. The copy should give no opportunity for misinterpretation by the compositor. If there is opportunity for confusion in only a few cases, the copy may be marked for lower-case or capital letters as shown. However, if the lower-case and capital letters appear repeatedly and close together, it is simpler to indicate the capitals by three underscores (the marking for a capital letter). Those not marked will be set in lower case.

Multi Sign	Cap	l.c.
5×10^{-10}	$X^2 = a^2 - x^2$	

"Subscripts (inferiors) and superscripts (superiors) should also be clarified for the typesetter by caret and inverted caret markings (made with a sharp black pencil), as shown. In particular, *subscripts* to subscripts must be distinguished from double subscripts, with corresponding care for double superscripts.

$x_r \quad x^r \qquad e_{ij} \quad e_{i_j} \qquad e^{xy} \quad e^{x^y}$

"Subscripts to subscripts and superscripts to superscripts pose a problem for machine setting. The subscripts can be set by machine only by using for the first subscript a smaller-sized type set on the line and then using the regular subscript matrix for the subscript to the subscript. This is satisfactory when the first inferior is a lower-case letter. In the example marked "bad" a better setting would require handwork.

satisfactory	bad
A_{a_b}	a_{A_b}

If authors can avoid notation such as the "bad" one in the example, mathematical papers will be much clearer to the reader. Often such notation can be improved by changing, for example, $\chi^{k}{}_{1}$ to χ_{k1}, $(\chi_k)_1$, $\chi_k{}'$, χ^{k1}, or $\chi(k_1)$.

"A subscript to a superscript can be set without any difficulty by using a regular superscript for the first superior and a smaller type piece for the inferior to it.

$$e^{b_a}$$

A superscript to a superscript cannot be set satisfactorily and should be avoided.

"In all mathematical copy care should be exercised that the alignment of symbols in an equation is clear and unambiguous. Many authors fail to draw the fraction bar long enough to extend under everything that is meant to be in the numerator. Note the example.

$$\text{Does} \quad \frac{a}{b}^{\!\!\ln} \ \text{mean}$$

$$\ln(a/b) \text{ or } \frac{\ln a}{b}?$$

Some authors find it helpful before writing exceptionally complicated equations to rule in with pencil horizontal guide lines which are later erased. This procedure may appear to be tedious, but it undoubtedly involves less labor than the possible insertion of an "Erratum" in a subsequent issue of the journal.

"Authors may help considerably in the matter of marking a manuscript for type. All *text* material will be set in roman (i.e., not italic) type unless otherwise marked; all characters in *displayed equations* will be set in italics unless otherwise marked. Hence, in *text* material, underscore once *words* to be italicized. Our printer is instructed to assume that all isolated letters appearing in the body of the text are to be italicized; hence they need not be marked explicitly unless they are letters easily mistaken for words, like an 'a' or 'I,' or unless they are combinations that may easily be confused with abbreviations, like Ne (when it means $N \times e$, not neon). Roman characters in equations and isolated roman letters in text should be indicated by encircling in pencil. Isolated letters also should be marginally noted if to be set in other type (e.g., script). Greek letters need only be underscored in red. German or other less familiar characters should be so indicated by a marginal note.

E.g., 'If one sets $a = \alpha$, only then is . . . $xe^x = \text{const.}$

"Symbols for the chemical elements are common examples of non-italicized letters, but the editorial staff of the Institute is trained to recognize these and the author need make no special marking of them.

"Marking for boldface type (heavy type) is indicated by underscoring with a single wavy line as shown in the table 'Signs Used in Correcting Proofs,' page 36.

"Vectors are set in boldface roman type, with their components in italic type. A unit vector in the k direction of a vector **k** is denoted by $\hat{k}$. Four-vectors are conventionally set in italic type. Tensors without indices are normally set in boldface roman type, except in rare cases when it is essential to distinguish between vectors and tensors. Here, if the tensor is an English capital letter, it can be set in sans serif, e.g., **T** (Section IV). Vectors should be marked with a single wavy line, which should not extend beyond the vector, that is, not under subscripts; arrows should not be placed above or below them. Such arrows are a bête noire to the editorial office since, if they are left in the copy, the printer may think that they are to be set in type.

$$\underset{\sim}{\mathrm{H}} \text{ not } \overrightarrow{\mathrm{H}} \text{ nor } \underrightarrow{\mathrm{H}} \qquad \underset{\sim}{\mathrm{A}} \cdot \underset{\sim}{\mathrm{B}} \text{ center dot} \qquad \dot{x} \text{ note dot}$$

"Obscure primes and dots over letters, etc., should be called to the compositor's attention. If the centered dot is used for a product, it is wise to call the printer's attention to it by a marginal note as shown.

"The large variety of type faces and symbols that the printer has is catalogued in Section IV. Any of these may be used when necessary, but it must be emphasized that, with due regard for accepted practice, mathematical symbols and notation should be kept as simple as possible. Most of the unusual signs and symbols are for emergency use only. In the monotype process, the compositor can handle a limited number of symbols at one time. Hence the use of an excessive number of symbols requires expensive hand insertion.

"When there is a particular demand for a new symbol, it can be specially made, but the cost is high. The Institute welcomes suggestions on new symbols, but a concerted demand for a symbol, rather than just the desire of a single author, is essential to make the ordering of new type worthwhile. For instance, the ordering of the Dirac $\hbar$, as well as $\sqrt{2}$ and $\sqrt{3}$, has been more than justified, but there are some special symbols for which the demand has not been as great as anticipated. Symbols which are on hand and which are at all unusual should be referred to by the numbers given in the list in Section IV. All the journals published by the American Institute of Physics are set in 10-point type with footnotes and other reduced material in 8-point type. However, symbols smaller than 10 points can be used in the machine work whenever this is essential.

"It is helpful to the printer if at the beginning of an article a memorandum is made of unusual symbols (e.g., cap Greek theta Θ, but not, for instance, ordinary π) which occur frequently in the article, so that he can arrange to have them in his matrix case for equations and text.

2. SPACING

"In the mathematical material in most of the journals of the American Institute of Physics, the conventional spacing that obtains elsewhere is not used. For example, signs of operation and trigonometric functions are set closed up. Among the reasons for which this is done is the saving of space. With a two-column journal whose mathematical material may include long and involved equations, it is wise to set as compactly as possible to avoid having part of an expression in one line and part (a turnover) in the next or the necessity of extending equations across two columns. The printer is instructed to set cos*x* without any space after 'cos.' In such instances, if the omission of the space leads to confusion, parentheses may be used.

"Under these circumstances spacing in equations can usually be left to the discretion of the printer, but occasionally special situations arise which require marking by the author. A hairline space cannot be set by the Monotype machine; the smallest one mechanically set is of five units, the width of the space between the letters *a b*. To indicate a space of this size, the conventional typewriter # is used.

#

$$3^{2\ 2p}$$

"Mathematical redundancy should be avoided. When a complicated mathematical expression occurs several times, a symbol for it should be used. Even with only one occurrence it is well to use a symbol for the numerator or denominator of a fraction which is too long to set on one line, as a fraction cannot feasibly overflow from one line to another.

3. MATHEMATICAL ENGLISH

"All numbered equations and all unnumbered but complicated equations should be typed on separate lines—not run into the text material. Equations should be punctuated in all cases. Good usage is violated and ambiguity often arises when the equality sign is made to do the work of a main verb in a sentence. Note, however, that an equality sign can properly be the verb of a subordinate clause. It is rather bad form to begin a sentence with a mathematical symbol or with a number, particularly when the preceding sentence ends in a numeral or in a symbol. Similar inelegance results from beginning a main clause with a number or symbol after just ending the subordinate clause similarly, and vice versa. The examples cited make the reasons for these statements obvious.

'When (7) is substituted in (8), one obtains $a = b$.'

not

'When (7) is substituted in (8), $a = b$.'

'If $a = b$, then c holds too.'
not
'If $a = b$, c holds too.'
nor
'If $a = 4.3$, c is infinite.'

"Another unnecessary (and perhaps misleading) overconcern is that involved with mathematical expressions in apposition. The rule is: Do not set them off by commas (and usually do not set them off by parentheses).

'By using the equation $a = b$'
not
'By using the equation, $a = b$'
or
'By using the equation ($a = b$)'

"Many writers seem to have a penchant for dangling participles. Avoid writing, 'Substituting Eq. (5) in (6), the thermal conductivity becomes $\frac{1}{2}kNVL$.' After all, the thermal conductivity does not do the substituting.

"Spectroscopic terms of odd parity should be written with the degree sign rather than a superscript zero. The number of fractional subscript matrices is limited. Hence the shilling mark fraction should be used.

Degree Sign	Zero
x° not x^{0}	${}^{2}P_{3/2}{}^{\circ}$ *not* ${}^{2}P_{\frac{3}{2}}{}^{0}$

4. RADICAL SIGNS

"In keeping with the announced purpose of eliminating all unnecessary handsetting, radical signs are not permitted unless they are without the vinculum (horizontal bar). Therefore, in manuscripts always use fractional superscripts. There are, however, two exceptions; the printer has special matrices for $\sqrt{2}$ and $\sqrt{3}$ in both 10- and 8-point sizes. He also has the exponent $^{\sqrt{2}}$ in 8- and 10-point sizes. Use should be made of these available matrices as well as of the radical sign without the vinculum.

$(a + bx)^{\frac{1}{2}}$ not $\sqrt{a + bx}$
Exception: $\sqrt{2}$ and $\sqrt{3}$ and $x^{\sqrt{2}}$
Thus: $\frac{1}{2}\sqrt{3}$ not $\frac{1}{2}3^{\frac{1}{2}}$

Beware, however, of this unroofed radical sign when its termination is not clear. In the example it is not clear whether what is meant is $a/\sqrt{bc}$

or $a/\sqrt{bc}$. The alternate $a/(bc)^{\frac{1}{2}}$ may be written even better as $ab^{-\frac{1}{2}}c^{-\frac{1}{2}}$ or $a(bc)^{-\frac{1}{2}}$.

$$(a+bx)/\sqrt{7} \text{ but avoid } a/\sqrt{bc}\text{: use instead } a/b^{\frac{1}{2}}c$$

"The rules regarding radical signs are quite stringent because they arise from definite limitations in the content of type cases and in the limitations in machine typesetting. In the following section other items are discussed on which there is more leeway.

5. WIDE VERSUS NARROW COLUMNS

"Normally the journals of the American Institute of Physics are printed in a two-column format. Occasionally a mathematical expression or equation is so long that it cannot be accommodated in one line in the narrow 19-pica column. In this case, part of the expression must be carried over to the following line or lines as a turnover or turnovers. If the situation occurs frequently in a portion of an article, it is usually better to print that portion so that both equations and text extend in a single wide column across the page, since equations with an excessive number of turnovers are inelegant and often difficult to read.

"Authors should leave to the publication office the decision as to setting in one- or two-column width.

6. BUILT-UP FRACTIONS VERSUS SOLIDUS

"One of the most constantly occurring typographical questions is when to use a built-up fraction and when the solidus (or shilling mark). In the text the solidus should invariably be the choice, for built-up fractions require breaking into the lines above and below, present an awkward appearance, and increase costs. In displayed equations, it is true, built-up fractions can be set by machine by using three lines, but this procedure does require more labor. (In the printer's bill, charges are made by the *inch* and 3 lines involve more inches of mathematical typography than 1 line.) Hence good judgment on the part of the author is a firm requisite. Obviously the solidus should be used for short expressions, but for longer expressions it may be ungainly—in which case the built-up fraction should be used instead.

$$\text{write } a/b = c/d \text{ but } \frac{a+by+cy^2}{c+ey+fy^2} = \frac{az+bz^2}{\cos x}$$

"It is important to note that the solidus *need not be used* in displayed equations whenever it has already been necessary to use extra lines for other parts of the equation (e.g., another fraction and limits above and below a summation). No particular economy ensues by using the solidus in these cases.

Do not write $a/b = \frac{c+dx}{e+mx}$ or $a/b = \sum_{i=1}^{\infty} a_i$; use $\frac{a}{b} = \frac{c+dx}{e+mx}$ and $\frac{a}{b} = \sum_{i=1}^{\infty} a_i$.

Note, however, that the limits on an integral sign do *not* involve additional space (they go to the right of the top and bottom of the sign, respectively). Similarly, if a *large* integral sign, breaking into other lines, is employed, a built-up fraction involves comparatively little extra labor. In fact, it usually is preferred. With a small integral sign, however, the above statement no longer applies, for with it simple expressions can still be set on one line. Cases will occasionally occur where the introduction of the solidus makes an equation so long that it can no longer be set in a single line on the page. It is particularly evident that such instances are to be distinctly avoided.

$$a/b = \textstyle\int_0^{\pi/2} dx \text{ but } \displaystyle\frac{\sin\theta}{nNG} R \int_0^{\infty}$$

"Many of the common fractions are available as single matrices (called 'case fractions'), and it is clearly advantageous to use them whenever possible, both for neat appearance and for economy's sake. Those available are listed in Section IV. These fractions are obviously about half the size of fractions formed from two ordinary pieces of type (such as a built-up fraction), and thus we interject a word of caution about using the former as exponents in expressions in footnotes, table legends, etc., where the main text is reduced to 8-point type. The solidus should always be used for fractional subscripts and superscripts. The case fraction permits many economies in typesetting, but *only if the author cooperates!* The compositor will set fractions exactly as they are marked. Thus in all doubtful cases the author should indicate the difference between built-up and case fractions by a marginal note as shown.

$\tfrac{1}{2}f(x)$ rather than $f(x)/2$ $\quad \tfrac{1}{4}\pi$ rather than $\pi/4$

Equally good:

$$\tfrac{1}{2} \textstyle\int e^{ax} dx \text{ or } \displaystyle\frac{1}{2}\int e^{ax} dx \frac{a}{b}$$

small large case built-up

"In using the solidus, be sure to use enough parentheses to make the meaning unambiguous. According to accepted convention, all factors appearing to the right of a solidus are to be construed as its denominator. To settle all doubt in cases of possible confusion, put in a superfluous pair of parentheses or two. The use of the unbracketed solidus for continued fractions should be avoided, as should the monstrosity $a/b/c$.

$a/b \cos x$ means $a(b \cos x)^{-1}$, but would *you* think it to be $(a/b) \cos x$?

$$a/b + c = (a/b) + c$$
$$= ab^{-1} + c$$
$$a/b + c \neq a/(b + c)$$

7. SOLIDUS VERSUS NEGATIVE POWERS

"Many authors do not seem to realize that proper introduction of a negative power often avoids the use of an inelegant solidus or an expensive built-up fraction. Furthermore, an equation can often be reduced to a simple typographical structure by clearing of fractions, unless some particular emphasis is desired. Whenever equations requiring complicated mathematical typography are involved, the author should consider whether their printing will not be facilitated by use of fractional powers, or by multiplying through by a common factor, or by any of the well-known tricks, without in any way sacrificing mathematical clarity or rigor.

$ax^{-1} \cos y + bz^{-1} \tan y$ is preferable to
$(a \cos y/x) + (b \tan y/z)$.
But $a = 14^{-1}x + 42^{-1}y$ is worse than
$a = x/14 + y/42$; but best is
$14a = x + \frac{1}{3}y$.

8. PARENTHESES

"Parentheses, brackets, etc., are cheap, easily printed, and often avoid ambiguities. We have already quoted examples where parentheses make the meaning of the solidus much clearer. Other examples are encountered in other connections where parsimony with parentheses, etc., may lead to the confusion of the reader. For instance, if one writes $\cos xy$, one may wonder whether $y \cos x$ or $\cos(xy)$ is meant. According to the accepted convention, one works outward with parentheses according to the scheme shown. Boldface brackets are also available, but usually should be avoided as they give an unpleasantly heavy appearance to the text (see Section IV, items 120-123). Of course, large-sized parentheses, brackets, and braces, extending over more than one line, are used with built-up fractions or other mathematical expressions occupying more than one line.

$$\Bigg\{ \quad \Bigg[\quad \Bigg(\quad \Bigg) \quad \Bigg] \quad \Bigg\}$$

9. BARS VERSUS AV

"The whole alphabet, as well as all lower case Greek letters, is available with single bars over the letters, but of course this type of one-piece character is helpful only for very simple kinds of averages. Note that in single barred characters, the bar is shorter than the width of the character and lies very close to it. Hence, for more complicated expressions, e.g., $\cos x$, xv, x^2, the bar must be set separately, either by hand or at the bottom of the line above, much as one would overscore on a typewriter. The one-piece barred letters can be used singly anywhere, but a bar over more than one character or over a superscript should be avoided where possible in displayed equations and in the text.

Do not use x^2, xv, $\cos x$ Satisfactory x, $\cos x$

It is also necessary to avoid bars over exponents or over expressions in the denominator of a built-up fraction, for complicated handwork is required. An "av" subscript is available, conveniently used with angular parentheses. In the majority of cases it works about as well as the bars; it often conveys the physical meaning more vividly. Often the use of bars can be avoided by employing asterisks, daggers, etc. (ψ^*, $\psi\dagger$).

$$\text{Avoid } e^{x+y} \text{ or } \frac{x+y}{z+a}. \qquad \text{Use } \langle x+y \rangle_{av} \text{ for } x+y.$$

10. DOTS

"Dots over letters are difficult for the printer to set unless he has type with the letters already dotted, as otherwise it is necessary to set the dot by hand. Fortunately he has the whole lightface alphabet (even in Greek) with dots over it, so that single dots do not cause trouble. Double dots, such as are used in double line differentiations, are available only for certain letters. Triple dots should never be used. Dots over letters, however, are hard to read and thus each author should consider carefully whether the same purpose could not be achieved by primes or similar symbolism.

11. EXP VERSUS *e*

"Another question is when to use "exp" and when *e*. It is clear that when the argument of the exponential is too complicated, the "exp" rather than *e* symbolism should be used, but with short arguments, *e* is usually better. The difficulty wih the use of *e* is that superscripts in the argument of the exponent (superscripts to superscripts) must be set by hand. Avoid anything involving more than one line in the argument of an exponent.

$$\exp\left[\tfrac{1}{8}\sqrt{3}\log(a+bx^{-1})\right] \text{ not } e^{\frac{1}{8}\sqrt{3}\log(a+bx^{-1})}$$

$$\exp\frac{(a+bx)}{(c+dy)} \text{ or } e^{(a+bx)/(c+dy)}$$

$$\text{not } e^{\frac{a+bx}{c+dy}}$$

12. BOLDFACE

"Most boldface characters are not available as superscripts or subscripts, but certain ones are available so as to permit writing familiar wave-mechanical expressions (see Section IV, items 22 and 23).

13. LIMITS TO SUMMATIONS

"If all the other characters come on one line, the variable of summations is printed *following* the summation, rather than *under* it, to avoid the expense of setting an extra line unless one is already needed for other purposes. When there is only one character attached to the summation, and all other characters come on one line, the publication office is instructed to make manuscripts read Σ_i rather than $\sum\limits_i$. Authors are asked to prepare their manuscripts accordingly. On the other hand, when the summation involves limits, the lower limit should be written under the summation sign, and the upper limit above, since no lone likes $\Sigma_{i=1}{}^{\infty}$. Whether to use $\Sigma_{j>i}$ or $\sum\limits_{j>i}$ is a borderline case. Oftentimes, however, the judicious author can eliminate the specification of the limits of summation every time if they are clear from context, juxtaposition, etc.

14. OVERHANGING SUPERSCRIPTS

"If a subscript and a superscript must be set to the same symbol, they can be set by machine if they are not aligned over one another. The rather more elegant appearing perfect alignment is more difficult, particularly in text, for the overhand must be eliminated either by handwork, as in f_a^b, or else by setting the superscripts as subscripts in the line above, as in $\overset{b}{f_a}$, and vice versa.

$$f_a{}^b, \overset{b}{f_a}, \underset{a}{f^b}$$

Editorial instructions are to bring the subscripts and superscripts into alignment when there is a complicated array of them, but not in the simpler forms. Superscripts ordinarily follow subscripts and will be set in that way unless special instructions to the contrary are given by the author.

$$\overset{def}{P_{abc}}, P_{ab}{}^{cd}, P_2{}^4$$

"Clearly, notations involving a multitude of subscripts and superscripts are to be avoided whenever possible, although admittedly much is asked of the author's ingenuity. For example, $A(nm; n'm')$ will usually serve as well as $A_{nm}{}^{n'm'}$.

15. MATHEMATICAL TERMS

"The Publication Board recommends that the following symbols, and no others, be used to represent the mathematical phrases printed below."

approximately equal to	$\approx$
proportional to	$\propto$
tends to	$\rightarrow$
asymptotically equal to	$\sim$
complex conjugate of A	A^*
Hermitian conjugate of matrix A	$A^\dagger$
transpose of matrix A	$\tilde{A}$

General Spherical Harmonic Tensors in the Boltzmann Equation

Tudor Wyatt Johnston
RCA Victor Company, Ltd., Research Laboratories, Montreal, Canada
(Received 6 January 1966)

The irreducible velocity space direction cosine tensors associated with velocity magnitude spherical harmonic expansion of the distribution function are manipulated in the Boltzmann–Vlasov flow terms to yield a linked chain of equations whose general (lth) equation is given explicitly. This generalizes earlier results for $l = 0, 1, 2, 3$.

INTRODUCTION

The object of this note is the presentation of a simple derivation from the Boltzmann equation of the general set of equations for the irreducible base tensors associated with the velocity space spherical harmonic expansion of the one-particle distribution function for charged particles.

To the author's knowledge, Wallace[1] was the first (in connection with neutron transport) to give a general explicit direction cosine tensor generalization of the spherical harmonics themselves. Ikenberry[2] also evolved an equivalent form for statistical mechanics problems. Unfortunately, this work, unknown to the author in 1960, was not mentioned before.[3] Delcroix[4] has hinted at the tensor application in the Boltzmann equation, based on the spherical harmonic work of Jancel and Kakan.[5]

The next step is to obtain the equations resulting from the substitution into the Boltzmann equation. Allis[6] had given the zero-order (scalar) and first-order (vector) equations and the general one-dimensional form in which the spherical harmonics and the tensors reduce to Legendre polynomials. The author[3] then derived the tensor equations up to the third order and Shkarofsky[7] included the intrinsic velocity effects up to the second order, but each case was calculated separately. The (successful) object of this work was to obtain the general form for the equations to all orders.

Using a bit of hindsight, together with the extremely useful approach developed by Wallace,[1] the general tensor equation including intrinsic velocity for any order is derived here in a manner much simpler than the brute force methods[3,7] previously used for the second-order and third-order results. The tensor equations are far more compact, symmetric and understandable than the clumsy spherical harmonic result.

SPHERICAL HARMONIC TENSORS

Owing to the habit of using powers of v with coefficients of one in the velocity moment equations, the tensor form used here differs by a numerical constant C_l from that of Wallace[1] but agrees with Ikenberry[2] in having the first term coefficient equal to 1. The fully symmetric lth-order tensor $\mathbf{T}_l$ is therefore defined as follows:

$$\mathbf{T}_l(\boldsymbol{\mu}) = T_l\left(\frac{\mathbf{v}}{v}\right) = \frac{(-1)^l}{l!}\, C_l \nabla_{\mathbf{v}}^l \left(\frac{1}{v}\right)_{v=1}, \qquad (1)$$

where

$$C_l \equiv \frac{2^l l!\, l!}{(2l)!} = \frac{l!}{1\cdot 3\cdot 5\,\cdots\,(2l-1)},$$

$\mathbf{v}$ is the velocity vector with magnitude v, $\nabla_{\mathbf{v}}$ is the gradient operator in velocity space, and $\boldsymbol{\mu} = \mathbf{v}/v$ is the velocity direction cosine vector of unit magnitude.

As Wallace[1] points out, $1/v$ is a solution of the Laplace equation in velocity space, i.e., $\nabla_{\mathbf{v}}^2(1/v)$ is zero, therefore $\mathbf{T}_l$ is an irreducible or base tensor, one for which any contraction gives zero ($\sum_i \mathbf{T}_l \cdots i \cdots i \cdots = 0$). Each of the $\frac{1}{2}(l+1)(l+2)$ elements of $\mathbf{T}_l$ is a linear combination of the $2l + 1$ spherical harmonics of order l, but the $\frac{1}{2}l(l - 1)$ conditions from the irreducibility feature leave just $2l + 1$ independent elements.[1,3] An equivalent situation exists in considering multipoles and spherical harmonics in electrostatic problems.[8]

The z^l element $T_{l(z)}$ is particularly simple, being

[1] P. R. Wallace, Can. J. Res. **A26**, 99 (1948).
[2] E. Ikenberry, Ann. Math. Monthly **62**, 719 (1955); E. Ikenberry and C. Truesdell, J. Ratl. Mech. Anal. **5**, 1 (1956); J. Math Anal. Appl. **3**, 355 (1961).
[3] T. W. Johnston, Phys. Rev. **120**, 1103, 2277 (1960).
[4] J. L. Delcroix, *Introduction à la théorie des gaz ionisés* (Dunod Cle., Paris, 1959), p. 69 [English transl.: *Introduction to the Theory of Ionized Gases* (Interscience Publishers, Inc., New York, 1960), p. 59].
[5] R. Jancel and T. Kahan, J. Phys. Radium **20**, 35, 804 (1959); later work [by C. A. Carpenter and F. W. Metzger, J. Math. Phys. **2**, 694 (1961)] appears to be very similar.
[6] W. P. Allis, in *Handbuch der Physik*, S. Flügge, Ed. (Springer-Verlag, Berlin, 1956), Vol. 21, pp. 404–408.
[7] I. P. Shkarofsky, Can. J. Phys. **41**, 1776 (1963).
[8] P. M. Morse and H. Feshbach, *Methods of Theoretical Physics* (McGraw-Hill Book Company, Inc., New York, 1953), pp. 1276–1283; M. H. Cohen, Phys. Rev. **95**, 674 (1954).

equal to the product of C_l and the lth-order Legendre polynomial in μ_z,

$$T_{l(z)} = C_l P_l(\mu_z) = \mu_z^l - \frac{l(l-1)}{2}\frac{\mu^{l-2}}{(2l-1)} + \frac{l(l-1)(l-2)(l-3)}{8(2l-1)(2l-3)}\mu^{l-4} \mp \cdots .$$

By unique extension the general element is then

$$\mathbf{T}_l(\boldsymbol{\mu}) = \boldsymbol{\mu}^l - \frac{l(l-1)}{2}[\mathbf{I}\boldsymbol{\mu}^{l-2}]_l + \frac{l(l-1)(l-2)(l-3)}{8(2l-1)(2l-3)}[\mathbf{I}\mathbf{I}\boldsymbol{\mu}^{l-4}]_l \mp \cdots . \quad (2)$$

Here $\boldsymbol{\mu}^l$ is the symmetric lth-order vector product tensor of $\boldsymbol{\mu}$ and $[\ ]_l$ denotes an lth-order symmetrization operation, adding all the $l!$ permutations and dividing by $l!$, $\mathbf{I}$ is the diagonal limit or identity tensor.

Wallace showed that the complete contraction or scalar product of two lth-order spherical tensors of different argument is simply given by the formula

$$\mathbf{T}_l(\boldsymbol{\mu})_l \cdot \mathbf{T}_l(\boldsymbol{\mu}') = C_l P_l(\boldsymbol{\mu}\cdot\boldsymbol{\mu}').$$

This means that $f(v)$ can be expanded as follows:

$$f(\mathbf{v}) = \frac{1}{4\pi}\sum_l (2l+1)\int f(v')P_l(\boldsymbol{\mu}\cdot\boldsymbol{\mu}')\, d^2\Omega'$$

$$= \frac{1}{4\pi}\sum_l \frac{(2l+1)}{C_l}\left\{\int f(v')\mathbf{T}_l(\boldsymbol{\mu}')\, d^2\Omega\right\}_l^{\cdot}\mathbf{T}_l(\boldsymbol{\mu}) \quad (3)$$

$$= \sum_l \mathbf{f}_l(v)_l^{\cdot}\mathbf{T}_l(\boldsymbol{\mu}),$$

where[9]

$$\mathbf{f}_l \equiv \frac{2l+1}{4\pi C_l}\int f(v)\mathbf{T}_l(\boldsymbol{\mu})\, d^2\Omega. \quad (4)$$

Note that the $f_{l(z)}$ element is just the coefficient of the lth-order $m = 0$ Legendre polynomial in the spherical harmonic expansion,[3] for from Eq. (4) we have

$$f_{l(z)} = \frac{2l+1}{4\pi C_l}\int f(v)T_{l(z)}\, d^2\Omega = \frac{2l+1}{4\pi}\int f(v)P_l(\mu_z)\, d^2\Omega \equiv f_{l00}.$$

Thus, the $f_{l(z)}$ tensor equation can be checked immediately with the polar spherical harmonic ($m = 0$) result given by Allis.[6]

Because $\mathbf{T}_l$ is irreducible, any contraction on $\mathbf{f}_l$ which gives a nonzero result cannot appear in the result and should be eliminated, and indeed this is the result of the definition of Eq. (4) for $\mathbf{f}_l$. Note that once we have made $\mathbf{f}_l$ irreducible by definition[10] then other $\boldsymbol{\mu}^l$ polynomials can be used and, in particular, using Eq. (2) and the fact that $\mathbf{f}_l\cdot[\mathbf{I}_2\boldsymbol{\mu}^{l-2}] = 0$, the following combinations are equivalent:

$$\mathbf{f}_{l l}^{\cdot}\, T_l = \mathbf{f}_{l l}^{\cdot}\, \boldsymbol{\mu}^l. \quad (5)$$

It is this equivalence that enables the simple derivation of the chain of tensor equations. Note that the only property required for $\mathbf{f}_l$ is irreducibility. The same result will hold if the tensor is not fully symmetric. We have (since $\boldsymbol{\mu}^l$ and $\mathbf{T}_l$ are fully symmetric) the following result $\{[\ ]_l$ defined after Eq. (2)$\}$:

$$\mathbf{g}_{l l}^{\cdot}\, \boldsymbol{\mu}^l = \mathbf{g}_{l l}^{\cdot}\, \mathbf{T}_l = [\mathbf{g}_l]_l^{\cdot}\, \mathbf{T}_l = [\mathbf{g}_l]_l^{\cdot}\, \boldsymbol{\mu}^l.$$

In that case, however, the tensor obtained by tensor multiplication by $\mathbf{T}_l$ and integration over angle is $[\mathbf{g}_l]_l$ the symmetrized version of the arbitrary irreducible tensor $\mathbf{g}_l$.

BOLTZMANN–VLASOV EQUATION

The collision terms are not discussed here. Shkarofsky[11] has treated the Fokker–Planck equation and spherical harmonic tensors in considerable detail and electron–neutron collision effects are well known.[6,12] Only the flow terms, those common to the Boltzmann and Vlasov equations, are discussed. To tackle the problem in two stages, the straightforward extrinsic flow terms referred to a rest frame are treated first, providing the generalization for the particular equations given previously.[3,6] The intrinsic velocity (velocity referred to some velocity $\mathbf{C}$) generalization of the particular results of Shkarofsky[12] (which introduce additional terms) are then derived.

EXTRINSIC VELOCITY EQUATION

The application to the Boltzmann–Vlasov flow terms is at first just like the previous work.[3] The flow terms $D(f)$ are as follows:

$$D(f) \equiv \frac{\partial f}{\partial t} + \mathbf{v}\cdot\nabla f + (\mathbf{a} + \mathbf{v}\times\boldsymbol{\omega}_b)\cdot\nabla_v f, \quad (6)$$

where ∇ is the configuration space gradient operator, ∇_v is the velocity space gradient operator, $\mathbf{a}$ is the velocity-independent acceleration; $\mathbf{a}$ =

[9] One should not leap to the incorrect conclusion that the $\mathbf{T}_l$ are orthogonal in angle integration. Contributions to an $\mathbf{f}_l$ element in (4) come from other elements as well.

[10] An indication of this is given in the bcok by A. Sommerfeld, *Lectures on Theoretical Physics, Vol. 5, Thermodynamics and Statistical Mechanics* (Academic Press, Inc., New York, 1964), p. 338. (This section was actually completed after the author's death by F. Bopp and J. Meixner.)

[11] T. W. Johnston, Can. J. Phys. **41**, 1208 (1962).

[12] I. P. Shkarofsky, Can. J. Phys. **41**, 1753 (1963).

$(q/m)\mathbf{E} - \nabla\psi$ with q, m the particle charge and mass, $\mathbf{E}$ the electric field, ψ the gravitational potential, and $\boldsymbol{\omega}_b$ is the magnetic cyclotron angular frequency vector $q\mathbf{B}/m$ so that $\mathbf{v} \times \boldsymbol{\omega}_b$ is the acceleration due to the magnetic field. Substituting the irreducible direction cosine expansion of Eq. (3) gives, as before,[3]

$$D(f) = \sum_l \frac{\partial \mathbf{f}_l}{\partial t} \underset{l}{\cdot} \boldsymbol{\mu}^l + v\nabla\mathbf{f}_l \underset{l+1}{\cdot} \boldsymbol{\mu}^{l+1} + \mathbf{a}v^l \frac{\partial}{\partial v}\left(\frac{\mathbf{f}_l}{v^l}\right) \underset{l+1}{\cdot} \boldsymbol{\mu}^{l+1} + \frac{l\mathbf{a}\cdot\mathbf{f}_l}{v} \underset{l-1}{\cdot} \boldsymbol{\mu}^{l-1} + l\boldsymbol{\omega}_b \times \mathbf{f}_l \underset{l}{\cdot} \boldsymbol{\mu}^l = 0.$$

As before we wish to group the terms by $\boldsymbol{\mu}^l$ rather than by $\mathbf{f}_l$. Now, however, we recognize the special value of arranging to have irreducible tensor forms multiplying $\boldsymbol{\mu}^l$, $\boldsymbol{\mu}^{l+1}\boldsymbol{\mu}^{l-1}$. We can symmetrize immediately but irreducibility must be contrived. It is evident that only the $\boldsymbol{\mu}^{l+1}$ terms require special treatment; the coefficients of $\boldsymbol{\mu}^l$ and $\boldsymbol{\mu}^{l-1}$ are already irreducible. Both $\boldsymbol{\mu}^{l+1}$ terms are of the form $\mathbf{Af}_l$. In order to form the irreducible tensors, we add and subtract the nonzero results of the contractions of $\mathbf{Af}_l$. The only nonzero contraction is $A \cdot f_l$, and the form we require is

$$[\mathbf{Af}_l]_{l+1} = [\mathbf{Af}_l - \alpha(\mathbf{A}\cdot\mathbf{f}_l)\mathsf{I}]_{l+1} + \alpha[(\mathbf{A}\cdot\mathbf{f}_l)\mathsf{I}]_{l+1}.$$

The required coefficient of α is that which will make the first tensor on the right zero on any contraction. There are l indices in $\mathbf{f}_l$ to choose in contracting $[\mathbf{Af}_l]_{l+1}$ and $2(l-1)+3 = 2l+1$ in $[\mathbf{A}\cdot\mathbf{f}_l\mathsf{I}]_{l+1}$, since contraction on I gives 3 and either index in I may be equated with the $l-1$ free indices in $\mathbf{A}\cdot\mathbf{f}_l$. Thus, α is $l/(2l+1)$ and the required form is

$$\mathbf{Af}_l \underset{l}{\cdot} \boldsymbol{\mu}^{l+1} = [\mathbf{Af}_l] \underset{l}{\cdot} \boldsymbol{\mu}^{l+1} = \left[\mathbf{Af}_l - \frac{l}{2l+1}\mathbf{A}\cdot\mathbf{f}_l\mathsf{I}\right]_{l+1} \cdot \boldsymbol{\mu}^{l+1} + \frac{l}{2l+1}[\mathbf{A}\cdot\mathbf{f}_l] \underset{l-1}{\cdot} \boldsymbol{\mu}^{l-1}.$$

(We have used $\mathsf{I}{:}\boldsymbol{\mu\mu} = \boldsymbol{\mu}\cdot\boldsymbol{\mu} = 1$.) Note that $\mathbf{A}\cdot\mathbf{f}_l$ is irreducible as well.

We can therefore write Eq. (7) as follows:

$$D(f) = \sum_l \frac{\partial \mathbf{f}_l}{\partial t} \underset{l}{\cdot} \boldsymbol{\mu}^l + \left\{v\left(\nabla\mathbf{f}_l - \frac{l}{2l+1}\mathsf{I}\nabla\cdot\mathbf{f}_l\right) + v^l \frac{\partial}{\partial v}\left[\frac{1}{v^l}\left(\mathbf{af}_l - \frac{l}{2l+1}\mathsf{I}\mathbf{a}\cdot\mathbf{f}_l\right)\right]\right\}_{l+1} \cdot \boldsymbol{\mu}^{l+1} + l[\boldsymbol{\omega}_b \times \mathbf{f}_l] \underset{l}{\cdot} \mu^l + \frac{l}{2l+1}\left[v\nabla\cdot\mathbf{f}_l + v^l \frac{\partial}{\partial v}\left(\frac{\mathbf{a}\cdot\mathbf{f}_l}{v^l}\right) + \frac{l\mathbf{a}\cdot\mathbf{f}_l}{v}\right]_{l-1} \cdot \boldsymbol{\mu}^{l-1}.$$

Grouping now by $\boldsymbol{\mu}^l$ and using the following identity on the $\mathbf{a}\cdot\mathbf{f}_l\dot{}_{l-1}\boldsymbol{\mu}^{l-1}$ terms,

$$\frac{1}{v^p}\frac{\partial v^p f}{\partial v} = \frac{pf}{v} + \frac{\partial f}{\partial v}, \tag{7}$$

we obtain

$$\begin{aligned} D(f) &= \sum_l \boldsymbol{\mu}^l \underset{l}{\cdot} \mathbf{D}_l \\ &\equiv \sum_l \boldsymbol{\mu}^l \underset{l}{\cdot} \left\{\frac{\partial \mathbf{f}_l}{\partial t} + v\left(\nabla_r\mathbf{f}_{l-1} - \frac{l-1}{2l-1}\mathsf{I}\nabla_r\cdot\mathbf{f}_{l-1}\right) + v^{l-1}\frac{\partial}{\partial v}\left(\frac{\mathbf{af}_{l-1}}{v^{l-1}} - \frac{l-1}{2l-1}\mathsf{I}\frac{\mathbf{a}\cdot\mathbf{f}_{l-1}}{v^{l-1}}\right) + l\boldsymbol{\omega}_b \times \mathbf{f}_l \right. \\ &\quad \left. + \frac{l+1}{2l+3}\left[v\nabla\cdot\mathbf{f}_{l+1} + \frac{1}{v^{l+2}}\frac{\partial}{\partial v}(v^{l+2}\mathbf{a}\cdot\mathbf{f}_{l+1})\right]\right\}_l. \end{aligned} \tag{8}$$

Now, because $\mathbf{D}_l$ is irreducible we can immediately apply Eq. (4), that is multiply by $\mathbf{T}_l$ and integrate over angle to obtain the $\mathbf{D}_l$ elements in a chain of equations which contain $\mathbf{f}_{l-1}$, $\mathbf{f}_l$, and $\mathbf{f}_{l+1}$. Provided one can do this for the collisions as well ($C = \sum_l \mathbf{C}_l \cdot \boldsymbol{\mu}^l$, with irreducible $\mathbf{C}_l$), then the result is the chain of equations

$$\mathbf{C}_l = \mathbf{D}_l = \left[\frac{\partial \mathbf{f}_l}{\partial t} + v\left(\nabla\mathbf{f}_{l-1} - \frac{l-1}{2l-1}\mathsf{I}\nabla\cdot\mathbf{f}_{l-1}\right) + v^{l-1}\frac{\partial}{\partial v}\left(\frac{\mathbf{af}_{l-1}}{v^{l-1}} - \frac{l-1}{2l-1}\mathsf{I}\frac{\mathbf{a}\cdot\mathbf{f}_{l-1}}{v^{l-1}}\right) + l\boldsymbol{\omega}_b \times \mathbf{f}_l + \frac{l+1}{2l+3}\left(v\nabla\cdot\mathbf{f}_{l+1} + \frac{1}{v^{l+2}}\frac{\partial}{\partial v}(v^{l+2}\mathbf{a}\cdot\mathbf{f}_{l+1})\right)\right]_l \tag{9}$$

instead of the original equation

$$C = D.$$

The $\mathbf{D}_l$ elements for $\mathbf{f}_l$ with l less than 2 were obtained by Allis[6] but irreducibility is not necessary. The equations for $\mathbf{D}_l$ and $\mathbf{f}_l$ with $l = 2, 3$ were given by Johnston[3] by direct calculation and agree with Eq. (9). The calculation shown here gives the result for all l with less labor than that required for $l = 2$ or $l = 3$ by the direct approach. The tensor form of $\mathbf{D}_l$ in Eq. (9) is far more symmetric and compact than the direct spherical harmonic form.[5]

Another check is the one-dimensional result, for which the magnetic field is along the z axis, say, and Eq. (8) becomes simply

$$D_l = \frac{\partial f_{l(z)}}{\partial t} + \frac{1}{2l-1} \times \left\{\frac{\partial}{\partial z}f_{l-1(z)} + v^{l-1}a_z\frac{\partial}{\partial v}\left[\frac{f_{l-1(z)}}{v^{l-1}}\right]\right\} + \frac{l+1}{2l+3}\left(\frac{v\,\partial f_{l+1(z)}}{\partial z} + \frac{a_z}{v^{l+2}}\frac{\partial v^{l+2}f_{l+1(z)}}{\partial z}\right).$$

Since, as pointed out above $f_{l(s)} = f_{l00}$, this can be compared directly with Allis'[6] one-dimensional result obtained from Legendre polynomial recursion relations; the two results are identical.

INTRINSIC VELOCITY EQUATION

The intrinsic velocity flow term $D^w(f)$ involves the intrinsic velocity $\mathbf{w} = \mathbf{v} - \mathbf{C}$, where $\mathbf{C}$ is the reference velocity. The intrinsic flow terms, as given by Bernstein and Trehan,[13] are

$$D^w(f) = \frac{df}{dt} + \mathbf{w}\cdot\nabla f + (\mathbf{h} + \mathbf{w}\times\omega_b)\cdot\nabla_w f - \mathbf{w}\cdot\nabla\mathbf{C}\cdot\nabla_w f, \qquad (10)$$

where

$$d/dt = \partial/\partial t + \mathbf{C}\cdot\nabla,$$

$$\mathbf{h} \equiv \mathbf{a} + \mathbf{C}\times\omega_b - d\mathbf{C}/dt.$$

Define

$$D^{wa}(f) \equiv \frac{df}{dt} + \mathbf{w}\cdot\nabla f + (\mathbf{h} + \mathbf{w}\times\omega_b)\cdot\nabla_w f,$$

$$D^{wb}(f) \equiv -\mathbf{w}\cdot\nabla\mathbf{C}\cdot\nabla_w f.$$

Thus

$$D^{wb}(f) = D^{wa}(f) + D^{wb}(f).$$

Evidently the $D^{wa}(f)$ term is just like $D(f)$ with d/dt replacing $\partial/\partial t$, $\mathbf{w}$ replacing $\mathbf{v}$ and $\mathbf{h}$ replacing $\mathbf{a}$, and the final result for $D_l^{wa}(f)$ can be obtained with these transformations in Eq. (8).

The term D^{wb} needs more treatment. Substitution of $\mathbf{f}_l(w)_l^{\cdot}\mathbf{w}^l$ in D^{wb} gives the result quoted by Shkarofsky[7] [his Eq. (9)]:

$$D^{wb}(f) = -\mathbf{w}\cdot\nabla\mathbf{C}\cdot\nabla_w f = \sum_l - l\nabla\mathbf{C}\cdot\mathbf{f}_l{}_l^{\cdot}\boldsymbol{\mu}^l - \nabla\mathbf{C}w^{l+1}\frac{\partial}{\partial w}\left(\frac{\mathbf{f}_l}{w^l}\right)_{l+2}^{\cdot}\boldsymbol{\mu}^{l+2},$$

where $\boldsymbol{\mu}$ now is given by $\boldsymbol{\mu} = \mathbf{w}/w$.

The $\nabla\mathbf{C}\cdot\mathbf{f}_l$ term is easily dealt with by the same type of reasoning as above. We see that it can be written as

$$-l\nabla\mathbf{C}\cdot\mathbf{f}_{l}{}_l^{\cdot}\boldsymbol{\mu}^l = -\left[\nabla\mathbf{C}\cdot\mathbf{f}_l - \frac{l-1}{2l-1}\,\mathsf{I}\nabla\mathbf{C}:\mathbf{f}_l\right]_l^{\cdot}\boldsymbol{\mu}^l + \frac{l-1}{2l-1}\nabla\mathbf{C}:\mathbf{f}_l\cdot\boldsymbol{\mu}^{l-2}.$$

[13] I. B. Bernstein and S. K. Trehan, Nucl. Fusion 1, 3 (1960), Chap. 1, Eq. (42).

The last term in D^{wb} is of the form $\nabla\mathbf{C}\mathbf{f}_l{}_{l+2}^{\cdot}\boldsymbol{\mu}^{l+2}$ and requires the subtraction and addition of two terms with I and II to reach the desired irreducible form.

If we contract and symmetrize $\nabla\mathbf{C}\mathbf{f}_l$, the result is $[\nabla\cdot\mathbf{C}\mathbf{f}_l + 2l\mathbf{f}_l\cdot[\nabla\mathbf{C}]_2]_l$; hence the terms to be subtracted and added are in the form

$$\pm[-\beta\mathsf{I}\nabla\cdot\mathbf{C}\mathbf{f}_l + 2\gamma\mathsf{I}\mathbf{f}_l\cdot[\nabla\mathbf{C}]_2 + \delta\mathsf{II}\mathbf{f}_l:\nabla\mathbf{C}]_{l+2}.$$

Contracting, we have

$$\begin{aligned}\mathsf{I}:[\nabla\mathbf{C}\mathbf{f}_l &- \beta\mathsf{I}\nabla\cdot\mathbf{C}\mathbf{f}_l - 2\gamma\mathsf{I}\mathbf{f}_l\cdot[\nabla\mathbf{C}]_2 - \delta\mathsf{II}\mathbf{f}_l:\nabla\mathbf{C}]_{l+2}\\ &= [\nabla\cdot\mathbf{C}\mathbf{f}_l + 2l[\nabla\mathbf{C}]_2\cdot\mathbf{f}_l - \beta(2l+3)\nabla\cdot\mathbf{C}\mathbf{f}_l\\ &\quad - 2\gamma(2l+3)[\nabla\mathbf{C}]_2\cdot\mathbf{f}_l - 2\gamma(l-1)\mathsf{I}\nabla\mathbf{C}:\mathbf{f}_l\\ &\quad - (2\times 3 + 4 + 4(l-2))\,\delta\nabla\mathbf{C}:\mathbf{f}_l\,\mathsf{I}]_l.\end{aligned}$$

Setting the coefficients of $\nabla\cdot\mathbf{C}\mathbf{f}_l$, $[\nabla\mathbf{C}]_2\cdot\mathbf{f}_l$, and $\nabla\mathbf{C}:\mathbf{f}_l$ equal to zero gives the following results:

$$\beta = \frac{1}{2l+3}, \qquad \gamma = \frac{l}{2l+3},$$

$$\delta = -\frac{(l-1)\gamma}{2l+1} = -\frac{l(l-1)}{(2l+1)(2l+3)}.$$

Thus, the expression

$$\left[\nabla\mathbf{C}\mathbf{f}_l - \frac{\mathsf{I}}{2l+3}(2l[\nabla\mathbf{C}]_2\cdot\mathbf{f}_l + \nabla\cdot\mathbf{C}\mathbf{f}_l) + \frac{l(l-1)}{(2l+1)(2l+3)}\,\mathsf{II}\nabla\mathbf{C}:\mathbf{f}_l\right]_{l+2}$$

is irreducible. We also require that the coefficients of the I and II terms are each irreducible. This is automatically true for $\mathbf{f}_l:\nabla\mathbf{C}$, the II coefficient and for $[\nabla\cdot\mathbf{C}\mathbf{f}_l]_l$ but not for the $[\nabla\mathbf{C}]_2\cdot\mathbf{f}_l$ combinations, to which a term of the form $\epsilon[\mathsf{I}\nabla\mathbf{C}:\mathbf{f}_l]_l$ must be added and subtracted.

We have already done this for the $\nabla\mathbf{C}\cdot\mathbf{f}_l$ coefficient earlier, the coefficient being $(l-1)/(2l-1)$ for each $\nabla\mathbf{C}\cdot\mathbf{f}_l$ term, so ϵ is then given by

$$-\epsilon = \frac{2l(l-1)}{(2l+3)(2l-1)}.$$

We have a term of the identical form already, so the final coefficient for the last II term is given below

$$\epsilon - \delta = \frac{l(l-1)}{2l+3}\left(\frac{1}{2l+1} - \frac{2}{2l-1}\right) = -\frac{l(l-1)}{(2l+1)(2l-1)}.$$

The final result is that we can write $\nabla\mathbf{C}\mathbf{f}_l{}_{l+2}^{\cdot}\boldsymbol{\mu}^{l+2}$ as follows:

$$\nabla \mathbf{C}\mathbf{f}_l \underset{l+2}{\cdot} \boldsymbol{\mu}^{l+2}$$

$$= \left(\nabla \mathbf{C}\mathbf{f}_l - \frac{\mathsf{I}}{2l+3}(2l[\nabla\mathbf{C}]_2\cdot\mathbf{f}_l + \nabla\cdot\mathbf{C}\mathbf{f}_l)\right.$$
$$\left. + \frac{l(l-1)}{(2l+1)(2l+3)}\,\mathsf{II}\nabla\mathbf{C}:\mathbf{f}_l\right)\underset{l+2}{\cdot}\boldsymbol{\mu}^{l+2} + \frac{\mathsf{I}}{2l+3}$$
$$\times\left(2l[\nabla\mathbf{C}]_2\cdot\mathbf{f}_l + \nabla\cdot\mathbf{C}\mathbf{f}_l - \frac{2(l-1)l}{2l-1}\,\mathsf{I}\nabla\mathbf{C}:\mathbf{f}_l\right)\underset{l}{\cdot}\boldsymbol{\mu}^{l}$$
$$+ \frac{l(l-1)}{(2l+1)(2l-1)}(\nabla\mathbf{C}:\mathbf{f}_l)\underset{l+2}{\cdot}\boldsymbol{\mu}^{l-2}.$$

All the $\boldsymbol{\mu}^l$, $\boldsymbol{\mu}^{l+2}$, $\boldsymbol{\mu}^{l-2}$ coefficients are now irreducible. As before, we group by $\boldsymbol{\mu}^l$ rather than by $\mathbf{f}_l$ and then isolate by multiplication by $\mathbf{T}_l(\boldsymbol{\mu})$ and integration over w angle, the result being

$$\mathbf{D}_l^{wb}(f) = -w^{l-1}\frac{\partial}{\partial w}\left[\frac{1}{w^{l-2}}\left(\nabla\mathbf{C}\mathbf{f}_{l-2} - \frac{\mathsf{I}}{2l-1}\right.\right.$$
$$\times(2(l-2)[\nabla\mathbf{C}]_2\cdot\mathbf{f}_{l-2} + \nabla\cdot\mathbf{C}\mathbf{f}_{l-2})$$
$$\left.\left.+\frac{(l-2)(l-3)}{(2l-3)(2l-1)}\,\mathsf{II}\mathbf{C}:\mathbf{f}_{l-2}\right)\right]_l$$
$$- l\left[\nabla\mathbf{C}\cdot\mathbf{f}_l - \frac{l-1}{2l-1}\,\mathsf{I}\nabla\mathbf{C}:\mathbf{f}_l\right]_l$$
$$-\frac{w^{l+1}}{2l+3}\frac{\partial}{\partial w}\left[\frac{1}{w^l}\left(2l[\nabla\mathbf{C}]_2\cdot\mathbf{f}_l\right.\right.$$
$$\left.\left.+\nabla\cdot\mathbf{C}\mathbf{f}_l - \frac{2(l-1)l}{2l-1}\,\mathsf{I}\nabla\mathbf{C}\cdot\mathbf{f}_l\right)\right]_l$$
$$-\frac{(l+2)(l+1)}{(2l+3)(2l+5)}\frac{1}{w^{l+2}}\frac{\partial}{\partial w}(w^{l+3}\nabla\mathbf{C}:\mathbf{f}_{l+2}). \quad (11)$$

The identity of Eq. (7) has been used on the $\nabla\mathbf{C}:\mathbf{f}_{l+2}$ terms to collapse them into one term.

The final form, with both D^{wa} and D^{wb} included, is

$$\mathbf{D}_l^w(f) = D_l^{wa} + D_l^{wb}$$
$$= \left[\frac{d\mathbf{f}_l}{dt} + l\omega_b\times\mathbf{f}_l + w\left(\nabla\mathbf{f}_{l-1} - \frac{l-1}{2l-1}\,\mathsf{I}\nabla\cdot\mathbf{f}_{l-1}\right)\right.$$
$$+ w^{l-1}\frac{\partial}{\partial w}\left(\frac{\mathbf{a}\mathbf{f}_{l-1}}{w^{l-1}} - \frac{l-1}{2l-1}\,\mathsf{I}\mathbf{a}\cdot\mathbf{f}_{l-1}\right)$$
$$+\frac{l+1}{2l+3}\left[w\nabla\cdot\mathbf{f}_{l+1} + \frac{1}{w^{l+2}}\frac{\partial(w^{l+2}\mathbf{a}\cdot\mathbf{f}_{l+1})}{\partial w}\right]$$
$$- w^{l-1}\frac{\partial}{\partial w}\left(\frac{1}{w^{l-2}}\right)\left\{\nabla\mathbf{C}\mathbf{f}_{l-2} - \frac{\mathsf{I}}{2l-1}\right.$$
$$\times\left[2(l-2)[\nabla\mathbf{C}]_2\cdot\mathbf{f}_{l-2} + \nabla\cdot\mathbf{C}\mathbf{f}_{l-2}\right]$$
$$\left.+\frac{(l-2)(l-3)}{(2l-3)(2l-1)}\,\mathsf{II}\nabla\mathbf{C}:\mathbf{f}_{l-2}\right\}$$
$$- l\left(\nabla\mathbf{C}\cdot\mathbf{f}_l - \frac{l-1}{2l-1}\,\mathsf{I}\nabla\mathbf{C}:\mathbf{f}_l\right)$$
$$-\frac{w^{l+1}}{2l+3}\frac{\partial}{\partial w}\left\{\frac{1}{w^l}\left[2l[\nabla\mathbf{C}]_2\cdot\mathbf{f}_l\right.\right.$$
$$\left.\left.+\nabla\cdot\mathbf{C}\mathbf{f}_l - \frac{2(l-1)l}{2l-1}\,\mathsf{I}\nabla\mathbf{C}:\mathbf{f}_l\right]\right\}$$
$$\left.-\frac{(l+1)(l+2)}{(2l+3)(2l+5)}\frac{1}{w^{l+2}}\frac{\partial}{\partial w}(w^{l+3}\nabla\mathbf{C}:\mathbf{f}_{l+2})\right]_l. \quad (12)$$

The result can be checked for the $D_{l(z)}$ element in the same way as for the extrinsic velocity case, by comparison with the Legendre polynomial result. Only the D_l^{wb} term needs checking. In the same way as before, we have in one dimension

$$-\mathbf{w}\cdot\nabla\mathbf{C}\cdot\nabla_w f = -\sum_l w\cos\theta$$
$$\times\frac{\partial C_z}{\partial z}\cdot\left(\cos\theta\frac{\partial f_l}{\partial w} - \frac{\sin\theta}{w}f_l\frac{\partial P_l}{\partial\theta}\right)$$

with $f = \sum_l f_l P_l$. Using Legendre polynomial recursion relations twice and the identity of Eq. (7), we have

$$-\mathbf{w}\cdot\nabla\mathbf{C}\cdot\nabla_w f = -\frac{\partial C_z}{\partial z}\sum_l\left(P_{l-2}\frac{l(l-1)}{(2l+1)(2l-1)}\right.$$
$$\times\frac{1}{w^{l+1}}\frac{\partial(f_l w^{l+1})}{\partial w} + P_l\left\{\frac{l^2}{2l-1}\frac{f_l}{w}\right.$$
$$\left.+\left[\frac{l^2}{2l-1} + \frac{(l+1)^2}{2l+3}\right]\frac{w^{l+1}}{2l+1}\frac{\partial}{\partial w}\left(\frac{f_l}{w^l}\right)\right\}$$
$$\left.+P_{l+2}\frac{(l+1)(l+2)}{(2l+1)(2l+3)}w^l\frac{\partial}{\partial w}\left(\frac{f_l}{w^l}\right)\right)$$

Regrouping by P_l results in the following:

$$-\mathbf{w}\cdot\nabla\mathbf{C}\cdot\nabla_w f = +\sum P_l D_{l(z)}^{wb}$$

with

$$-D_{l(z)}^{wb} = \frac{\partial C_z}{\partial z}\left[\frac{(l-1)l}{(2l-3)(2l-1)}w^{l-1}\frac{\partial}{\partial w}\left(\frac{f_{l-2}}{w^{l-2}}\right)\right.$$
$$+\frac{l^2}{2l-1}f_l + \frac{2l(l+1)-1}{(2l-1)(2l+3)}w^{l+1}\frac{\partial}{\partial w}\left(\frac{f_l}{w^l}\right)$$
$$\left.+\frac{(l+1)(l+2)}{(2l+3)(2l+5)}\frac{1}{w^{l+2}}\frac{\partial}{\partial w}(f_{l+2}w^{l+3})\right] \quad (13)$$

This agrees with the z^l element of $D_{l(z)}^{wb}$ in Eq. (11).

SUMMARY

The general intrinsic velocity (w) spherical harmonic equation has been derived [Eq. (12)] with the extrinsic case as a particular case [Eq. (8)] obtained when the reference velocity $\mathbf{C}$ and its derivatives are zero so that $\mathbf{v}$ and $\mathbf{w}$ are equal. Legendre polynomial recursion relations have been used to check the result. The worth of this general result can best be appreciated by those who have laboured through the piecemeal derivation of several particular cases.

With this general expression for $\mathbf{D}_l$ available, it may now be worthwhile to examine more general expressions for collisions in order to extend Shkarofsky's[12] work on the Fokker–Planck terms on effects on irreducible anisotropic tensor pressure to other effects of higher order.

ACKNOWLEDGMENTS

It is a pleasure to acknowledge the helpfulness of several authors (P. R. Wallace, E. Ikenberry, T. Kahan) in sending me copies of their work and of Dr. Suchy and Dr. Ziering in drawing my attention to the works of A. Sommerfeld and E. Ikenberry, respectively. My thanks are due to Dr. I. P. Shkarofsky, who urged the extension of the work to obtain the intrinsic results and who was always available for invaluable and stimulating discussion.

Comments on the Article

The article is an excellent example of how content can determine form. While the style and practice follow the *Style Manual* of the AIP, the organization is determined by the logic of a mathematical proof as discussed in Chapter 3. Analysis of the article, however, reveals the style expected in an article in the field of physics.

The *title* is made up of keywords whose arrangement specifically limits the topic. Cross-referencing is thus made easy, and the meaning is clear.

The *by-lines* or *author credits* cite the author, the name and address of the corporate laboratory to which he is attached, and the date (inverted) upon which he submitted the article for publication. Had the laboratory supported the work in any fashion other than paying the publication fee of the journal, that fact would have been noted in either a footnote or in the introduction.

The *abstract* is informative rather than descriptive. Because it contains fewer than fifty words, it may be collated verbatim by a variety of abstracting services. As required by the journal, the abstract is printed in italics to differentiate it from the rest of the article.

The *introduction* to this article serves several purposes. The first paragraph states the purpose or object of the article. The second paragraph reviews the literature on the topic and explains an oversight in an earlier article by the author. The third paragraph begins with a statement of the methodology to be followed and ends with a statement of the object of this methodology. The parenthetical inclusion of the word "successful" is good practice. While the work would be of value if only to show an unsuccessful approach, the fact that the approach was successful should be shown early in the article. The last paragraph of the introduction shows the value and application of the results.

The first subsection, "Spherical Harmonic Tensors," synthesizes the end results of the work of others to show the rational basis upon which the author's work rests.

The next three subsections derive and apply equations. Of importance here is the printing and positioning of the equations. They meet the requirements reprinted above from the *Style Manual* of the American Institute of Physics. The progression is mathematically logical, and the last sentence is essentially a QED.

The *summary* indicates the significance of the work and suggests applications of it to extend the work of others. (The variant spelling of the word "laboured" indicates the Canadian/English background of the writer. Most editors will pass such variants because their use lends an international tone to a professional article.)

The *acknowledgements* are more than a list of names taken from the footnotes. The contribution of each person listed is specifically stated.

Geological Journals

As stated earlier, the literature of geology follows either the *Handbook for Authors* of the American Chemical Society or the style guides of the public and private agencies which sponsored the work being discussed. The geological article reprinted below with the kind permission of the editors of *Geological Magazine* illustrates practice in the professional journals of geology. The use of an interlinked bibliography permits a footnoting system limited to content notes.

Zones of Progressive Regional Metamorphism Across the Western Margin of the Mozambique Belt in Rhodesia and Mozambique

By J. R. Vail

Abstract

Investigations of the sedimentary rocks of the Umkondo System have enabled six zones of eastwardly increasing metamorphic grade to be delineated on the basis of index mineral assemblages in pelite horizons.

Chlorite, garnet (almandine), staurolite, kyanite, and sillimanite isograds have been recognized over a distance of 500 km. along the eastern border of Southern Rhodesia and are plotted on a map. More detailed reconnaissance work has been confined to the Barué highlands north-east of Inyanga. In addition, a " zone of intermediate isotopic age measurements " is proposed to the west of the lowermost chlorite zone. The isograds lie parallel to the structural edge of the north–south trending Mozambique orogenic belt and can be joined with those of similar metamorphic rocks in the east–west trending Zambesi orogenic belt, suggesting contemporaneity of the *c.* 500 m.y. regional tectono-thermal event in these two belts.

I. Introduction

The part of eastern southern Africa, with which this paper is concerned, may be considered to comprise three main units (Text-fig. 2). The oldest group of rocks concerned is the Archaean of Rhodesia, which forms a cratonic nucleus upon which later events were superimposed. It comprises mainly batholithic granitic masses which enclose narrow belts of meta-sedimentary and meta-volcanic rocks. The grade of metamorphism in the sediments is variable, but in many cases the rocks remain in the greenschist facies. To the north and east the craton is bounded by highly foliated granitic gneisses belonging to the Zambesi and Mozambique belts. These belts include remobilized basement material of the craton which had reached a high grade of metamorphism prior to the deposition of the overlying Umkondo sediments. The gneisses also came under the influence of the regional metamorphism that affects the Umkondo cover and it is consequently not certain to what extent the gneisses may represent highly metamorphosed Umkondo material.

Recent investigations by the Geological Surveys and independent workers in Southern Rhodesia and Mozambique have indicated that the sedimentary group, known as the Umkondo System, which is developed along the frontier between the two countries, covers a much wider area than was previously known (e.g., Swift, 1961). The sediments in question suffer increasing folding and metamorphism in an easterly direction (Johnson, Slater, Vail, 1963), and when this fact was recognized, it was then possible to identify them in areas where their characteristics were very different from those in the unmetamorphosed and unfolded regions to the west. It was further recognized that

GEOL. MAG. VOL. 103 NO. 3

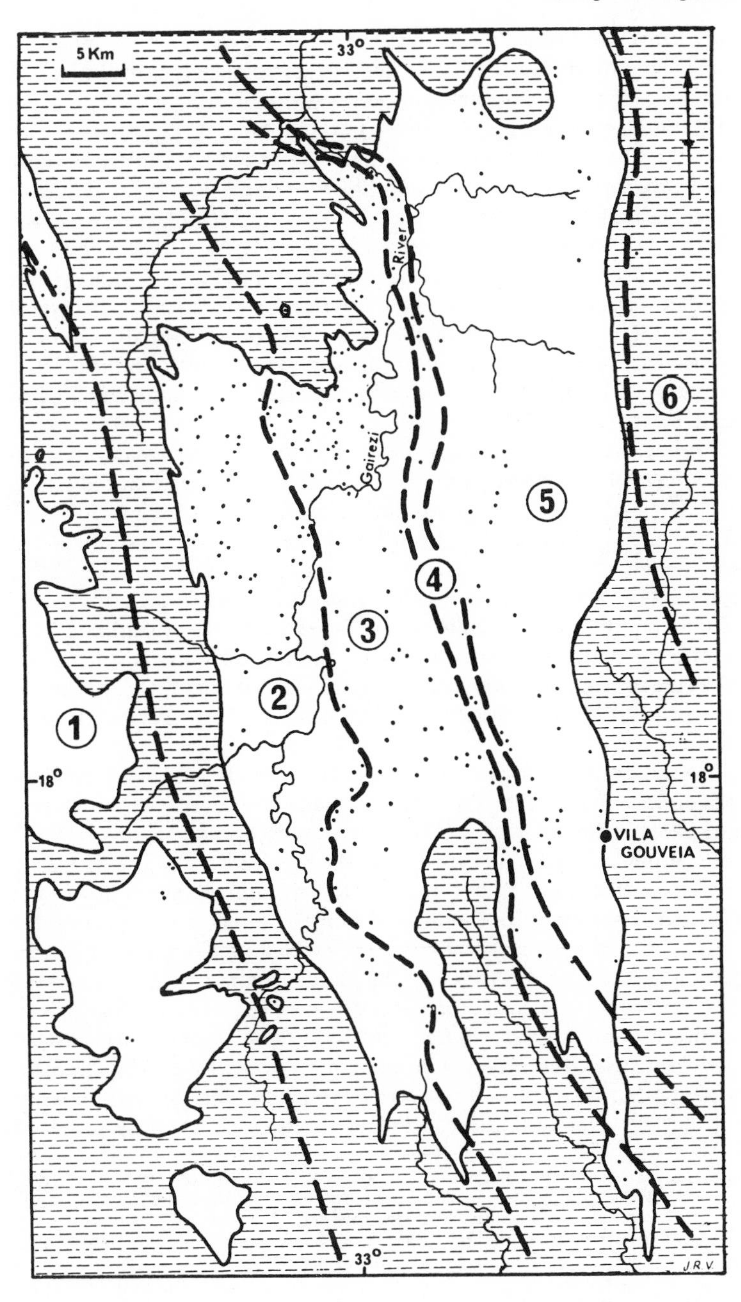
5 Km
33°
River
Gairezi
1
2
3
4
5
6
18°
18°
VILA
GOUVEIA
33°
J.R.V.

the change from an almost unmetamorphosed sequence to relatively highly metamorphosed and deformed rocks coincided closely with the contact between the Archaean Rhodesia craton in the west and the Mozambique mobile belt (Holmes, 1951) in the east.

Within the western part of the Mozambique belt the Umkondo rocks show progressive regional metamorphism and it is possible to delineate a sequence of metamorphic zones utilizing the principles employed by Barrow (1912), Tilley (1925), and Kennedy (1949). Except for brief mention (Teale, 1924, Bond, 1951, Borges and Pinto Coelho, 1957, Tyndale-Biscoe, 1957) no previous systematic work has been carried out on the metamorphism of the area. The purpose of this paper is to provide a preliminary account of the regional metamorphism. The petrography of the Umkondo rocks adds little to the understanding of fundamental metamorphic petrology.

The nature of the regional metamorphism is most clearly seen in the rocks of the Umkondo system, which have been traced along the Anglo-Portuguese border for a distance of about 500 km. (Text-fig. 2). In the Melsetter-Chipinga area, where the rocks are non-metamorphosed, the sequence consists of about 1,200 m. of argillaceous, arenaceous and calcareous rocks, intruded by thick sills of Umkondo dolerite. Sedimentary facies changes are common and, despite their age (at least Cambrian, but probably much older), the rocks are unmetamorphosed and almost unfolded (Swift, 1962, Watson, in press). A similar situation occurs near Inyanga (Tyndale-Biscoe, 1957).

Traced from the undisturbed areas towards the east, two changes appear together at the edge of the basement craton. The first is a noticeable sedimentary facies variation and the second involves the onset of regional metamorphism. In the eastern part of the Melsetter district conspicuous white quartzites with minor pelites form the spectacular Chimanimani mountain range. In the central part of the Umkondo belt the change in sedimentary facies tends towards a thick development of pelitic rocks in the Barué Highlands, and it is here that the progressive metamorphism can best be seen (see Text-fig. 1). Eastwards and northwards, across the Mazoe river in the eastern Darwin district, the proportion of psammites increases with decrease in pelitic horizons. In all the areas calcareous horizons are present, but they are usually thickest in the north and west and discontinuous

---

TEXT-FIG. 1.—Metamorphic zones in the Umkondo System in the Gairezi–Barué highlands. Localities of diagnostic minerals shown by dots.

1. Non-regionally metamorphosed Umkondo sediments and dolerites, 2. Chlorite Zone, 3. Garnet Zone, 4. Staurolite Zone, 5. Kyanite Zone, 6. Sillimanite Zone.

Ornamented areas are pre-Umkondo, unornamented areas are Umkondo sediments and dolerites.

and poorly developed in the east. At the same time, the Umkondo dolerites, which attain their maximum development in the unfolded western areas, are almost absent east of the Mozambique front.

II. Zones of Progressive Regional Metamorphism

Two types of metamorphism affect the Umkondo and older rocks, but are not always distinguishable. The first is a mild, localized, thermal effect produced at the contacts of the Umkondo dolerites mainly with adjacent calcareous rocks that are converted to calcsilicate hornfels in which epidote is very common and grossular garnet may be present. (Slater, private communication.) Other rock types are usually unaffected. The second is a conspicuous regional metamorphism that obliterates any effect of the contact metamorphism and is most readily observable in the Umkondo pelites.

Six zones, based on the first appearance of certain diagnostic metamorphic minerals in the pelites of the Umkondo System have been recognized, particularly in the Gairezi-Barué highlands (Text-fig. 1) and allow isograds to be drawn. As little is known of the chemical composition of the pelites, the zones need not everywhere imply changing tectono-thermal conditions but may also reflect geochemical differences. Mineral paragenesis as discussed, for example, by Francis (1956), has not been studied in detail.

The following six zones of regional metamorphism are distinguished in the pelites:—

(1) *Regionally non-metamorphosed pelites*.—The unmetamorphosed Umkondo pelites are usually fine-grained shales, mudstones and siltstones which show no cleavage or mineral orientation. Small shreds of sericite and an interstitial mass of chlorite occur between angular grains of quartz and rare feldspar. Larger sericite flakes of clastic origin lie parallel to the bedding. Other detrital minerals include tourmaline and zircon. Magnetite bands are not uncommon and are composed of small euhedral grains, in places associated with hematite.

(2) *Chlorite Zone*.—With the onset of regional metamorphism a poor axial plane cleavage develops and becomes more pronounced until it predominates over the bedding. Concurrently the bedding becomes contorted and folded on a small scale. In the higher grade parts of the zone this first cleavage is marked by oriented muscovite aligned close to the bedding surface. A second, strain-slip cleavage intersects the bedding plane cleavage and produces a conspicuous mineral crenulation.

White mica predominates in lustrous grey pelites, but where chlorite is the dominant mineral the rocks are green in colour. Chlorite forms small laths and crystalloblasts which become larger with increasing metamorphism. They lie in the plane of the first cleavage, and are associated with muscovite formed from original sericite. Angular, detrital quartz grains recrystallize and are orientated in the cleavage between micas. Detrital tourmaline also recrystallizes and, in places, forms radiating needles. Magnetite attains its maximum development in the chlorite zone, and perfect octohedra are common. Rare, pale-brown to green biotite forms at the expense of chlorite and muscovite, but is never prominent.

(2*a*) *Chlorite-Chloritoid Sub-Zone.*—In the Barué area chloritoid is found in the upper portions of the chlorite zone and in the lower parts of the garnet zone. The mineral is usually present as a felted mass of unoriented laths associated with chlorite. With the development of a strong cleavage the chloritoid crystals take the form of large porphyroblasts several millimetres in length within which a relic structure of the first cleavage may be clearly preserved in the form of inclusions. These crystals are in turn affected by the second, strain-slip cleavage. In the Chimanimani area, on the other hand, chloritoid occurs in low-grade phyllites in the chlorite zone.

(3) *Garnet Zone.*—The first development of garnets in the pelites marks the lower limit of the garnet zone. In the lowermost part of the zone the garnets are usually small and widely scattered. With advance in temperatures, however, they increase in size to a maximum of about 2 cm. in diameter. Many are disc-shaped and in thin-section they show a variety of internal structures suggesting a complex growth history. They are usually poikiloblastic, with quartz inclusions arranged along relic cleavages. Retrograde effects are not common, but there is some alteration to chlorite and magnetite.

X-ray fluorescence analyses (by Dr. G. Hornung) show that the garnets are iron-rich and contain only very small amounts of manganese, indicating that they are normal almandine garnets. This is confirmed by optical studies by Slater.

The pelitic schists in which the garnets occur consist of varying proportions of muscovite, chlorite and, less commonly, pale green or yellow to almost black, pleochroic biotite. Quartz is present as scattered, recrystallized grains or as augen centred on garnet nuclei. Recrystallized detrital tourmaline is a minor accessory and magnetite is common in some of the garnetiferous pelites.

(3*a*) *Garnet-Chloritoid Sub-Zone.*—Within the lower parts of the garnet zone some pelites contain chloritoid. Nowhere, however, was a chloritoid-bearing rock seen to contain garnets. The pelites in this sub-zone consist of the usual assemblage of muscovite, chlorite, magnetite, and quartz; no tourmaline has been recognized. The chloritoid occurs as stubby crystals, often with conspicuous simple twinning; the crystals are generally parallel to the bedding plane cleavage and are bent by the strain-slip cleavage.

(4) *Staurolite Zone.*—The appearance of staurolite in the pelites

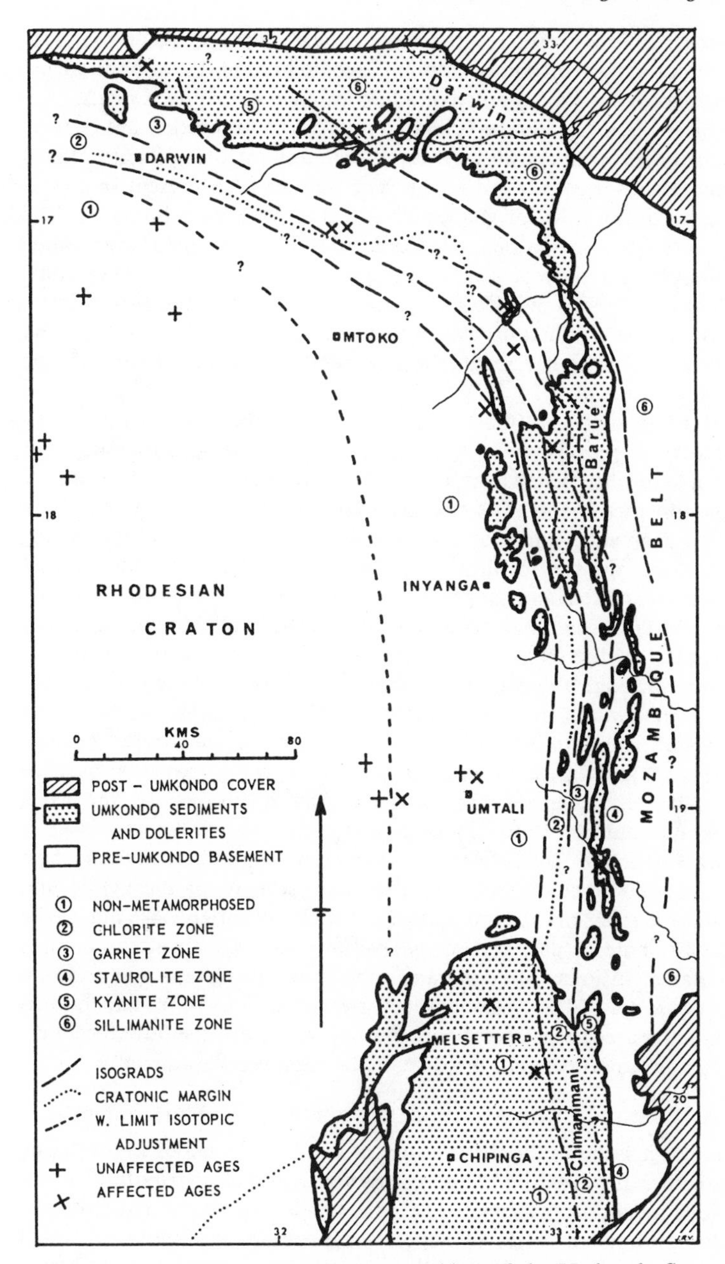

TEXT-FIG. 2.—Progressive regional metamorphism of the Umkondo System across the margin of the Mozambique Belt.

defines the lower limit of the next zone. There is otherwise little difference in mineral composition between schists of this and the garnet zone, except for the presence of black, tabular staurolite crystals.

Muscovite is usually the principal mineral and the flakes are bent by the strain-slip cleavage into crenulations and small folds. Recrystallized quartz grains are scattered in the rock and are also present in strain-shadows in the foliation planes around staurolite crystalloblasts. The latter are frequently rolled, bent and full of oriented inclusions. Some of the crystals show a clear outer zone and a poikilitic inner zone. Accessory minerals include small amounts of chlorite and brown mica, which may be products of retrogression. Small tourmaline crystals, pink garnets, and irregular magnetite fragments are scattered through the rocks.

(5) *Kyanite Zone.*—Kyanite occurs in two distinct associations. First, as scattered porphyroblasts, of about 3 cm. in length, in micaceous garnet-staurolite schists, and secondly, in irregular quartz veins or pods, in which the kyanite crystals attain 20 cm. or more in length, in bladed aggregates intergrown with glassy quartz. The kyanite-bearing pelites are similar to those on the garnet and staurolite grades. Muscovite, quartz and magnetite are the common minerals; tourmaline, green or brown biotite, garnet and rare staurolite are all accessory minerals. In the semi-pelites and psammitic rocks kyanite is usually absent, although the massive muscovite quartzites of the eastern Chimanimani contain small amounts of kyanite. Dr. J. W. Wiles reports (personal communication) the development of muscovite pseudomorphing andalusite in pegmatite, and kyanite, possibly pseudomorphing andalusite, in pelitic rocks just north-west of the main Umkondo outcrop shown in Text-fig. 1.

(6) *Sillimanite Zone.*—As in the case of kyanite, sillimanite also occurs both in quartz veins and as a constituent of the pelitic and psammitic rocks of the Umkondo System. In the latter rocks muscovite is subordinate to deep-red or pale-yellow mica, which often contains acicular inclusions. Ilmenite and sphene are also present in addition to magnetite. Clear, pink garnets, subordinate kyanite, and fibrous sillimanite are the principal metamorphic minerals, the latter accounting for up to 30 per cent of some of the specimens examined.

III. Discussion of Age and Distribution of Metamorphism

Text-fig. 1 shows the distribution of the index minerals in the Barué highlands, and Text-fig. 2 represents the regional distribution of the various metamorphic zones. There is progression outwards from the craton through a series of zones, the position and width of which are probably influenced not only by temperature-pressure conditions, but also by the inclination of the isograd surfaces and the chemical composition of the rocks. The presence of chloritoid may be linked with compositional factors indicated by the absence of a biotite zone and the sparse development of the mineral, even in the garnet zone and higher metamorphic grades.

It is of interest that the isograds follow, with remarkable regularity, the craton-orogen boundary, as delineated from structural considerations (Johnson and Vail, 1965). It is also significant that the isograds in the northerly-trending Mozambique belt can be joined to those in the westerly-trending Zambesi belt, in the north, suggesting that the metamorphism of the two belts was continuous and coeval.

Apart from the five isograds, an additional line can be drawn to mark a western limit of " intermediate isotopic age measurements ". East of this line seventeen determinations, taken from Holmes and Cahen (1957) and Vail (1965) have yielded, with only one exception, radiometric ages lower than might have been expected from the nature of the rocks concerned. Seven of these lie west of the chlorite zone, a discrepancy that might perhaps be related to the effects of the regional metamorphism. If this is indeed the case it may be suggested that isotopic adjustment can take place in areas up to about 50 km. from the assumed mineralogical limit of regional metamorphism. However, the width of such a zone is dependent upon the analytical methods used, rock textures, and the minerals analyzed.

The geological age of the metamorphism is later than the intrusion of the post-Umkondo dolerites and prior to the deposition of the Karroo System. Radiometric (K-Ar, Rb-Sr, and U-Th-Pb) analyses give a minimum age of about 450 m.y. for the metamorphism although ages in the order of 500 and 600 m.y. have also been recorded (see, for example, Snelling *et al.*, 1964). These ages are in accordance with ages delimited for similar tectono-thermal events in Mozambique, north-west Southern Rhodesia, Katanga, and elsewhere (Cahen, 1961).

The pattern of metamorphism within the western part of the Mozambique belt, as outlined above, is based on an initial survey, and further study is called for. In particular, the geochemistry of the pelites requires investigation, having regard to the identification of a chloritoid sub-zone, the absence of a biotite zone and the limited width of the staurolite zone. There is an opportunity to study the response of basement granitic rocks to progressive metamorphism as identified on the basis of the response in pelitic rocks.

Finally, an attempt might be made to establish the extent of migration of radiogenic isotopes between various minerals in an ancient terrain adjacent to a belt of regional metamorphism.

IV. Acknowledgments

The author is grateful to the Director of the Research Institute of African Geology, Professor W. Q. Kennedy, and to colleagues, especially Dr. R. L. Johnson, Dr. D. Slater, and Mr. C. J. Talbot for critical discussion and freely supplying field information. Mr. J. R. F. Araujo of the Mozambique Geological Survey and Mr. I. Ralston of Anglo American Prospecting (Rhodesia) also generously provided information about mineral occurrences. Dr. G. Hornung kindly undertook the X-ray fluorescence analyses.

V. REFERENCES

BARROW, G., 1912. On the geology of lower Dee-side and the Southern Highland border. *Proc. Geol. Ass., Lond.*, **23,** 268–284.

BOND, G., 1952. Structures and metamorphism in the Chimanimani Mountains, Southern Rhodesia. *Trans. geol. Soc. S. Africa*, **54,** for 1951, 69–83.

BORGES, A., and A. V. PINTO COELHO, 1957. Primeiro reconhecimento petrografico du circonscricão du Barué. *Surv. Geol. e Minas, Mozambique*, Bol. 21.

CAHEN, L., 1961. Review of geochronological knowledge in middle and northern Africa. *Ann. N.Y. Acad. Sci.*, **91,** 159–594.

FRANCIS, G. H., 1956. Facies boundaries in pelites at the middle grades of regional metamorphism. *Geol. Mag.*, **93,** 5, 353–368.

HOLMES, A., 1951. The sequence of pre-Cambrian orogenic belts in south and central Africa. *XVIII Intern. Geol. Congr.*, **14,** 254–269.

HOLMES, A., and L. CAHEN, 1957. Géochronologie africain. *Mem. Acad. roy. Sci. Col., Sci. nat. med.*, **8,** 51, 169 pp.

JOHNSON, R. L., D. SLATER, and J. R. VAIL, 1963. Progress report on work on the Manica Belt and adjacent areas of Central Africa. *7th Ann. Report Res. Inst. African Geol., Univ. Leeds*, 49–50.

JOHNSON, R. L., and J. R. VAIL (1965). The junction between the Mozambique and Zambesi Orogenic Belts; north-east Southern Rhodesia. *Geol. Mag.* **102** (6), 489–495.

KENNEDY, W. Q., 1949. Zones of progressive regional metamorphism in the Moine Schists of the Western Highlands of Scotland. *Geol. Mag.*, **86,** (1), 43–56.

SNELLING, N. J., E. HAMILTON, D. REX, G. HORNUNG, R. L. JOHNSON, D. SLATER, J. R. VAIL, 1964. Age determinations from the Zambesi and Mozambique Orogenic belts, central Africa. *Nature, Lond.*, **201,** 4918, 463–464.

SWIFT, W. H., 1961. An outline of the geology of Southern Rhodesia. *Geol. Surv. S. Rhodesia, Bull.* 50.

—— 1962. The geology of the Middle Sabi Valley. *Geol. Surv. S. Rhodesia, Bull.* 52.

TEALE, E. O., 1924. The geology of Portuguese East Africa between the Zambesi and Sabi Rivers. *Trans. geol. Soc. S. Africa*, **26,** for 1923, 103–129.

TILLEY, C. E., 1925. A preliminary survey of metamorphic zones in the southern Highlands of Scotland. *Quart. J. geol. Soc., Lond.*, **81,** 100–110.

TYNDALE-BISCOE, R. M., 1957. The geology of a portion of the Inyanga District. *Geol. Surv. S. Rhodesia, Short Report* 37.

VAIL, J. R., 1965. An outline of the geochronology of the late Precambrian formations of eastern central Africa. *Proc. Roy. Soc.*, A, **284,** 354–369.

WATSON, R. L. (in press). The geology of the Cashel, Melsetter and Chipinga Districts. *Geol. Surv. S. Rhodesia, Bull.* 60.

RESEARCH INSTITUTE OF AFRICAN GEOLOGY,
THE UNIVERSITY OF LEEDS,
ENGLAND.

Comments on the Article

The *title* might be faulted for length. However, it so precisely pinpoints both a topic and an area that most editors would not revise it for brevity. This is particularly true of British journals, such as *Geological Magazine,* which "edit" articles less heavily than do their North American counterparts.

The *by-line* is uniquely British. The author's name and not his job title or employer is important. Should the reader be interested, however, he may find the information in the end note.

The *abstract* is informative; it discusses the investigation and not the article.

The *introduction* to a geological article, like the introduction to a philosophical article, is preoccupied with defining precise areas. This definition is achieved in the article above by references to men and to maps and by a review of the literature. The result is a lengthy introduction, but it furnishes the reader with factual data to which he may refer while reading the rest of the article. The introduction reprinted above is well handled; it contains only information essential to an understanding of the findings discussed.

The remainder of the article is divided into two parts. First the zones and then the age and distribution of metamorphism are discussed. Five zones are introduced with numbered subheadings. Within two of these zones, numbered sub-subheadings are used. These headings serve as a table of contents for the section, and they permit the author to omit transitional phrases, sentences, and paragraphs that might otherwise be required.

The text ends with suggestions as to the direction future work might take. This is fair comment, for the author of a professional article usually knows more about the topic than his reader and is thus qualified to make recommendations.

The *acknowledgements* are for specific services.

The *references* constitute an alphabetized bibliography arranged according to the practice recommended by the American Chemical Society.

Journals of Government and Industry

Chemical and physical science articles written for the journals of agencies other than the professional societies usually follow practice recommended by the Government Printing Office *Style Manual,* the University of Chicago *Style Manual,* or the manual of a specific agency such as the

National Aeronautics and Space Administration. Recommended practices are similar. The excerpts reprinted by permission below illustrate the minor variations in referencing recommended by the Government Printing Office and the National Aeronautics and Space Administration.

Footnotes and references

(See also Reference marks and footnotes, p. 160.)

180. Figures are used for footnote references, beginning with 1 in each table, but if figures might lead to ambiguity (for example, in connection with a chemical formula), asterisks, daggers, or italic superior letters, etc., may be used.

(*a*) When an item carries several reference marks, the superior-figure reference precedes an asterisk, dagger, or similar character used for reference. These, in the same sequence, precede mathematical signs. A thin space is not used to bear off an asterisk, dagger, or similar character.

181. If a reference is repeated on another page it should carry the original footnote; but, to prevent repetition, especially of a long note, it may carry instead, as a cross-reference, the words "See footnote 1 [*or* 2, 3, etc.], p. —."

(*a*) Footnote references are repeated in box heads or in continued lines over tables unless special orders are given not to do so.

182. References to footnotes are numbered consecutively across the page from left to right and across both pages in a parallel table. Footnotes to a parallel table begin on the even page unless there are no references on that page. (For sample, see par. 209, p. 134.)

183. Footnote references are placed at the right in reading columns and date columns and at the left in figure columns (also at the left of such words as *None* in figure columns) and in symbol columns and are borne off. If a date column is the last column, however, the references are placed at the left. (See also par. 158, p. 127.) Two footnote references occurring together are separated by a space, not a comma. (For sample, see par. 209, p. 134.)

184. In a figure column or date column a footnote reference standing alone is set in parentheses and centered; in a reading column it is set at the left in parentheses and is followed by leaders, but in the last column by a period and quads, as if it were a word.

185. The numbered footnotes are placed immediately beneath the table. Should it be requested that a sign or letter reference in the heading to a table be followed, it is not changed to become the first numbered reference mark, and the footnote to it precedes all other footnotes. If the table runs over more than one page, the appropriate footnotes go with each page.

(*a*) If for better make-up or other reason all footnotes are placed at end of a table making more than one page, it is necessary to supply at bottom of each page "See footnotes at end of table, p. —."

186. If the footnotes to both table and text fall together at the bottom of a page, the footnotes to the table are placed above the footnotes to the text, and the two groups are separated by a 50-point rule flush on left. If there are footnotes to the text and none to the table, the 50-point rule is omitted.

(*a*) Footnotes to tables in rules that are centered are set full measure; footnotes to tables that are cut in are set in the same measure as the tables.

187. Footnotes are set as paragraphs, but two or more short footnotes may be combined by the maker-up in one line, with the blank spaces equalized, provided the spaces are not less than 2 ems. In a series of short footnotes the reference numbers are alined on the right.

188. Footnotes in measures 30 picas or wider will be set half measure and doubled up.

189. The footnotes and notes referring to a table are set solid if the table is solid and leaded if the table is leaded.

190. Footnotes and notes referring to tables are usually set in type 2 points smaller than the table but not smaller than 6-point.

191. Footnotes to tables follow tabular style in the use of abbreviations, figures, etc.

192. In footnotes numbers are expressed in figures, even at the beginning of a note or sentence. Fractions standing alone will be spelled at the beginning of footnotes.

193. If a footnote consists entirely or partly of a table, the footnote table is indented 3 ems on left. It should always be preceded by introductory matter carrying the reference number; if necessary copy preparer should add an introductory line, such as "[1] See the following table:"

Some typical reference forms used in NASA papers are as follows:

Books. —

One edition:	Dodge, Russell A.; and Thompson, Milton J.: Fluid Mechanics. McGraw-Hill Book Co., Inc., 1937.
Revised edition:	Boyd, James E.: Strength of Materials. Fourth ed., McGraw-Hill Book Co., Inc., 1935.
One volume of a series:	Fuller, Charles E.; and Johnston, William A.: Applied Mechanics. Vol. II. John Wiley & Sons, Inc., 1919.
More than one volume:	Chapman, Sydney; and Bartels, Julius: Geomagnetism. Clarendon Press (Oxford), 1940. Vol. I — Geomagnetic and Related Phenomena. Vol. II — Analysis of the Data, and Physical Theories.

Foreign book:	Flügge, W.: Statik und Dynamik der Schalen. Julius Springer (Berlin), 1934.
Foreign book distributed by U.S. publisher:	Mitra, S. K.: The Upper Atmosphere. Second ed., The Asiatic Soc. (Calcutta, India), 1952. (Available from Hafner Pub. Co., New York.)
Translation:	Jost, Wilhelm (Huber O. Croft, trans.): Explosion and Combustion Processes in Gases. McGraw-Hill Book Co., Inc., 1945.
Edited book:	Eshbach, Ovid W., ed.: Handbook of Engineering Fundamentals. Vol. I. John Wiley & Sons, Inc., 1936.
One section of an edited collection:	Betz, A.: Applied Airfoil Theory. Airfoils or Wings of Finite Span. Vol. IV of Aerodynamic Theory, div. J, ch. III, sec. 5, W. F. Durand, ed., Julius Springer (Berlin), 1935, pp. 56-62.
Book compiled by a staff:	Staff of Battelle Memorial Institute: Prevention of the Failure of Metals Under Repeated Stress. John Wiley & Sons, Inc., 1941.
Book of anonymous authorship:	Anon.: SAE Handbook. Soc. Automotive Engrs., 1949.

Periodicals.—

Foreign:	Gebelein, H.: Über die Integralgleichung der Prandtlschen Tragflügeltheorie. Ingr.-Arch., Bd. VII, Heft 5, Oct. 1936, pp. 297-325.
Foreign in non-Latin alphabet (use transliteration from GPO manual or translated title):	Author (use transliteration): Title in English (or transliteration). Bull. Acad. Sci. USSR (or Izv. Akad. Nauk SSSR), no. , year, pp.
American:	Evans, Thomas H.: Tables of Moments and Deflections for a Rectangular Plate Fixed on All Edges and Carrying a Uniformly Distributed Load. J. Appl. Mech., vol. 6, no. 1, Mar. 1939, pp. A-7 — A-10.
	Albini, Frank A.; and Jahn, Robert G.: Reflection and Transmission of Electromagnetic Waves at Electron Density Gradients. J. Appl. Phys., vol. 32, no. 1, Jan. 1961, pp. 75-82.
Paper with discussion in same issue:	Beitler, S. R.: The Effect of Pulsations on Orifice Meters. Trans. ASME, vol. 61, no. 4, May 1939, pp. 309-312; Discussion, pp. 312-314.

Paper with discussion in different issue:	Goland, Martin; and Luke, Y. L.: The Flutter of a Uniform Wing With Tip Weights. J. Appl. Mech., vol. 15, no. 1, Mar. 1948, pp. 13-20. (Discussion by R. H. Scanlan, J. Appl. Mech., vol. 15, no. 4, Dec. 1948, pp. 387-388.)

NACA publications. —

Report:	Reissner, Eric: On the Theory of Oscillating Airfoils of Finite Span in Subsonic Compressible Flow. NACA Rept. 1002, 1950. (Supersedes NACA TN 1953.)
Technical Note:	Neurath, Peter W.; and Koehler, J. S.: Creep of Lead at Various Temperatures. NACA TN 2322, 1951.
Technical Memorandum:	Ringleb, F.: Some Aerodynamic Relations for an Airfoil in Oblique Flow. NACA TM 1158, 1947.
Research Memorandum:	Wier, John E.; Pons, Dorothy C.; and Axilrod, Benjamin M.: Effects of Molding Conditions on Some Physical Properties of Glass-Fabric Unsaturated-Polyester Laminates. NACA RM 50J19, 1950.

NASA publications. —

Technical Note:	Mayo, Alton P.; Hamer, Harold A.; and Hannah, Margery E.: Equations for Determining Vehicle Position in Earth-Moon Space From Simultaneous Onboard Optical Measurements. NASA TN D-1604, 1963.
Technical Report:	Swift, Calvin T.; and Evans, John S.: Generalized Treatment of Plane Electromagnetic Waves Passing Through an Isotropic Inhomogeneous Plasma Slab at Arbitrary Angles of Incidence. NASA TR R-172, 1963.
Technical Report superseding another publication:	Roberts, Leonard: Mass Transfer Cooling Near the Stagnation Point. NASA TR R-8, 1959. (Supersedes NACA TN 4391.)
Technical Memorandum:	Hastings, Earl C., Jr., compiler: The Explorer XVI Micrometeoroid Satellite. Supplement II, Preliminary Results for the Period March 3, 1963, Through May 26, 1963. NASA TM X-899, 1963.
Special Publications:	
Single report:	Martz, C. William: Tables of the Complex Fresnel Integral. NASA SP-3010, 1964.
Proceedings:	Huber, Paul W.; and Nelson, Clifford H.: Plasma Frequency and Radio Attenuation. Proceedings of the NASA-University Conference on the Science and Technology of Space Exploration, Vol. 2, NASA SP-11, 1962, pp. 347-360. (Also available as NASA SP-25.)

Contractor Report: Price, Harold E.; Smith, Ewart E.; and Gartner, Walter B.: A Study of Pilot Acceptance Factors in the Development of All-Weather Landing Systems. NASA CR-34, 1964.

Technical Translation: Vlasov, V. Z.: The Theory of Momentless Shells of Revolution. NASA TT F-6, 1960.

Technical Translation published by outside source: Mushtari, Kh. M.; and Galimov, K. Z.: Non-Linear Theory of Thin Elastic Shells. NASA TT F-62, The Israel Program for Sci. Transl., 1961. (Available from OTS, U.S. Dept. Com.)

NASA Paper with appendix by different author: James, Robert L., Jr. (With appendix B by Norman L. Crabill): A Three-Dimensional Trajectory Simulation Using Six Degrees of Freedom With Arbitrary Wind. NASA TN D-641, 1961.

NASA paper presented at AGARD meeting and not published as AGARDograph, etc.: Campbell, John P.: Techniques for Testing Models of VTOL and STOL Airplanes. Presented to Wind Tunnel and Model Testing Panel of AGARD (Brussels, Belgium), Aug. 27-31, 1956.

Government publication.—

British R. & M.: Squire, H. B.; and Trouncer, J.: Round Jets in a General Stream. R. & M. No. 1974, British A.R.C., 1944.

University and Industry Reports done under government contract: Eckert, E. R. G.; Schneider, P. J.; Hayday, A. A.; and Larson, R. M.: Mass Transfer Cooling of a Laminar Boundary Layer by Injection of a Lightweight Foreign Gas. Tech. Note 14 (AFOSR TN 57-323, ASTIA No. AD 132395), Inst. Technol., Mech. Eng. Dept., Univ. of Minnesota, June 1957.

Note: Use Contract number within parentheses if no agency number given. In some reports DDC may have replaced ASTIA. Where ASTIA No. is stamped on cover only, insert parenthetical statement at end: (Available from ASTIA (or DDC) as .)

School publications.—

Bulletin: Fried, Bernard; and Weller, Royal: Photoelastic Analysis of Two- and Three-Dimensional Stress Systems. Bull. No. 106, Eng. Expt. Sta., Ohio State Univ. Studies, Eng. Ser., vol. IX, no. 4, July 1940.

Foreign school report: Tanaka, Keikiti: Air Flow Through Suction Valve of Conical Seat. Part I. Experimental Research. Rept. No. 50 (vol. IV, no. 9), Aeron. Res. Inst., Tokyo Imperial Univ., Oct. 1929, pp. 259-360.

Thesis for a degree:	Krebs, Charles V.: Determination of Stress Concentration Factors for Hyperbolically Notched Tension Members. M. S. Thesis, Univ. of Notre Dame, 1950.
Commercial publication. —	
Industrial report:	Howell, F. M.: Tensile Properties of XB18S-T at Elevated Temperatures. Rep. 9-45-2, Aluminum Res. Lab., Aluminum Co. of Am., Mar. 23, 1945.
Papers presented at meetings. —	
With no preprint number:	Jensen, Duane C.; and Whitaker, W. A.: Spheric Harmonic Analysis of the Geomagnetic Field. Paper presented at 41st Annual Meeting, Am. Geophys. Union (Washington, D.C.), Apr. 1960.
With preprint number:	Author: Title of paper. Preprint ____, Am. Inst. Aeron. and Astronaut., month and year of meeting. (If only number appears, insert preprint within brackets before number.)

D. HUMANITIES JOURNALS

More than thirty university presses and almost all the journals of language and literature follow the usage recommended in the Modern Language Association *Style Sheet*. Because articles in the fields of language and literature usually contain references and citations, the documentation sections of the *Style Sheet* are complex but all-inclusive. The section upon bibliographic usage has been reprinted, by permission, in Chapter 4 on p. 133. The sections upon footnotes and other forms of documentation are reprinted by permission below.

Documentation

18 **In general,** citation of sources for statements of fact or opinion, or for quoted matter, should be kept as concise as the demands of clarity and complete accuracy permit. If the reference is brief, insert it, within parentheses, in the text itself (see Sec. 13*f*); if it is lengthy, put it in a note. Let the test be whether or not it interferes seriously with ease in reading, remembering that the

[9] The problems of capitalization of titles in foreign languages are too complicated to be treated adequately here. Many journals follow the Univ. of Chicago Press's *Manual of Style*, 11th ed. (Chicago, 1949).

[10] *Ital.* accents all vowel caps in Italian. *RPh.* and *NRFH* accent all vowel caps in Spanish; *Hisp.* and *HR* do not.

footnote number, which teases the reader to look at the bottom of the page,[11] may be more of an interruption than such a simple reference in your text as (II, 241) or (page 72). Your first, full reference to a work from which you intend to quote a number of times should usually be in a note, but you may say there that subsequent references to this edition will appear in your text. Avoid large numbers of very short notes. Except for glosses, definitions, and incomplete quotations introduced by three periods, the first word in a footnote should always be capitalized. Footnotes normally end with a period.

19 **Footnote logic.** The conventions of documentation are largely means to an end—enabling the reader to check you with ease—and any practice which ignores this end may result in pedantry. For example, it is rarely necessary to give references for proverbs or familiar quotations (e.g., "stilus virum arguit"), or to give line references for short poems (e.g., sonnets), or to spell out the full names of familiar authors, or to give page references to works alphabetically arranged.[12] To do such things—and to omit *dates* of works or events discussed—is to forget the reader and think only of the machinery of scholarship. Information given in your text need not be repeated in a footnote; hence many notes can easily be shortened by taking the trouble to give *complete* titles or dates or names of authors in the text itself. Successive quotations in one paragraph may usually be documented in a single note. In references to classics of which many editions are available and the nature of which makes line citations impractical, it is often helpful to give the reader more information than the page number of the edition used, e.g., "p. 271 (Bk. IV, Ch. ii)."

20 **Footnote numbers.** Footnotes should be numbered consecutively, starting from 1, throughout an article or review or chapter of a book unless a special section, such as an annotated text or numerical tables, requires a separate series. Never number by pages, for this necessitates renumbering when the work is set in type. Do not use asterisks or other symbols; use Arabic numerals, and type them exactly as they are to appear in print, without such embellishments (which would have to be deleted by your editor) as periods, parentheses, or slashes. Footnote numbers are "superior figures"; in your text type them slightly above the line, always *after* the punctuation, if any,[13] and always after the quotation—not after the author's name or the introductory verb or the colon preceding quoted matter. As for the footnote itself, indent the first line several spaces and, again, type the number without punctuation slightly above the line. The footnote numbers as well as the references themselves should always be verified before the manuscript is submitted anywhere.

[11] Like this. And suppose you had found only "Ibid."

[12] An adequate reference: Sidney Lee in *DNB* s.v. "Wither, George." Since this is a familiar reference work, alphabetically arranged, a full reference to *The Dictionary of National Biography*, ed. Sir Leslie Stephen and Sir Sidney Lee (London: Oxford Univ. Press, 1937–38), XXI, 730–739, would be not only pedantic but possibly confusing: this volume originally had a different number and it has been several times reprinted.

[13] In German text (e.g., in *GR*) the footnote number may precede the punctuation.

21 First references: published books.[14] Use the following order, subject to abridgment by omission of unnecessary items:

Exceptions: Many presses and some journals use considerably less documentation in footnotes when a "bibliography" is included at the end. A number of scientific journals simply insert in the text reference numbers to a list at the end.

a. Author's or authors' names in normal order, not as though they were being alphabetized, followed by a comma. By always giving names in the fullest form known to you, or at least the most usual form,[15] you may save your reader many minutes of searching in a library catalogue, e.g., for "H. M. Jones."

b. Title of the chapter or part of the book cited, enclosed in quotation marks (not underlined), followed by a comma inside the final quotes. This detail is rarely necessary except in references to articles in collections, Festschriften, etc.

c. Title of the work, underlined, followed by a comma unless the next detail is enclosed in parentheses. Some abbreviation of the title is permissible in cases of books with unusually long titles; but the first few words should always be cited intact, and any later omissions *within* the portion cited should be indicated by three periods (. . .). Always take the title from the title page, not from the cover or running title. If there is a subtitle, underline it as well and, if necessary, supply appropriate punctuation (usually a colon). If there are typographical peculiarities in the title (e.g., an italicized name of a play), you may normalize them, using quotation marks or roman for something there italicized.

d. Editor's or translator's name in normal order, preceded by "ed." or "trans." (without parentheses), followed by a comma unless the next detail is enclosed in parentheses. If the editor's or translator's work rather than the text is under discussion, give his name *first* in your reference (followed by a comma, followed by "ed." or "trans." without punctuation) and the author's name *after* the title, with a comma and "by."

e. Edition used, whenever the edition is not the first, in Arabic numerals (e.g., "4th ed."), followed by a comma unless the next detail is enclosed in parentheses. Unless you are concerned with your author's changes of opinion, or with differences in text, you will of course have used the latest *revised* edition or will inform your reader of your inability to do so.

f. The series (if unnamed on the title page), not underlined and not in quotation marks, followed by a comma, followed by the number of this work in the series (e.g., "VII" or "Vol. VII" or "No. 7"), followed by a comma unless the next detail is enclosed in parentheses. If, however, your reference is to part of a

[14] These instructions apply, of course, to notes, not necessarily to the full lists of references ("bibliographies") given at the end of some articles and many books. Moreover, the full first reference is very different from the elaborate and technical *bibliographical description;* for a thorough analysis of the problems involved in the latter, see Fredson Bowers' *Principles of Bibliographical Description* (Princeton, 1949).

[15] Common sense will have to guide the author in applying this rule; consider: Thomas Stearns Eliot, Herbert George Wells, Ulrich von Wilamowitz-Moellendorff. *AHR* requires that the first name always be given in full. Some editors believe that square brackets should be used to indicate parts of a name supplied, i.e., not found in the work cited.

collection or section of works named on the title page, the more general title is also underlined and may be introduced by "in."

g. The number of volumes with this particular title, if more than one (e.g., "3 vols.") and if the information is pertinent. It is usually not pertinent when your reference is to a specific passage rather than to the book as a whole.

h. Place(s) and date(s) of publication, within parentheses, the place followed by a comma; but if the publisher's or bookseller's name is also supplied, it follows the place of publication, preceded by a *colon,* and followed by a comma. Except in articles with a bibliographical slant or purpose, or except when acknowledgment for permission to quote must be made, it is usually pointless to add the name of the publisher, for this information rarely aids in identification of the book.[16] Some scholars, however, regularly include the information for all works still in the copyright period (i.e., published within the last 56 years), assuming a legitimate interest by the reader. In listing places avoid ambiguity (e.g., Cambridge—Mass. or Eng.) or vagueness.

i. Volume number, if one of two or more, in capital Roman numerals, preceded and followed by a comma, unless it is necessary to give the date of a single volume (in parentheses, followed by a comma).[17] Use the volume number alone (without "Vol.") if the page number follows, e.g., "III, 248–251."

j. Page number(s) in Arabic numerals (unless the original has small Roman numerals), preceded by a comma, followed by a period unless an additional reference is required, e.g., "p. 47, n. 3." The numerals are preceded by "p." or "pp." only for works of a single volume.

Sample Footnotes

[1] Archer Taylor, *Problems in German Literary History of the Fifteenth and Sixteenth Centuries* (New York, 1939), p. 213.

[The simplest form of reference; the publisher (the MLA) might have been reported also.]

[1] *Problems in German Literary History of the Fifteenth and Sixteenth Centuries* (New York, 1939), p. 213.

[See above; use this form if the author's full name has been given in the text.]

[1] (New York, 1939), p. 213.

[See above; use this form if the work's title and the author's full name have been given in the text.]

[2] René Wellek and Austin Warren, *Theory of Literature* (New York, 1949), pp. 289–290. Quoted by permission of the publishers, Harcourt, Brace and Co.

[A form of acknowledgment in articles. In books such acknowledgments are usually made in a preface or otherwise collected in one place.]

[16] But see note 7, above, especially Art. 2. The attitudes of presses and journals necessarily differ on this point.

[17] This necessity rarely arises. It may do so, however, when a work appears over a very long period of time or when a single volume has been revised after the completed publication of the work of which it is a part, e.g., David Masson, *The Life of John Milton* (London, 1859–80), I (rev. ed., 1881), 56. Note incidentally that this 6-volume work has a long title, 18 words, here legitimately shortened.

[3] Taylor, p. 12.

[To add "op. cit." to this clear reference—see note 1, above—would be quite superfluous. A short title (*Problems*) might have been added.]

[4] Gottfried von Strassburg, *The "Tristan and Isolde,"* trans. Edwin H. Zeydel (Princeton, 1948), pp. 201–209 (Notes).

[5] Thomas Mossman, trans. *The Great Commentary*, by Cornelius à Lapide, 4th ed. (London, 1890), p. 84 (on St. Luke).

[6] Joannes Caius, *Of English Dogs*, trans. Abraham Fleming (1576), in *An English Garner*, ed. Edward Arber (London, 1877–83), III, 225–268.

["Of English Dogs" may be either italicized (as separately published) or enclosed in quotation marks (as here part of a volume).]

[7] Ernest A. Baker, "Fanny Burney," *The History of the English Novel*, V (London, 1934), 156.

[Note that the volumes of this work by Baker were published in different years. Cf. the less precise treatment in note 6, above, and the different position of the volume number.]

[8] Baker, IV (1932), 190–193.

[9] Baldwin Maxwell, "Middleton's *The Phoenix*," *Joseph Quincy Adams Memorial Studies*, ed. James G. McManaway et al. (Washington, D. C., 1948), pp. 750–752.

[Since the reference is to an article, not to the collection, which had three editors, the name of one editor (the first listed) suffices. The word "in" might have been added before "*Joseph*"; cf. note 6, above.]

[10] Karl Young, *The Origin and Development of the Story of Troilus and Criseyde*, Chaucer Soc., 2nd Ser., No. 40 (London, 1908 [for 1904]), p. 106.

[11] Ruth Wallerstein, *Richard Crashaw: A Study in Style and Poetic Development*, Univ. of Wis. Stud. in Lang. and Lit., No. 37 (Madison, 1935), p. 52.

[12] Edward G. Cox, *A Reference Guide to the Literature of Travel*, Vol. III: *Great Britain*, Univ. of Wash. Pubs. in Lang. and Lit., XII (Seattle, 1949), p. 33.

[Since a series number is involved and the volume number is made part of the title, the abbreviation "p." is used.]

[13] *The Complete Works of Chaucer*, ed. Fred N. Robinson (Cambridge, Mass., 1933)—hereafter cited as *Works*.

[14] Citations from Chaucer in my text are to *The Complete Works*, ed. Fred N. Robinson (Cambridge, Mass., 1933).

[15] See Hans Koch, *Die lyrische Gestaltung und die Sprachform Stefan Georges* (diss. Bonn, 1929), p. 12.

[16] See the unpubl. diss. (Columbia, 1958) by U. Fuller Schmaltz, "The *Weltschmerz* of Charles Addams," p. 7.

[Cf. note 15. Titles of unpublished works, regardless of length, are not underlined but enclosed in quotation marks.]

[17] Thomas Dekker, *His Thirde Childehood* (London, n.d.), sig. B4^{v}.

[18] See *Barbour's Bruce*, ed. Walter W. Skeat, Scottish Text Soc., O.S. 31–33 (Edinburgh, 1893–95), Bk. IV, lines 99–109.

[Note that with this form of reference the reader can easily find the passage in a different edition.]

[19] *Bruce* IV.111–115.

[The brief title is unambiguous and is as simple as "Ibid."]

[20] Fernando de Herrera, *Relación de la guerra de Chipre*, in *Colección de documentos inéditos para la historia de España* (Madrid, 1852), XXI, 242–382.

[21] Gilbert Chinard, "Montesquieu's Historical Pessimism," in *Studies in the History of Culture: The Disciplines of the Humanities* (Waldo G. Leland Presentation Volume), ed. [Percy W. Long] (Menasha, Wis., 1942), pp. 161–172.

[An unusually elaborate form of reference, identifying the book and editor, whose name is hidden. Cf. note 9, above.]

22 **First references: articles in periodicals.** Use the following order, subject to abridgment by omission of unnecessary items:

a. Author's name in normal order, followed by a comma. Here it is not so important as in the case of books (cf. 21*a*) to give the name in the fullest possible form, but if only initials are given, give them all (and, in typing, leave a space between them).

b. Title in full, enclosed in quotation marks (not underlined), followed by a comma inside the second quotation marks.

c. Name of the periodical, abbreviated in accord with good usage (see Sec. 26 and Sec. 27, below), underlined, followed by a comma.

d. Volume number (without "Vol." preceding)) in capital Roman numerals, followed by a comma unless the next detail is enclosed in parentheses. Volume numbers of newspapers and weekly or monthly magazines may be omitted and the complete date given instead—with commas, not parentheses.

e. Issue number or name (e.g., "Autumn") if the pagination of the issue is separate and if, then, the month of publication is not also given.

f. The year (preceded by the month, if needed; see *e*), enclosed in parentheses, followed by a comma. The year should always be given unless it has been noted in your text, for it tells the reader at once how recent the study is and serves also as a useful check on the volume number in the act of location. If the volume covers more than one year, give only the year of the number involved.

g. Page number(s) in Arabic numerals, without "p." or "pp." preceding, followed by a period unless an additional reference to a footnote is needed.

Sample Footnotes

[22] R. B. McKerrow, "Form and Matter in the Publication of Research," *RES*, XVI (1940), 116–121; reprinted in *PMLA*, LXV (April 1950), 3–8.

[In the latter reference the mention of either month or issue is essential, for this particular issue of *PMLA* has separate pagination. In the first reference the year alone suffices. On the other hand, since the *object* of documentation is to enable the critical reader to check you with ease, it is often a courtesy to mention the month in references to periodicals likely to be in the reader's possession or so recently published that they may still be unbound in libraries.]

[23] Hyder E. Rollins, "A New Holograph Letter of Keats," *K-SJ*, I (1952), 37–39.

[Since this article appears in a comparatively new journal, a reference to *Keats-Shelley Jour.* would have been clearer to many readers.]

[24] W. H. Bond, "The Bibliographical Jungle," *TLS*, Sept. 23, 1949, p. 624.

[25] H. C. Lancaster, "The Decline of French Classical Tragedy," *Bull. Modern Humanities Research Assn.*, No. 20 (March 1949), pp. 9–17.

[For some audiences this might have been abbreviated further: many scholars speak familiarly of the MHRA. Notice that a volume number is not involved here; hence the use of "pp."]

[26] Lancaster, p. 12.

[To add "op. cit." to this clear reference would be quite superfluous; to use "ibid." would save but little space.]

[27] See, e.g., R. S. Crane, "Cleanth Brooks; or, The Bankruptcy of Critical Monism," *MP*, XLV (1948), 227.

[Vol. XLV of *MP* covered two years, but the date of the article is specified.]

[28] Crane, loc. cit.; cf. McKerrow, *RES*, XVI, 117.

[The reference is to the particular passage in Crane's article already noted, and comparison is invited with a passage in McKerrow's article cited in note 22, above.]

[29] W. W. Greg, rev. of Percy Simpson, *Proof-Reading in the Sixteenth, Seventeenth, and Eighteenth Centuries* (London, 1935), *RES*, XIII (1937), 190–205.

[30] William Bridgwater, "Who Writes on the Campus?" *Sat. Rev. of Lit.*, June 17, 1944, pp. 8–10.

[The volume number, XXVII, might have been cited also; see Sec. 22*d*. Note that a comma is not used after a question or exclamation mark ending a title. Note, incidentally, that this particular periodical has since changed its name.]

[31] "The Laureate and His 'Arthuriad' " (anon. rev.), *London Quarterly Review*, XXXIV (April 1870), 154–168.

[32] "Problems of Atomic Energy" (editorial), New York *Times*, March 4, 1951, Sec. 4, p. 8.

[The section number is given because the paper is divided into separately paged sections.]

[33] D[erwent] C[oleridge], "An Essay on the Poetic Character of Percy Bysshe Shelley . . . ," *Metropolitan Quarterly Magazine*, II (1826), 191–203.

[34] Information in a letter to the author from Professor John Harold Wilson of Ohio State University, April 1, 1958.

[35] Morgan Library MS. 819, fol. 17. I wish here to express my thanks to the authorities of the J. Pierpont Morgan Library for permission to consult this MS., and particularly to Dr. Curt Bühler for supplying me with helpful information.

23 **Parenthetical documentation.** References within sentences in notes or text may be handled in any one of three or more different ways. Consider these sentences, for example: (1) Henry Watson Fowler, *A Dictionary of Modern English Usage* (Oxford, 1926), p. 10, makes this point. (2) Henry Watson Fowler and F. G. Fowler—*The King's English* (Oxford, 1906), p. 10—make this point. (3) Herbert W. Horwill (*A Dictionary of Modern American Usage* [Oxford, 1944], p. 10) makes this point. Notice that the first two of these methods avoid the necessity for square brackets within parentheses. (Some journals, e.g., *PMLA*, *HLB*, *MLR*, would simply use commas instead of square brackets in the third example.) When any of these methods involve awkwardness, as when the reference itself is unusually long or complicated, try recasting the sentence to make the reference come at the end, e.g., "This point is clearly made by George B. Ives, *Text, Type and Style: A Compendium of Atlantic Usage* (Boston, 1922), p. 10."

24 **Subsequent references.** Be brief. Be clear. If only one work by a given author is being referred to, if the full reference is readily found in a *recent* note (i.e., not buried in the middle of a long note or lost somewhere in the text), and if no other writer with an identical surname is being cited, your author's surname, never abbreviated, suffices for the reference, e.g., "Taylor, p. 14." Adding "op. cit." to such a note contributes nothing. But in the absence of any of the above-mentioned conditions, enlarge your note accordingly, consistently using intelligible short titles or identifying the author more fully or helping the reader to locate the full reference, e.g., "Taylor, *Problems*, p. 14" or "Taylor, op. cit. (above, note 1), p. 14." Always avoid repeating long titles. Use Latin reference words or abbreviations only when they are perfectly clear and really save space. Never abuse your reader's patience; remember that the note "Ibid." may turn up in page proof as the first footnote on a left-hand page. Make brief subreferences always point to something near at hand.[18]

25 **Plays and long poems,** etc. After your first, full reference specifying the edition you are using, refer to them in documentation by short titles or familiar abbreviations[19] and by main divisions and lines separated by periods (not commas) without spacing. Most such references can be inserted within parentheses in your text immediately after quotations; see Sec. 13*f*, above. Examples: *Romeo* III.ii.83; *Merch.* II.i.17–18; *F.Q.* III.iii.53.3; *Iliad* XI.119; Luke xiv.5 and I Chron. xxv.8. Notice that commas are not used after the titles in these references. If the title has been mentioned in your text, or is clearly implied, it need not, of course, be repeated in the documentation.

[18] COLUMBIA states that "using *op. cit.* for the first citation in a chapter to refer to a work cited in a previous chapter we will not tolerate; we insist, if full bibliographical material is being given in the notes, that it be given at the first occurrence in every chapter. This ruling is one of our few rigid rules."

[19] Learned journals occasionally publish lists of accepted abbreviations; one for the classical field appeared in *AJA*, LIV (July 1950), 269–272. *SQ* recommends the use (in documentation only) of the following abbreviations of titles of Shakespearian plays: *All's W.*, *Antony*, *A.Y.L.*, *Errors*, *Cor.*, *Cym.*, *Ham.*, *1 H.IV*, *Caesar*, *John*, *Lear*, *L.L.L.*, *Macb.*, *Meas.*, *Merch.*, *Wives*, *Dream*, *Much*, *Oth.*, *Per.*, *R.II*, *R.III*, *Romeo*, *Shrew*, *Temp.*, *Tim.*, *Tit.*, *Troi.*, *Twel.*, *T.G.V.*; also *Venus*, *Lucr.*, *Sonn.*, *Lov. Com.*, *Pass. Pil.*, *Phoenix*. Accepted abbreviations of the names of books of the Bible (Gen., Exod., Lev., Num., Deut., etc.) are given in many dictionaries.

The humanistic article reprinted below by permission of *English Language Notes* illustrates MLA documentation procedures. It also shows random idiosyncracies of the *Style Sheet* such as the use of italics for quotations in Latin but not in other languages.

English Language Notes

Volume III June, 1966 Number 4

GOOD AND BAD FRIDAYS AND MAY 3 IN CHAUCER

In telling of the misfortunes of Arcite, Chaucer's Knight comments (KT, 1534-1539),

> Right as the Friday, soothly for to telle,
> Now it shyneth, now it reyneth faste,
> Right so kan geery Venus overcaste
> The hertes of hir folk; right as hir day
> Is gereful, right so chaungeth she array.
> Selde is the Friday al the wowke ylike.

An experienced reader of medieval literature is struck immediately by the conjunction of suggestive words like "Friday," "Venus," and "array." But notes to Chaucer's work are seldom satisfying; Robinson, for instance, cites examples to show that "Friday is an off-day,"[1] and hence special; but oddly enough, he does not make reference to the Nun's Priest's Tale, 3341-3346, in his note to the KT. In the NPT "on a Friday fil al this meschaunce," namely that "Venus," whose "servaunt was this Chauntecleer," let her servant fall prey to the fox on her day. Clearly, Venus in NPT is as tricky as she is in the Knight's Tale.

Robinson notes in his commentary on the Priest's story that the tale is in part a parody of Geoffrey of Vinsauf's lament on the death of Richard I, *"O Veneris lacrimosa dies,"* and that Friday is Venus' day;[2] but again Robinson fails to cross-reference the two tales. More importantly, he misses important points about

[1]*The Works of Geoffrey Chaucer,* ed. F. N. Robinson (2nd. ed.; Cambridge, Mass., 1957), p. 674, note 1539. All citations in our text are to this edition.

[2]Robinson, p. 754, note 3347.

the interrelationships between Chaucerian works which allude to Venus, Friday, and May 3.

He notes of KT, 1462 ff., that "The reason is not certain for the selection of May 3 as a starting point" for the escape of Palamon from prison, but he remarks further, "Curiously enough the same date is given in NPT (VII, 3187 ff.) for the tragic seizure of Chanticleer, and in *Troilus* (ii, 55 ff.) it is the day on which Pandarus suffers from a misfortune (*teene*) in love."[3] The reader of course makes an instant connection between ill luck in love on Friday, the day of fickle Venus, and ill luck in love and in general on May 3. In his discussion of May 3 as an unlucky day in the year, Robinson in passing mentions a comment by Root that is valuable, i.e., that May 3 is the date in the Church calendar for the Invention of the Cross.[4] He points out also a fact noted by Eleanor Hamilton, that May 3 is part of the Maytime festival season,[5] a fact made clear by Chaucer in his comment that Emily had risen to "doon honour to May" (KT, 1047). Again the perceptive reader notes that there ought to be a connection between Friday, May 3, Venus, the May festival season, and the Invention of the Cross, and that Chaucer indeed intended such a connection.

A cursory examination of some of Chaucer's likely sources reveals the meaning of the dates. As Genesis I: 26-31 makes clear, the sixth day, Friday, is the day on which Adam and the garden were created. According to a tradition used by Dante in *Paradiso,* 26: 139-142, Adam was credited and fell within a six-hour period, from "la prim' ora a quella che seconda / (Come 'l sol muta quadra) l'ora sesta," and the journey of the pilgrim Dante begins on Good Friday. The *Legenda Aurea* notes that *"Adam factus fuit et peccavit in mense Martio, feria sexta*

[3]Robinson, p. 674, note 1462 ff. In a note to NPT, p. 754, note 3190, Robinson discusses the difficulty of the astrological dating in lines 3190 ff., and decides for May 3, cross-referencing KT, I, 1462 and *Tr.* ii, 55.

[4]Jacobus a Voragine, *Legenda Aurea,* ed. Th. Graesse (Leipzig, 1850), p. 229. D. W. Robertson, in "Chaucerian Tragedy," *ELH,* XIX (1952), 19, briefly notes the relevance of the Invention; however, in *A Preface to Chaucer* (Princeton, N. J., 1962), p. 482, he suggests that McCall (note 5, below) has found a more relevant source in Ovid.

[5]Robinson, p. 674, note 1462 ff. The relevance of the May festival season is pointed out by John Halverson, "Aspects of Order in the Knight's Tale," *SP,* LVII (1960), 606-611, in his discussion of popular *ludi* and the combat of winter and summer, and John P. McCall, "Chaucer's May 3," *MLN,* LXXVI (1961), 201-205, who discusses the relevance of the erotic festivals described in Ovid's *Fasti.*

et hora sexta,"[6] a tradition echoed in NPT, 3187 ff., which speaks of March as "the month in which the world bigan," and then goes on (3258 ff.) to allude to "Adam" and his fall from "Paradys." Clearly, the narrative and thematic pattern of the tale is based upon the Adam and Eve story; Chauntecleer's fall even takes place according to tradition, for the time between the end of his argument with Pertelote, in which she misleads him, to the time he is caught, is some time (a six-hour period, no doubt) between dawn and the hot part of the day. And, as lines 3341 ff. make clear, the fall comes on a bad Friday, Venus' day.

The Old Law, Genesis, points out that the Old Adam was made on Friday; the New Law, John 19: 25-30, points out just as clearly that the New Adam, Christ, was crucified on Friday. Tradition, as exemplified in the *Legenda Aurea,* points out that the tree of Adam and the Cross are the same wood,[7] and that *"quia Adam factus fuit et peccavit"* in March, so Christ *"pati voluit in Martio, quia in die, qua fuit annuntiatus, fuit et passus."*[8] Durandus remarks that Christ "a été conçu le vendredi" and "il a été crucifié le vendredi" because Adam was born and fell on Friday.[9] Moreover, Durandus comments, Christ was aged "trente-deux ans et trois mois" or "trente-trois ans et demi,"[10] phrases echoed in the Nun's Priest's Tale, 3190: "thritty dayes and two." Further in his discussion, Durandus comments on Christ as Light, a concept not original with him, but important in NPT in that it recalls the fact that on the fateful Friday Chauntecleer "Caste up his eyen to the brighte sonne," symbol of Him Who died on Good Friday. Such conjunctions of symbols in NPT indicate that Chauntecleer, the Adam-figure, succumbs to a woman because of "al his pryde," the sin of Adam, and in plain sight of the sun repeats the loss of Eden on a day dedicated to Venus and Christ.[11]

The theme of the Nun's Priest's Tale brings together several points in the preceding analysis. May 3, the time of pagan festival, is a time of erotic pleasure and rebirth ritual, a fusion of

[6]*Legenda,* p. 229.
[7]*Ibid.*
[8]*Ibid.*
[9]Guillaume Durand, *Rational ou Manuel des Divins Offices,* trans. C. Barthélemy (Paris, 1854), vol. 3, V:13, p. 218, and vol. 4, VI:77, p. 128.
[10]Durandus, vol. 4, VI:77, p. 128.
[11]For a discussion of the Adam and Eve parallels in NPT, see John Speirs, *Chaucer of the Maker* (2nd. ed.; London, 1960), pp. 185-193.

human impulse and natural forces. The time is therefore not sinful in itself, and is part of God's plan. Thus both lusty Chauntecleer and chaste Emily respond to it, as does skeptical Pandarus in *Troilus,* ii. But to make Friday, May 3, into a festival of praise for Venus is to misplace the emphasis on Love, and to concentrate on the erotic, as do Chauntecleer, Pandarus, and the two lovers of Emily. The story of the Invention of the Cross, available to readers in the Breviary, makes clear the condemnation of Venus. When the Cross was found by Queen Helen on May 3, a church was dedicated to it on the spot where a temple of Venus stood, "overcaste" when the church was built. The Invention of the Cross is thus a concrete reenactment of the replacement of the Old Law by the New, in terms of a Venus-Christ conflict.

We are now able to specify Chaucer's use of Friday and May 3. As the Knight's Tale shows, the right view of Love means a movement away from the courtly-erotic devotion to Venus, through a ritualized expression, the tournament, to a view of Love that sees it as the force behind the amity of man and man, nation and nation, and the marriage vows of man and woman. *Troilus,* of course, is a statement about the wrong use of the will and a false definition of Love, again in the pagan May season. Chauntecleer's story is the repetition of the fall from Paradise, the victim being a servant of Venus and hence subservient to a woman. In direct opposition to the pagan celebration of Venus, who can change her "array," is the Friday celebration of Christ and the Invention of the Cross, and of the New Adam Who overthrew Venus and Who offers a New Garment to all lovers.

GEORGE R. ADAMS, BERNARD S. LEVY

Harpur College, SUNY

Comments on the Article

The *title* of the article seems at first reading to be a glib attempt to attract reader interest. Yet the specificity of "May 3" implies that the article is not written on a popular level. The result is a suggestion of sophistication essential to the level of *English Language Notes.* On such a level the exact terms—or perhaps the jargon terms—of literary research are essential. The title is thus both promise and warning; it is also sufficiently specific for a reference librarian's use.

There is no headnote. The authors' *by-lines* appear as endnotes in which their names and their affiliation are given. The sequence of names implies that Adams is senior author and that he may be reached for reprints at Harpur College. The latter implication is incorrect; George R. Adams is now at Wisconsin State University at Whitewater. Both authors, however, were at Harpur College *when the article was submitted* for publication. Thus the usage is correct. As discussed on p. 96, an author must list his affiliation at the time he submits the article.

The structure of the article is not standardized by formal subheadings such as those required for articles in science journals. The organization, however, follows the concepts of logic discussed in Chapter 3. First, an example of Chaucer's usage is given. Then, a gap in knowledge is pointed out—"But notes to Chaucer's work are seldom satisfying. . . ." Then the solution is introduced—"A cursory examination of some of Chaucer's likely sources reveals the meanings of the dates." This is followed by references to world literature documented even to the line. The final paragraph contains the conclusion. Beginning "We are now able to specify Chaucer's use of Friday and May 3", that paragraph states precisely the conclusions necessary to fill the knowledge gap pointed out in the first paragraph.

The *internal documentation* follows the MLA *Style Sheet.* There are both content notes and reference notes. They are numbered consecutively throughout the article. Significantly, the first footnote identifies the edition of the original source and states that all subsequent citations will be to that edition. This procedure permits the parenthetical identification of lines from that source. For texts whose lines have been fixed, for example, *Old Testament, New Testament,* and *Paradiso,* the primary note is unnecessary. Parenthetical reference is adequate.

E. FINE ARTS JOURNALS

The "Notes for Contributors" of *The Art Bulletin* refers writers to the MLA *Style Sheet.* This, in effect, indicates that the divisions of art, art criticism, and art history tend to follow usage common to the disciplines of history, literature, and language. An example of approved bibliographic practice in these devisions is reprinted on pp. 134–135 by permission of *Architectural History.* In addition to mechanics, however, there are unique problems posed by the visual nature of art. Some of these problems are solved by the "Notes for Contributors" reprinted below by kind permission of the editor of *The Art Bulletin.*

The illustrative article, reprinted by permission of the author and of the editors of *The Art Bulletin,* combines the requirements of the journal and of the MLA *Style Sheet.*

These notes supersede those printed in the March, 1965 issue.

Manuscripts should be submitted to the Editor-in-Chief of *The Art Bulletin.*

The funds of *The Art Bulletin* do not admit of an expenditure of over ten per cent (10%) of the cost of composition for alterations in articles once set up in galley proof. In order that contributors may be spared the expense of exceeding this allowance, they are urged to prepare their manuscripts as nearly as possible in conformity with the following rules. In cases of doubt as to form, contributors are referred to the *MLA Style Sheet,* obtainable from the Modern Language Association, 4 Washington Place, New York, N.Y. 10003.

Articles improperly prepared, even though accepted for publication, may be returned to the author for retyping.

Form of Manuscripts

1. Contributors should retain a carbon copy of their manuscripts, since it is undesirable to return the original with the galley proofs.

   Page and plate proofs are usually not submitted to the author.

2. All manuscripts must be typewritten and double-spaced, on one side of the paper only, on sheets of 8½" x 11" and regular weight, numbered consecutively.

   In order to effect maximum economy in time and cost involved in typesetting, contributors are *urgently requested* to observe the following rules in typing their manuscripts:

TEXT	LEFT MARGIN	RIGHT MARGIN
Pica type	15	74
Elite type	10	81
FOOTNOTES		
Pica type	20	60
Elite type	20	66
REVIEWS		
Pica type	15	72
Elite type	10	78

   No more than 22 lines of copy per page.

3. The name of the institution with which the author is connected should be typed at the end of his contribution; brackets will be used to denote that the author is a student at that institution.

4. Words, phrases, passages, or titles intended to be printed in italics should be underlined in the typescript. This includes titles of works of art, titles of books, poems and periodical publications, and technical terms or phrases in a foreign language; but does not include direct quotations in a foreign language, foreign titles preceding proper names, place names, names of buildings, or words anglicized by usage.

Footnotes

1. Footnote references in the text should be clearly designated by means of superior figures, placed after punctuation.

2. Footnotes should be numbered consecutively and typed *double-spaced* on separate pages subjoining to the article.

3. All references should be verified before the manuscript is submitted for publication. Contributions that are incomplete in this respect will be returned to the author for completion.

4. In the case of certain books frequently cited in our field, titles and bibliographical data should be abbreviated in the first and all succeeding references. A list of abbreviations will be found on the following page.

5. In all references to books cited frequently in the text the shortest intelligible form of author and title should be used. Such abbreviations as *op. cit.* and *loc. cit.* should not be employed. Thus:

   Swindler, *Ancient Painting,* 65.
   Diehl, *Manuel,* I, 482.

   A bibliography consisting of an alphabetical listing by author of those books so cited should be appended to the footnotes. Full bibliographical data should be given.

The form of references to books should be as follows: (1) author's name, preceded by his given name or initials, and followed by a comma; (2) title, italicized, followed by a comma; (3) the edition, where necessary, followed by a comma; (4) place of publication, followed by a comma; (5) date of publication, followed by a comma; (6) reference to volume in Roman numerals without preceding "Vol.," and followed by a comma; (7) page or column number (using "f") without preceding "p.," "pp.," "col.," or "cols."

Mary H. Swindler, Ancient Painting, New Haven, 1929, 60f.

Charles Diehl, Manuel d'art byzantin, 2nd ed., Paris, 1925, II, 324–45.

In the case of periodicals frequently cited in our field, titles should be abbreviated in the first and all succeeding references. A list of abbreviations will be found on the following page.

In all references to articles in periodicals cited frequently in the text the shortest intelligible form of author and title should be used. Such abbreviations as *op. cit.* and *loc. cit.* should not be employed. Thus:

Sauerlander, "Westportale," 21–33.

A bibliography consisting of an alphabetical listing by author of those articles so cited should be appended to the footnotes. Full bibliographical data should be given.

The form of references to periodical literature should be as follows: (1) author's name, preceded by his given name or initials, and followed by a comma; (2) title of article in roman, followed by a comma and enclosed within double quotation marks; (3) full or abbreviated title of periodical in italics, followed by a comma; (4) reference to volume in Arabic numerals without preceding "Vol."; (5) date, followed by a comma; (6) page number (using "f") without preceding "p.," or "pp." Thus:

Willibald Sauerlander, "Die kunstgeschichtliche Stellung der Westportale von Notre-Dame in Paris," *MarbJb*, 17, 1959, 1–55.

Illustrations

1. Good photographs made directly from the work to be reproduced are essential and wherever negatives of these are available, they should be submitted also. Only when the work to be reproduced is lost, destroyed, or completely inaccessible will photographs made from reproductions be considered. In such cases the author should supply a photograph and a full bibliographical reference for the best possible reproduction. All photographs and negatives will be returned. Contributors are urged to consult the Note by Clarence Kennedy in the March, 1961 issue. Offprints are available on request from the Assistant Editor.
2. Drawings should be in India ink on white drawing paper. Unsatisfactory copy, such as photostats, will be returned to the author for redrawing according to these specifications.
3. Each photograph or drawing should be clearly marked with the name of the author and the figure number on the reverse. Marking should be done lightly with a soft pencil. Do not type or write heavily; the marks will show through on the finished plate. If only part of the illustration is required, the area to be reproduced should be outlined lightly in pencil on the reverse side.
4. Permission for the use of Alinari, Anderson, Brogi, and Mannelli photographs will be cleared by *The Art Bulletin*, but contributors are responsible for obtaining permission, whenever necessary, for the reproduction of other photographs.
5. A separate list of illustrations must accompany the manuscript when submitted. Captions should be stated as briefly as possible. For example:
   1. *Enemy herdsman on horseback*. Rome, Vat. MS gr. 549, IX cent. (Courtesy Museum of Art, Rhode Island School of Design)
   2. Roman Sarcophagus, *Death of Meleager* (detail). Paris, Louvre (photo: Anderson)
   3. Castiglione, *Crucifixion*. Genoa, Palazzo Bianco (photo: Frick Art Reference Library)

Book Reviews

Book reviews represent solely the opinions of the reviewers, who have complete freedom, within the limits of scholarly discourse, to set forth their own evaluations of the publications assigned to them for review. In order to ensure maximum usefulness, it is desirable that every review include a factual description of the contents of the work under review, as distinguished from the reviewer's critical estimate of its merits. (*The Art Bulletin* will not consider unsolicited manuscripts for book reviews.)

Book reviews should be prepared in the same style as other contributions to *The Art Bulletin* except that footnotes should be incorporated into the text in parentheses.

Letters to the Editor

Any letter to the Editor which comments on a contribution to *The Art Bulletin* is submitted to the author of the contribution in question in order to permit a letter in reply to be published concurrently if desired. Writers of letters to the Editor are therefore requested to enclose a carbon copy.

Offprints

Authors of Articles and Notes will receive thirty offprints free of charge. Fifteen copies of the whole back section will be sent free of charge to authors of Reviews of Books and Exhibitions, and ten to writers of Letters. Additional offprints of Articles, Notes, and Reviews may be obtained at cost.

Offprints of these Notes for Contributors and the list of Abbreviations are available on request to the Assistant Editor, Lucy Freeman Sandler, Washington Square College, New York University, New York, New York 10003.

NOTES

EL GRECO'S ROMAN CATHOLICISM: A DOCUMENT RECONSIDERED

PÁL KELEMEN

The sole legal document directly referring to El Greco's faith is a power of attorney which he signed a week before his death. Instead of writing out the twenty-three letters his name contains, El Greco affixed as his signature only a short illegible scrawl, thereby reflecting what the text of a preamble to the document informs us to have been his physical condition: "for the gravity of my illness I cannot make nor execute nor enact my testament as is meet for the service of Our Lord and the salvation of my soul. . . ." This statement and the wording of the rest of the preamble (given below) have been taken as evidence for El Greco's belief in the Holy Trinity and the Mother Church of Rome. This author and others, however, have regarded the opening phrases of this power of attorney as a standard text form with which notaries and scribes of that age introduced the individual sections of the document.[1] This belief can now be substantiated for on a recent trip to Latin America I came upon confirmation in two books which prove that the text for such legal documents was rigorously prescribed.

In a vademecum of legal phraseology, published in Mexico in 1605, the scribe was given as a model for writing a will the following introductory text:

"En caso de testamento, la redacción variaba un tanto. La invocación en un testamento habia de ser mas o menos así:

"En el nombre de la Santísima Trinidad, Padre, Hijo y Espíritu Santo, tres personas y un solo Dios. Sepan quienes esta carta vieren como yo. . . . vecino de . . . , estando sano (enfermo), y en mi acuerdo y entendimiento, y creyendo como creo el misterio de la Santísima Trinidad y todo aquello que cree, tiene, y confiesa nuestra madre la Santa Iglesia Romana como todo fiel cristiano le debe tener y creer; y protestando como protesto vivir y morir en esta y por esta catolica fe y creencia, y deseando poner mi anima en carrera de salvación, y tomado para ello por mi abogada a la Virgen Santa María señora nuestra: otorgo que hago mi testamento en la forma y manera siguente. . . ."[2]

The other document to be brought into evidence here is the preamble to the testament of the sculptor, Mateo de Zúñiga, who died in the city of Santiago de Guatemala in 1678. It reads as follows:

"En el nombre de Dios, Amén, todo poderoso y de la Virgen Santa María Madre suya concebida sin mancha de pecado original, Amén. Sepan cuantos esta carta de mi testamento, última y final voluntad vieren, como yo Mateo de Zúñiga, vecino que soy de esta ciudad de Santiago de Guatemala y natural. . . . hijo natural de Juan del Castillo y de doña Francisca de Zúñiga, maestro que soy del arte de escultura, estando como estoy enfermo en cama de la enfermedad que Nuestro Senor Jesucristo ha sido servido darme, mas en mi entero juicio y entendimiento natural creyendo, como fiel y católicamente creo, en el Misterio inefable de la Santísima Trinidad, Padre, Hijo y Espíritu Santo, tres personas distintas y una esencia divina y en todo lo demas que tiene, cree y confiesa nuestra santa madre Iglesia Católica Romana, en cuya fé y creencia me huelgo haber vivido y protesto vivir y morir y deseando poner mi anima en carrera de salvacion . . . hago y ordeno mi testamento y última voluntad en la manera siguiente: . . . ."[3]

The complete text of the preamble to the power of attorney signed by El Greco in 1614 reads as follows:

"In dei nomine amen. Sepan q^tos [quantos] esta carta de poder para testam^to [testamento] bieren como yo dominico Teotocopuli pintor becino desta ciu^d [ciudad] de T^do [Toledo] estando echado en una cama enfermo de enfermedad que dios nuestro señor fue serb^do [serbido] de me dar y en mi buen seso juicio y entendimiento natural teniendo creyendo e confesando como tengo creo y confieso todo aquello que cree y confiesa la santa madre iglesia de roma y en el misterio de la santisima trenidad en cuya fe y crehenzia protesto bibir y morir como bueno fiel y catolico cristiano—digo que por q^to [quanto] por la gravedad de mi enfermedad yo no puedo hazer ni ctogar ni hordenar mi testam^to [testamento] como conbiene a el serbicio de dios nuestro s^r [señor] e salbacion de mi alma. . . ."[4]

By comparing the preambles of the three documents, it becomes evident beyond doubt that the texts used in Mexico, Guatemala, and Toledo all go back to an identical formula. Not only are the strict dogmatic ideas identical but even the sentence "ill of the sickness which the Lord our God has deemed fit to give me. . . ." In the light of what has been presented above, the claims based on the wording of the power of attorney are no longer tenable.

NORFOLK, CONNECTICUT

1. Pál Kelemen, *El Greco Revisited, His Byzantine Heritage*, New York, 1961-1962, p. 102.
2. Nicolás de Irolo Calar, *La Política de Escripturas*, Mexico, 1605, fol. 75.
3. Heinrich Berlin, *Historia de la imaginería colonial en Guatemala*, Guatemala, 1952, pp. 196-197.
4. Francisco de Borja de San Román y Fernández, *El Greco en Toledo*, Madrid, 1910, pp. 185-186.

Comments on the Article

The Kelemen article was written with great brevity to permit printing it as a "Note." The compression, however, results in the attainment and restraint essential to a good article. There are no prolixities or digressions. There is a straight-line progression of logic leading to a QED.

The tone of the article is authoritative, partially because the logic is apparent. Concealing the logic would make the tone authoritarian and arbitrary. Even a writer of Kelemen's stature will offend readers by simply being contentious and opinionated. He must justify his position in logic although he has, in the words of the "Notes for Contributors," "complete freedom, within the limits of scholarly discourse, to set forth their own evaluations. . . ."

The *title* achieves two things; it establishes a narrow topic and it indicates the approach of the article. This last can often be shown by the use of a colon in the title.

The first paragraph is both introduction and abstract. It states the problem, reviews the literature, introduces new evidence, and draws a conclusion.

The text proper contains direct quotations in seventeenth-century Spanish. The level is never compromised; it is assumed that the intended reader can translate classical Spanish.

The last paragraph is both summary and conclusion. It recapitulates the similarity of documents from Mexico, Guatemala, and Toledo. It summarizes that all had a common source. It concludes that a standard form is not valid indication of an artist's religion.

The "Notes for Contributors" of this journal require the listing of "the name of the institution with which the author is connected." Kelemen indicates his free-lance and independent position by giving his address. This is the most effective technique for a writer whose affiliations are many, conflicting, or professionally competitive.

F. MATHEMATICAL JOURNALS

Articles written for the journals of mathematics are often a printer's nightmare. Lengthy equations and complicated formulae (the American Mathematical Society prefers "formulas") can only be printed with accuracy when they are submitted in a standard form. That form has been established by the American Mathematical Society in their publication "A Manual for Authors of Mathematical Papers." Because the "Manual" discusses writing and editing as well as printing problems, it is reprinted below by permission in its entirety.

The illustrative article which is reprinted after the "Manual" presupposes a highly trained reader. Its complexity, however, is in the content and not in the form. The form follows the standards of the society and capitalizes upon a tenet of information theory—a message in an expected pattern will communicate more effectively than one in a random pattern.

Reprinted from
Bulletin of the American Mathematical Society
Volume 68, Number 5, September, 1962

Sixth Printing, 1966

MANUAL FOR AUTHORS OF MATHEMATICAL PAPERS[1]

1. **Introduction.** Our purpose is to advise mathematicians about preparing papers for publication so as to improve the readability and appearance of the printed article, and to eliminate unnecessary delay, trouble, and expense in the printing.

The crucial difficulty is that an uninformed author frequently makes severe demands on the skill of the compositor to achieve trivial effects. Notations convenient for handwriting are often troublesome and costly when printed (e.g. a tilde over several symbols); and some of the advantages of printing are used too little (e.g. boldface). This is a serious problem. The cost of setting a page of mathematics in type varies from $10 to $30 depending on the difficulties posed by the manuscript. The difference between an informed, careful author and his opposite can average $15 per page. An editor is very conscious of this problem. In a marginal case, his judgment of a paper may be swayed by this non-mathematical feature.

The information in this manual should enable authors to adapt to the medium of print, and to make only intelligent use of the compositor's skill. The most important section is headed: Selecting notations.

2. **Organizing a paper (for beginners).** The usual journal article is aimed at experts and near-experts, who are the people most likely to read it. Your viewpoint should be to say quickly what you have done that is good, and why it works. Avoid lengthy summaries of known results, and minimize the preliminaries to the statements of your main results. There are many good ways of organizing a paper which can be learned by studying papers of the better expositors. The following suggestions describe a standard acceptable style.

Choose a title which helps the reader place the paper in the body of mathematics. A useless title: *Concerning some applications of a theorem of J. Doe.* A good title contains several well-known key words, e.g. *Algebraic solutions of linear partial differential equations.* Make the title as informative as possible; but avoid redundancy, and eschew the medieval practice of letting the title serve as an inflated advertisement. A title of more than ten or twelve words is likely to be miscopied, misquoted, distorted, and cursed.

---

[1] This manual was prepared by a committee whose members were Professors J. L. Doob, Leonard Carlitz, F. A. Ficken and George Piranian, with Professor Norman E. Steenrod as chairman.

The first paragraph of the introduction should be comprehensible to any mathematician, and it should pinpoint the location of the subject matter. The main purpose of the introduction is to present a rough statement of the principal results. This should be done as soon as feasible, although it is sometimes well to set the stage with a preliminary paragraph. The remainder of the introduction can discuss the connections with other results.

It is sometimes useful to follow the introduction with a brief section that establishes notation and refers to standard sources for basic concepts and results. Normally this section should be less than a page in length. Some authors weave this information unobtrusively into their introductions, avoiding thereby a dull section.

The next section should contain the statement of one or more principal results. The rule that the statement of a theorem should precede its proof applies equally well to the main results. It is usually poor practice to postpone the statement of a main result until a series of partial results makes its proof a triviality. A reader wants to know the objective toward which he is heading, and the relevance of each section as he reads it. In the case of a major theorem whose proof is long, its statement can be followed by an outline of the proof with references to subsequent sections for proofs of the various parts.

Strive for proofs that are conceptual rather than computational. For an example of the difference, see A Mathematician's Miscellany, by J. E. Littlewood*, in which the contrast between barbaric and civilized proofs is beautifully and amusingly portrayed. To achieve conceptual proofs, it is often helpful for the author to adopt initially the attitude he would take if he could only communicate his mathematics orally (as when walking with a friend). Decide how to state your results with a minimum of symbols, and how to express the ideas of the proof without computations. Then add to this framework the details needed to clinch the results.

Omit any computation which is routine (i.e. does not depend on unexpected tricks). Merely indicate the starting point, describe the procedure, and state the outcome.

It is good research practice to analyze an argument by breaking it into a succession of lemmas, each stated with maximum generality. It is usually bad practice to try to publish such an analysis, since it is likely to be long and uninteresting. The reader wants to see the path—not examine it with a microscope. A part of the argument is worth isolating as a lemma if it is used at least twice later on.

The rudiments of grammar are important. The few lines that you write on the blackboard during an hour's lecture are augmented by spoken commentary, and at the end of the day they are washed away

* Published by Methuen Co. Ltd., London, 1953; see page 30.

by a merciful janitor. Since your published paper will forever speak for you, without benefit of the cleansing sponge, careful attention to sentence structure is worthwhile. Each author must develop a style that suits him; a few general suggestions are nevertheless appropriate.

The barbarism called the dangling participle has recently become more prevalent, but not less loathsome. "Differentiating both sides with respect to x, the equation becomes . . . " is wrong, because "the equation" cannot be the subject that does the differentiation. Write instead "Differentiating both sides with respect to x, we get the equation . . . ," or "Differentiation of both sides with respect to x leads to the equation . . . ."

Although the notion has gained some currency, it is absurd to claim that the informal "we" has no proper place in mathematical exposition. Strict formality is appropriate in the statement of a theorem, and casual chatting should indeed be banished from those parts of a paper which will be printed in italics. But fifteen consecutive pages of formality are altogether foreign to the spirit of the twentieth century, and nearly all authors who try to sustain an impersonal dignified text of such length succeed merely in erecting elaborate monuments to clumsiness.

A sentence of the form "if P, Q" can be understood. However "if P, Q, R, S, T" is not so good, even if it can be deduced from the context that the third comma is the one that serves the role of "then." The reader is looking at your paper to inform himself, and not with a desire for mental calisthenics.

3. **Typesetting.** Most mathematics is set on monotype machines. The compositor works at a keyboard. The pressing of a key results in the casting from a matrix of a single symbol in type which is then placed on a line, next to its predecessor. If the entire job can be done by pressing keys, the happy state of minimum cost is achieved.

Unfortunately the monotype has various limitations. When a place is reached where the machine is inadequate, the compositor inserts a blank and goes on. Subsequently, the blank must be located, removed, and replaced by a hand-cast symbol. The pressing of a key costs a fraction of a penny. Handwork is much more expensive.

One limitation is that the compositor cannot backspace, as on a typewriter, and place one symbol above another on the same line. Thus fractions cause difficulty, also subscripts under superscripts, and any inflections over subscripts.

Another limitation is that the monotype cannot use the matrix of a symbol whose height exceeds that of the standard line, e.g. oversize integral and summation signs, and also large parentheses, braces, and brackets.

The most important limitation is that the machine has only 255 keys. Compare this with the fact that there are about 1000 different symbols which an article in mathematics may use. Prior to composition, each manuscript is examined to determine which 255 symbols are used most frequently; and the matrices for these are placed in the monotype. All other symbols must then be hand-set for each occurrence.

4. Selecting notations. There are numerous ways in which the limitations described above can be avoided without sacrifice of clarity, and in most cases with an increase of clarity. The first is to strive for conceptual proofs as opposed to computational ones (see §2). This will prevent an unnecessary proliferation of special symbols, and it surely makes for easier reading.

Obtain a list of the symbols available on the typewriter to be used in typing the manuscript, and select notations from this list as much as possible. This has the added advantage of minimizing the filling in by hand of blanks in the typed copy.

It is not necessary to represent symbolically each concept which appears. If a concept occurs rarely, it is usually better to refer to it by name. For example, the expression "the upper derivative on the right of f at a" compared with

$$\limsup_{x \to a+} \frac{f(x) - f(a)}{x - a},$$

is just as clear, takes less space, and is very much less trouble to set in type. The purpose of a symbolism is to provide ready reference to objects so as to reveal the structure of an argument in a clear and compact fashion. If the symbols do not contribute to this purpose, use words instead.

It is a common tendency to use distinct symbols for distinct objects throughout the same paper. This is sometimes not necessary. If in §5 it is stated that α represents a complex number, this cancels all previous usages. Consistency of notation is necessary only for the important concepts recurring throughout the paper.

Subsubscripts, sub-to-superscripts, etc., are called indices of the second order. They should be used sparingly and with care. In the first place, they are more difficult to read, especially by those who need stronger glasses. The typist of the manuscript is frequently uncertain where to put them; and the compositor can't be more certain. How is he to know that a typewritten $x^b h$ means x^{b_h}? Also, there are certain combinations which, when properly printed, do not give the intended effect. For example, A-sub-γ-sub-b looks like this: A_{γ_b}; and A-super-b-super-γ looks like A^{b^γ}. At first glance one can be misled until it is recalled that b and γ on the same level have the appearance $b\gamma$. *Third order indices are not available.*

Any large-scale consistent use of second-order indices should be avoided, and it usually can be, to the advantage of all concerned. For example, the monstrosity

$$\sum_{i_1, i_2, \cdots, i_k = 1}^{n} a_{i_1 i_2 \cdots i_k} \Phi^{i_1 i_2 \cdots i_k}$$

can be replaced by

$$\sum_{\alpha \in S(n,k)} a_\alpha \Phi^\alpha,$$

where $S(n, k)$ denotes the set of sequences of length k from the integers $1, \cdots, n$. It is even preferable to write

$$\sum a_\alpha \Phi^\alpha \qquad (\alpha \in S(n, k)).$$

In the period 1900–1930, differential geometers developed and popularized an elaborate notation in which most of the information is communicated through systematic use of indices. The virtue of the notation is that it reduces many calculations to routine inspection and control of indices. Its disadvantages are many. It is expensive to print, hard to see, and prone to typographical errors. Also, it tends to emphasize technical details of computation at the expense of conceptual notions. Some of the more recent writers in the subject have successfully avoided the worst features of the old notation. It would be well if others would study their efforts, and follow suit.

Table I presents a list of alternatives to expensive notations. Any large-scale use of the expensive notations should be avoided. In most cases the trouble is caused by placing one symbol above another. In any displayed formula where much handwork is needed for other reasons, a bit of additional handwork is a minor matter. *Thus the left-hand notations of Examples* 6 *through* 10 *are acceptable in a line which must be handset, but should be avoided in text or in displayed lines not otherwise requiring handsetting.*

Example 5 is an instance of the general rule: Write e^u if u is simple, but write $\exp(u)$ if u is complicated.

In connection with Example 6, $\sqrt{3}$ is preferable to $3^{1/2}$.

Most printers have special matrices for Latin letters with bars or tildes. Thus if one barred letter, say $\overline{C}$, occurs frequently enough, its matrix can be included among the 255 used in the monotype. In contrast, if each barred letter of the alphabet were used just once, then all would be handset.

Bars, tildes, and carets on subscripts or superscripts must be handset. Composite matrices are not available. These should therefore be avoided, or used very rarely.

A bar which covers two or more letters must always be handset. If this occurs in the text, then the space between lines must be increased to accommodate the bar. A few of these would be of no concern; but, in case of a great many, the closure symbol Cl, as in Example 12, is recommended. A tilde or caret covering two or more symbols is too costly to be considered.

Boldface characters can be called for by underscoring the typed letter with a black wiggly line (or tilde). However, boldface is not usually available in subscripts and superscripts.

TABLE I

	Expensive notations	*Alternatives*
1.	$\bar{A}, \bar{b}, \check{\gamma}, \hat{g}, \mathring{\Lambda}$	$A', b'', \gamma^*, g_*, \Lambda^\#$
2.	$\vec{v}$	$\mathbf{v}$ (boldface)
3.	$\overline{\lim}, \underline{\lim}$	lim sup, lim inf
4.	$\underrightarrow{\lim}, \underleftarrow{\lim}$	inj lim, proj lim
5.	$e^{-\frac{x^2+y^2}{a^2}}$	$\exp(-(x^2 + y^2)/a^2)$
6.	$\sqrt{a^2 + b^2}$	$\sqrt{}(a^2 + b^2)$ or $(a^2 + b^2)^{1/2}$
7.	$\frac{7}{8}, \frac{a+b}{c}$	$7/8, (a + b)/c$
8.	$\sum\limits_{i=0}^{n}, \prod\limits_{i=1}^{\infty}$	$\sum_{i=0}^{n}, \prod_{i=1}^{\infty}$
9.	$\overline{A \cap B}$	$\mathrm{Cl}(A \cap B)$ (Cl = closure)
10.	$A \xrightarrow{f} B$	$f: A \rightarrow B$
11.	$\dfrac{\cos \frac{1}{x}}{\sqrt{a + \frac{b}{x}}}$	$\dfrac{\cos(1/x)}{(a + b/x)^{1/2}}$
12.	$\overset{-1}{f}$	f^{-1}
13.	$e_{i_1 i_2 \cdots i_n}$	$e(i_1, i_2, \cdots, i_n)$
14.	$d_{\check{a}}, d_{\bar{c}}$	$d_{a'}, d_{c''}$

5. **Displays, diagrams, figures, and tables.** A short and simple formula should be left in the line of text and not displayed unless it must be numbered for later reference. *A formula whose length is nearly half a line should be displayed*; because, otherwise, it is likely to be broken with part on one line and part on the next, and may thus be difficult to decipher. Compositors are requested, when breaking a formula, to try to do so immediately after an equality sign or one of the other relation symbols. Display any formula which is $1\frac{1}{2}$ inches long or more and has no convenient breaking point. Authors are urged to avoid long paragraphs with many formulas but no displays. The reason for this is explained in the second paragraph of §10.

Some authors number every displayed line of their typescript. The practice is unfortunate for two reasons: If all displayed lines are numbered, the numbers lose their value as clear signals and signposts (they are reduced to the status of section markers); and if the printer finds it necessary to display a few of the formulas that the author had left in the text, the few exceptions stand out as singularities and spoil the author's meticulous routine.

Diagrams of groups and homomorphisms in which the homomorphisms are represented by arrows can be set by the compositor if they are properly arranged. The compositor can readily build a long vertical or horizontal line with an arrowhead at one end; and with considerable difficulty he can piece together a 45° line. But he cannot construct lines at other angles or make curved lines. Since a diagram can usually be arranged in many ways, the author should choose the one which is simplest to set in type. If the diagram cannot be arranged simply, then it must be treated as an illustration or figure.

Illustrations and figures should be carefully drawn in *black* ink (india ink) on strong white paper so that photo-engravings can be made from them (blue ink does not photograph successfully). It is customary to reduce the figure to half its size in the process; and the author should take this into account in making the drawing—especially the lettering.

In matrices and tables, vertical alignment is important; and *it is essential that the manuscript exhibit the desired alignment clearly.* A poorly-typed manuscript can easily mislead the compositor.

It is expensive and wasteful to enclose matrices in long curving lines (oversize parentheses). It is recommended that the matrix be enclosed by a pair of vertical lines, and that a determinant be indicated by prefixing "det"; thus $|a_{ij}|$ denotes the matrix, and det $|a_{ij}|$ its determinant. A second good method is to enclose the matrix in brackets $[a_{ij}]$, and to use $|a_{ij}|$ for its determinant. A third acceptable method uses $\|a_{ij}\|$ for the matrix, and $|a_{ij}|$ for the determinant.

There is a serious difficulty that may arise with any matrix, diagram, figure, table or displayed formula *which occupies more than four lines.* If the author specifies its exact location in the text, then the compositor, while cutting the galley into pages, may find it necessary to cut it in two parts with the first part at the bottom of one page and the second at the top of the next. This will destroy vertical alignment, and may render the material almost unreadable. To avoid this difficulty, the author should label the diagram as a figure, and refer to it as such in the text. This allows the compositor to move it to a nearby place to avoid breaking it.

6. **Bibliography.** Items in the bibliography are usually ordered alphabetically by name of author, and they are numbered consecutively. The names of journals should be abbreviated. Standard abbreviations for some of the principal American journals are

Amer. J. Math.	J. Math. Mech.
Amer. Math. Monthly	J. Symbolic Logic
Ann. of Math.	Math. Reviews
Bull. Amer. Math. Soc.	Michigan Math. J.
Canad. J. Math.	Pacific J. Math.
Comm. Pure Appl. Math.	Proc. Amer. Math. Soc.
Duke Math. J.	Proc. Nat. Acad. Sci. U.S.A.
Illinois J. Math.	Trans. Amer. Math. Soc.

A full list of standard abbreviations can be found in the annual index of Math. Reviews. The name of the journal is followed in order by the volume number, the year, and the first and last page numbers, thus:

3. J. Doe, *Summability of Fourier series*, Pacific J. Math. **12** (1960), 232–257.

Observe that the month of issue and the issue number are *not* included. Some journals have discontinued an older numbering of volumes and started afresh with volume one, adding a serial number to distinguish the new block of volumes. In such cases, the serial number should be listed in parentheses before the volume number: Ann. of Math. (2) **52** (1950), 127–139.

A reference to a book should give in order author, title, edition (if not the first), name of series and number (if one of a series), publisher (or distributor), city, and year:

4. F. Hausdorff, *Mengenlehre*, 3rd ed., Göschens Lehrbuch. 7, W. de Gruyter, Berlin, 1935.

A reference in the text to the bibliography is customarily of the form: "It is well known [5, pp. 32–34] that . . . ," or "According to a result of J. Doe [3, Th. 7, p. 252], ..." Authors are urged to make references specific by including page numbers. The extra space and trouble is fully justified by the saving to the readers.

7. **Footnotes.** These are more expensive than equivalent material embodied in the text because, being in smaller type, all footnotes are set together in a separate operation. Subsequently, during the paging, they must be separated and inserted in their proper locations. Because of the smaller type, complicated formulas in footnotes must be avoided.

It is therefore best to have no footnotes. The usual literary excuse for a footnote (to avoid interrupting the continuity) does not apply here because the reading of mathematics is usually painfully discontinuous. It is increasingly the custom (and editors and their staffs applaud this warmly) to incorporate a footnote into the text, perhaps parenthetically, perhaps as a "Remark." Acknowledgements can be given in a paragraph at the end of the introduction.

If footnotes are deemed necessary, then all footnotes should be typed together on a single page at the end of the manuscript, numbered consecutively. The reference number in the text should be signalled by writing e.g. "footnote 2" in the margin. An initial footnote beginning with "Presented to . . . " or "Work done under contract . . . " need not be numbered.

8. **The typing of the manuscript.** It is important that the manuscript be typed with at least double spacing, and with generous margins on both sides ($1\frac{1}{4}$ inches). The double spacing provides clear separation of subscripts and superscripts, and also permits the author to insert changes directly above portions crossed out. The margins allow the technical editor to write instructions to the printer.

The pages of the manuscript should be numbered consecutively in the upper right corners. This insures that the absence of a misplaced page will be noticed.

The manuscript that goes to the printer must be of good quality paper. It is subjected to handling of sufficient roughness to cause thin paper to tear. A carbon copy is very unsatisfactory for two reasons: The paper is usually too thin, and moisture from the hand causes the letters to smear.

It is increasingly the custom to submit manuscripts produced by mimeographing or some other duplicating process. This is advantageous to an author, for it permits him to circulate his paper to a sizable audience a year before publication. Such a manuscript is acceptable for printing if it satisfies the above requirements: *at least double spacing, generous margins, good quality paper, clear letters, and non-smear ink.* A well-executed *mimeographed* or *multilithed* manuscript is satisfactory. Other methods are almost always unsatisfactory.

The time for deciding on the precise arrangement of the displayed formulas comes just before the typing. Sloppiness in the handwritten formulas tends to perpetuate itself throughout the typed and printed versions because no one but the author is competent to make alterations. Compositors and typists can only endeavor to reproduce the copy before them. Thus, for each display, the author should visualize exactly what he wants in the printed version, and then he should produce a reasonable facsimile which exhibits clearly the alignment and spacing of all its parts, both horizontally and vertically.

It is very helpful if the typed manuscript follows the standard format of journal articles in the matter of indentations of section headings and theorems. If the handwritten copy does not do this, then the typist should be asked to follow standard practice as exemplified by a reprint.

9. **Preparing the manuscript.** Because typewriters do not have enough symbols, it is necessary for authors to fill in quite a few formulas by hand. This must be done carefully, in ink, so that the compositor will know what is wanted. The first requirement is that the typist leave enough space, horizontally and vertically, for the handwritten symbols. (Inexperienced typists tend to leave too little space.) Secondly, the author should give extra thought and care to the placing of the symbols. The compositor does not understand mathematics, and he is guided solely by what he sees in the manuscript. He works along lines, and he must know what is on the line, what is a subscript, a superscript, and a subsubscript. He cannot guess that a hastily scribbled *gij* means g_{ij}. Also, the manuscript should distinguish clearly between similar symbols such as

$$\cup, \text{u}, \text{U}; \quad o, O, 0; \quad x, \times; \quad \phi, \varnothing; \quad l, 1; \quad \epsilon, \in.$$

It is very helpful if, in any doubtful case, the author explains his wishes by means of a pencilled note in the margin.

Small changes in the typed copy (a few words here and there) can be carefully handwritten *in black or blue-black ink* in the spaces between lines. Do not write lengthwise along the margin. Do not use proofreading symbols as if the typed copy were galley proof. Larger alterations should be typed. For example, the lower half of page 5 can be crossed out, and a marginal note can say: Insert here the material on page 5a. In deleting material, cross it out, and do not tear off part of a page. A half page may not stay in the paper clip and may be lost. Do not use scotch tape on the top side of a page, for nobody can write on it.

The author can call for special type by underlining according to the following conventions:

Italic —a straight underline in black,
Boldface—a wavy underline in black,
Greek —a straight underline in red,
German —a straight underline in green,
Script —encircle the letter in blue.

The author should leave to the editorial assistant the underlining of the title, section headings, and the statements of theorems, lemmas, and corollaries. Also, do not underline mathematical symbols which are to appear in the usual italic. The only italic underlining the author should do is that of the occasional word or sentence in the text he may wish to emphasize. Do not underline the special symbols

$\sum$ (sum), $\prod$ (product), $\in$ (is an element of), $\varnothing$ (empty set).

German letters and script capitals are obtained most conveniently by typing the letter and marking as indicated above. A second method is to fill in a fair approximation by hand, and then, on a separate page of instructions to the printer, make a table of the handwritten symbols adjacent to the corresponding typed symbols.

The page of instructions to the printer should also list alternatives for any special symbols the printer might not have.

The author should submit a manuscript which he believes to be complete. If errors are discovered after submission, typewritten replacement pages should be sent to the editor as soon as possible. Once a manuscript has been set in type, over half the cost of publication has been expended. If, at this stage, the author insists on major changes, or on the withdrawal of the manuscript, he may be charged the printing costs.

10. **Proofreading.** Signs used in correcting proof and examples of their use will be found on the next 3 pages. Alterations in type which has been set are done by hand. Authors should call for only such changes as are essential. The insertion or deletion of a comma, for example, should be based on a compelling reason and not a whim.

If a number of consecutive words or symbols must be deleted or altered, they should be replaced if possible by material occupying nearly the same amount of space. If this is not done, the compositor will have to reset a number of neighboring lines to make the change. For example, if an author adds a twenty-letter phrase to the first line of a paragraph, the entire paragraph may have to be reset to make room for it. Such changes are costly and may lead to new errors.

It is important that, for each correction, two marks appear, one in the margin showing what the correction is, and one in the text showing where it is to go. The signs in the margins must have the order of the corresponding errors in the line, and they must be separated by a vertical bar |.

TABLE II

SIGNS USED IN CORRECTING PROOFS

Sign	Meaning	Sign	Meaning
₰	Delete; take out	x	Change broken letter
⁀	Close up	stet	Let it stand as set
^	Insert	......	Let it stand as set
#	Insert space	w.f.	Wrong font, size or style
⊓	Raise	l.c.	Lower case, not capitals
⊔	Lower	rom.	Use Roman letter
[	Move to left	bf.	Use black type letters
]	Move to right	⊙	Period
‖	Straighten type line at side of page	^,	Comma
//	Straighten lines	∨'	Apostrophe
¶	Paragraph	∨2	Superior figure
center	Put in middle of page or line	^2	Inferior figure
∽	Transpose	=/	Hyphen
Tr	Transpose	sc.	Use small capitals
ɑ	Turn inverted letter right side up	caps	Use capitals
		ital.	Use italics

EXAMPLE SHOWING THE USE OF PROOF READING SIGNS

of B_l and $d_1 d'_p$ divides $d_1 d'_p + 1$. Hence **B** is reducible to the from (11.5) with diagonal terms $d_1, d_1 d'_2, \cdots, d_1 d'_p$ which proves (11.4).

12. Groups with a finite number of generators. We shall discuss certain properties of these groups culminating in the basic product decomposition (12.5).

(12.1) DEFINITION. *Let* $B = \{g_1, \cdots, g_n\}$, $B' = \{g'_1, \cdots, g'_n\}$ *be two sets of elementsof G containing the same number n of elements. By a unimodular transformation* $\tau: B \to B'$ *is meant a system of elations*

$$(13.2) \qquad g'_i = \sum a_{ij} g_j, \qquad \|a_{ij}\| \text{ unimodular.}$$

The following proposition shows in how natural a manner unimodular transformations make their appearance in the theory of groups with finite bases.

(12.3) *Let G be a group with a finite base* $B = \{g_1, \cdots, g_n$. *In order that* $B' = \{g'_1, \cdots, g_n\}$ *be a base for G it is necessary and sufficient that B′ be obtainable from B by a unimodula r transformation.*

For any given set $B' = \{g'_1, \cdots, g'_n\}$ of elements of G there exist relations

$$(12.4) \qquad g_i = \sum c_{ij} g_j, \qquad C = \|c_{ij}\|$$

A necessary and sufficient condition in order that $\{g'_i\}$ be as base is that the G_j be expressible as linear combinations of the g'_i, or that there exist relations

$$(12.5) \qquad g_i = \sum d_{ij} g'_j, \qquad D = \|d_{ij}\|.$$

From this follows

$$g_i = \sum d_{ij} c_{jk} g_k.$$

Hence since B is a base we must have $DC = 1.$
This matrix relation yields $|D||C| = 1$, and since the determinants are integers we must have $|C| = \pm 1$. Thus in order that B' be a base C must be unimodular, or the condition of (12.3) must be fulfilled. Conversely, if it is fulfilled, C is unimodular and (12.5) holds with $D = C - 1$, from which it follows readily that b' is a base.

THE PRECEDING PASSAGE PRINTED WITH ALL CORRECTIONS MADE

of B_1 and d_1d_p' divides d_1d_{p+1}'. Hence B is reducible to the form (11.5) with diagonal terms $d_1, d_1d_2', \cdots, d_1d_p'$ which proves (11.4).

12. **Groups with a finite number of generators.** We shall discuss certain properties of these groups culminating in the basic product decomposition (12.5).

(12.1) DEFINITION. *Let* $B=\{g_1, \cdots, g_n\}$, $B'=\{g_1', \cdots, g_n'\}$ *be two sets of elements of* G *containing the same number* n *of elements. By a unimodular transformation* $\tau: B \rightarrow B'$ *is meant a system of relations*

$$g_i' = \sum a_{ij}g_j, \qquad \|a_{ij}\| \text{ unimodular.} \tag{12.2}$$

The following proposition shows in how natural a manner unimodular transformations make their appearance in the theory of groups with finite bases.

(12.3) *Let* G *be a group with a finite base* $B=\{g_1, \cdots, g_n\}$. *In order that* $B'=\{g_1', \cdots, g_n'\}$ *be a base for* G *it is necessary and sufficient that* B' *be obtainable from* B *by a unimodular transformation.*

For any given set $B'=\{g_1', \cdots, g_n'\}$ of elements of G there exist relations.

$$g_i' = \sum c_{ij}g_j, \qquad C = \|c_{ij}\|. \tag{12.4}$$

A necessary and sufficient condition that $B'=\{g_i'\}$ be a base is that the g_j be expressible as linear combinations of the g_i', or that there exist relations

$$g_i = \sum d_{ij}g_j', \qquad D = \|d_{ij}\|. \tag{12.5}$$

From this follows

$$g_i = \sum d_{ij}c_{jk}g_k.$$

Hence since B is a base we must have $DC=1$. This matrix relation yields $|D|\cdot|C|=1$, and since the determinants are integers we must have $|C|=\pm 1$. Thus in order that B' be a base C must be unimodular, or the condition of (12.3) must be fulfilled. Conversely, if (12.3) is fulfilled, C is unimodular and (12.5) holds with $D=C^{-1}$, from which it follows readily that B' is a base.

TABLE III
A LIST OF SIGNS AND SPECIAL CHARACTERS

1.	+	*36.	≙	71.	|	106.	↣
2.	−	37.	<	72.	‖	107.	**→**
3.	×	38.	>	*73.	⫆	*108.	⇄
4.	=	*39.	≺	*74.	⫅	*109.	⇄
5.	÷	*40.	⋚	75.	卐	*110.	⇆
6.	**+**	*41.	⋛	76.	▥	111.	$\sqrt{}$
7.	**−**	42.	≤	*77.	**∨**	112.	$\sqrt[2]{}$
8.	**×**	43.	≥	78.	**⊃**	113.	$\sqrt[3]{}$
9.	**=**	44.	≦	79.	**⊂**	114.	$\sqrt[4]{}$
10.	**÷**	45.	≧	*80.	**⊅**	115.	$\sqrt[5]{}$
*11.	∔	*46.	≱	81.	**∪**	116.	$\sqrt[6]{}$
12.	**±**	*47.	≰	82.	**∩**	117.	$\sqrt[7]{}$
13.	±	48.	≯	*83.	∑	118.	$\sqrt[8]{}$
*14.	∓	49.	≮	*84.	∏	119.	$\sqrt[9]{}$
*15.	≬	*50.	⊀	85.	Γ	120.	$\sqrt[n]{}$
16.	≐	*51.	⊁	86.	→	121.	∨
17.	∽	52.	∧	87.	←	122.	⊼
18.	∼	53.	∨	88.	⇄	123.	▽
19.	≈	*54.	≫	89.	⇆	124.	△
20.	∾	*55.	≪	90.	↷	*125.	△S
21.	≃	56.	⊂	91.	↶	126.	◬
22.	≅	57.	⊃	92.	↔	127.	▲
23.	≊	58.	⊆	93.	↖	*128.	∠
24.	≌	59.	⊇	94.	↗	*129.	⦤
25.	≁	60.	⊄	95.	↓	130.	□
26.	≄	*61.	⋑	96.	↘	131.	⊡
27.	≠	62.	∈	97.	⤴	132.	■
28.	ǂ	63.	∩	98.	↰	133.	▭
29.	=	64.	∪	99.	↳	*134.	[S]
30.	≡	65.	∉	100.	⇌	135.	▱
31.	≣	66.	∪	101.	⇋	*136.	▱S
32.	≢	*67.	⊈	102.	⇸	137.	◇
*33.	⋕	*68.	⊉	103.	↑	138.	○
34.	≎	69.	∝	104.	⇒	139.	⊙
35.	≏	70.	∞	105.	⇔	*140.	Ⓢ

* An asterisk indicates symbols which are not available in a smaller font for use in inferior and superior position.

141.	ō	162.	|	183.	′	204.	♭
142.	⍜	163.	‖	184.	′	205.	∗
143.	ө	164.	╱	185.	′	206.	✱
144.	⊖	165.	╲	186.	″	207.	∗
145.	⊕	166.	⑊	187.	‵	208.	♏
146.	⊗	167.	⫽	188.	°	209.	♂
147.	∅	168.	╲	189.	⸰	210.	♀
148.	♊	169.	⋰	190.	·	211.	⚥
149.	♈	170.	⋰	191.	•	212.	☿
*150.	m	171.	⫽	192.	:	213.	ȯ
151.	℘	172.	……	193.	⋮	214.	∂
152.	⌢	173.	╱	194.	∴	215.	∇
153.	⌣	174.	┐	195.	✯	*216.	Δ
154.	↷	175.	⊥	196.	☆	*217.	∆
*155.	⤙	176.	⊢	197.	★	218.	*
*156.	⤚	177.	⊨	198.	★	219.	†
157.	⟨	*178.	∃	199.	⋆	220.	‡
158.	⟩	*179.	ẋ	200.	#	221.	§
159.	/	180.	∫	201.	♯	222.	¶
160.	\	181.	∫	202.	♮	223.	ℂ
161.	∤	182.	∮	203.	♭	*224.	ℓ

GREEK ALPHABET

Alpha	Α α	Iota	Ι ι	Rho	Ρ ρ
Beta	Β β	Kappa	Κ κ	Sigma	Σ σ
Gamma	Γ γ	Lambda	Λ λ	Tau	Τ τ
Delta	Δ δ ∂	Mu	Μ μ	Upsilon	ϒ υ
Epsilon	Ε ϵ	Nu	Ν ν	Phi	Φ ϕ φ
Zeta	Ζ ζ	Xi	Ξ ξ	Chi	Χ χ
Eta	Η η	Omicron	Ο ο	Psi	Ψ ψ
Theta	Θ θ ϑ	Pi	Π π	Omega	Ω ω

RESEARCH ANNOUNCEMENTS

The purpose of this department is to provide early announcement of significant new results, with some indications of proof. Although ordinarily a research announcement should be a brief summary of a paper to be published in full elsewhere, papers giving complete proofs of results of exceptional interest are also solicited. Manuscripts more than eight typewritten double spaced pages long will not be considered as acceptable.

MULTIPLIERS OF p-INTEGRABLE FUNCTIONS[1]

BY ALESSANDRO FIGÀ-TALAMANCA

Communicated by A. E. Taylor, April 9, 1964

1. **Introduction.** Let G be a locally compact Abelian group. Let $L^p(G)$ $(1 \leq p < \infty)$ be the space of p-integrable functions (with respect to the Haar measure) with the usual norm. A *multiplier* of $L^p(G)$ is a bounded linear operator T of $L^p(G)$ into $L^p(G)$ which commutes with the translation operators; that is, $\tau_y T = T\tau_y$ for all $y \in G$, where $\tau_y f(x) = f(x+y)$. The space of multipliers will be denoted by $M_p = M_p(G)$. It is known that M_1 is isomorphic and isometric to the space of bounded regular Baire measures on G and that M_2 is isomorphic and isometric to $L^\infty(\Gamma)$, where Γ is the character group of G, and thus M_2 is the conjugate space of the space $A(G)$ of continuous functions on G which are Fourier transforms of elements of $L^1(\Gamma)$. Theorem 1 below asserts that, for $1 < p < \infty$, M_p is also the conjugate space of a space A_p of continuous functions on G. A corollary of this fact is that M_p is the closure in the weak operator topology of the linear span of the translation operators. A theorem due to Hörmander relating tempered distributions on $\boldsymbol{R}^n$ to $M_p(\boldsymbol{R}^n)$ [2], is also an easy consequence of Theorem 1. In view of the fact that a multiplier T can be identified with an element $T^\frown \in L^\infty(\Gamma)$ (Γ being the character group of G), another consequence of Theorem 1 is that if $T \in M_p$, $T^\frown * \mu = U^\frown$ with $U \in M_p$, where μ is a bounded regular Baire measure on Γ. If G is a noncommutative unimodular group, a proposition analogous to Theorem 1 holds for operators commuting with right (respec-

[1] This research was supported by the Office of Naval Research, U. S. Navy. Reproduction of this paper in whole or in part is permitted for any purpose of the United States Government.

This note summarizes in part results contained in a dissertation submitted in partial satisfaction of the requirements for the Ph.D. degree at the University of California, Los Angeles, prepared under the direction of Professor Philip C. Curtis, Jr., to whom I am greatly indebted for much valuable advice. Proofs of the results contained in this paper will appear in full elsewhere.

tively, left) translations; the specialization to the case $p=2$ yields results established by Segal in [7]. Theorem 7 relates lacunary subsets of a discrete Abelian group Γ to multipliers on $L^p(G)$ where G is the character group of Γ.

2. **The space of multipliers as a conjugate space.** Hereafter p will be a fixed real number, $1<p<\infty$, and q will be such that $1/p+1/q=1$. $C_{00}(G)$ and $C_0(G)$ will be, respectively, the space of continuous functions with compact support and the space of continuous functions vanishing at infinity on G. The convolution with respect to the Haar measure of two measurable functions f and g will be denoted by $f * g$ whenever it is well defined.

DEFINITION 1. *Let A_p be the space of functions $h \in C_0(G)$ which can be written as $h=\sum_{i=1}^{\infty} c_i f_i * g_i$, where $f_i \in L^p(G)$, $g_i \in L^q(G)$ and $\sum |c_i| \, \|f_i\|_p \|g_i\|_q < \infty$; for $h \in A_p$ define*

$$\|h\|_{A_p} = \inf \left\{ \sum_{i=1}^{\infty} |c_i| \, \|f_i\|_p \|g_i\|_q : h = \sum_{i=1}^{\infty} c_i f_i * g_i \right\}.$$

THEOREM 1. *M_p is isometric and isomorphic to the conjugate space of A_p, the element $T \in M_p$ corresponding to the functional $\phi_T(h) = \sum c_i (Tf_i * g_i)(0)$, where $h=\sum c_i f_i * g_i$. The weak operator topology of M_p coincides, on the unit sphere of M_p, with the weak-star topology induced by A_p.*

One notices that if $f, g \in C_{00}(G)$, $T(f * g) = Tf * g$, so that $T(f * g)$ is, after correction on a set of Haar measure zero, a continuous function. One can then define the functional $\phi_T(h) = Th(0)$, on the space S of linear combinations of functions of the type $f * g$ with $f, g \in C_{00}(G)$. Thus the proof of Theorem 1 consists essentially of showing that the completion of S, under the norm

$$\|h\| = \sup\{ |Th(0)| : T \in M_p, \|T\|_{M_p} \leqq 1 \},$$

is A_p. This is accomplished using the fact that the BX topology on the conjugate space X^* of a Banach space X has the same continuous linear functionals as the X topology (weak star topology) (cf. [1, V, 5.6]).

COROLLARY 2. *M_p is the closure, in the weak operator topology, of the span of the translation operators.*

REMARK 3. *As a consequence of the Riesz convexity theorem [1, V, 10.11], and in view of the duality between $L^p(G)$ and $L^q(G)$, the restriction to $C_{00}(G)$ of an element $T \in M_p$ can be extended to an element of M_r for $p \leqq r \leqq q$. Furthermore $\|T\|_{M_p} = \|T\|_{M_q}$ and, if $p \leqq r \leqq s \leqq 2$, $\|T\|_{M_r} \leqq \|T\|_{M_s}$. Dually one has $A_p = A_q$ and, if $p \leqq r \leqq s \leqq 2$, $A_s \subseteq A_r$,* *with* $\|\cdot\|_{A_r} \leq \|\cdot\|_{A_s}$. One should also notice that $A_2(G) = A(G)$ is the space of Fourier transforms of elements of $L^1(\Gamma)$, where Γ is the character group of G. Thus, since A_2 is dense in A_p, each $T \in M_p$ *corresponds biuniquely to an element* $T^\frown(\gamma)$ of $L^\infty(\Gamma)$. $T^\frown$ will be called the *transform* of T.

REMARK 4. A simple consequence of Theorem 1 and of the fact that $A_2 = A(G)$ is continuously and densely embedded in A_p, is the result proved by Hörmander in [2], stating that for $G = \boldsymbol{R}^n$, M_p can be identified with a subspace of the space of tempered distributions on $\boldsymbol{R}^n$. It suffices to notice that the space $\mathcal{S}$ of rapidly decreasing functions on $\boldsymbol{R}^n$ is continuously and densely embedded in $A(\boldsymbol{R}^n)$ and therefore in A_p (cf., e.g., [**3**, I, 1.7]).

COROLLARY 5. *Let* $T \in M_p$ *and let* $T^\frown$ *be its transform in the sense of Remark* 3; *then, if* μ *is a bounded regular Baire measure on* Γ *(the character group of* G*),* $T^\frown * \mu$ *is also the transform of a multiplier in* M_p.

One shows that if $\hat{\mu}$ is the Fourier-Stieltjes transform of μ, $\hat{\mu}h \in A_p$ for every $h \in A_p$; thus the functional $\phi(h) = T(\hat{\mu}h)(0)$ on A_p defines a multiplier whose transform is $T^\frown * \mu$.

REMARK 6. Theorem 1 and Corollary 2 are valid for not necessarily commutative unimodular groups in the following sense: Let $\mathfrak{L}_p$ (respectively, $\mathfrak{R}_p$) be the space of bounded linear operators on $L^p(G)$ which commute with right (respectively, left) translations by elements of G. Let A_p^1 (respectively, A_p^2) be the space functions on G which can be written as $\sum_{i=1}^{\infty} c_i f_i * g_i$, with $f_i \in L^p(G)$, $g_i \in L^q(G)$ (respectively, $f_i \in L^q(G)$, $g_i \in L^p(G)$) with f_i, g_i, c_i satisfying the conditions of Definition 1 and with norms analogously defined; then $\mathfrak{L}_p$ (respectively, $\mathfrak{R}_p$) is the conjugate space of A_p^1 (respectively, A_p^2). Moreover, $\mathfrak{L}_p$ (respectively, $\mathfrak{R}_p$) is the closure, in the weak operator topology, of the space of the left (respectively, right) translations. Thus the space $\mathfrak{R}_p'$ of the operators commuting with elements of $\mathfrak{R}_p$ is $\mathfrak{L}_p$ and conversely, so that $\mathfrak{R}_p \cap \mathfrak{L}_p$ is the center of both $\mathfrak{L}_p$ and $\mathfrak{R}_p$. For the case $p = 2$ these results specialize to known results due to Segal [**7**]. One should also notice that $A_p^1 = A_q^2$ and therefore $\mathfrak{L}_p$ is isometric and linearly isomorphic to $\mathfrak{R}_q$.

3. **Multipliers and lacunary sets.** Let G be a compact Abelian group, Γ its discrete character group. A set $E \subseteq \Gamma$ is called a *Sidon set* (cf. [**5**, 5.7.2]) if every $f \in C(G)$ with $\hat{f}(\gamma) = 0$ for $\gamma \in E$ satisfies $\sum |\hat{f}(\gamma)| < \infty$, or equivalently if for every bounded function $\lambda(\gamma)$ on E, there exists a Baire measure μ on G such that $\hat{\mu}(\gamma) = \lambda(\gamma)$ for $\gamma \in E$ ($\hat{f}$ and $\hat{\mu}$ denote, respectively, the Fourier transform and the Fourier-

Stieltjes transform of f and μ). As measures (operating by convolution) are exactly the operators on $L^1(G)$ which commute with translations, it is natural to define an analogous concept for multipliers of $L^p(G)$, recalling that each $T\in M_p$ corresponds biuniquely to an element $T^\frown$ of $L^\infty(\Gamma)$ (cf. Remark 3).

DEFINITION 2. *A set $E\subseteq\Gamma$ is called a p-Sidon set if every $f\in A_p$ such that $\hat{f}(\gamma)=0$ for $\gamma\notin E$ satisfies $\sum|\hat{f}(\gamma)|<\infty$.*

THEOREM 7. *Let $p\neq 2$, then the following properties are equivalent for a subset E of Γ:*

(i) *E is a p-Sidon set;*

(ii) *if $\lambda\in L^\infty(\Gamma)$, there exists $T\in M_p$ such that $T^\frown(\gamma)=\lambda(\gamma)$ for $\gamma\in E$;*

(iii) *if $\lambda\in L^\infty(\Gamma)$, there exists $T\in M_p$ satisfying* (ii) *and moreover such that $T^\frown(\gamma)=0$ for $\gamma\notin E$;*

(iv) *if $f\in L^1(G)$ and $\hat{f}(\gamma)=0$ for $\gamma\notin E$, then $f\in L^r(G)$ where $r=\max(p, q)$.*

It should be noted that condition (iv) above is the defining property for what is called a *lacunary set of order r* or a $\Lambda(r)$ set. Properties of these sets are investigated in [6] and [4, Chapter VIII]. In particular, it is known that a Sidon set is a lacunary set of order r for every r and hence a p-Sidon set for every p. One should also notice that condition (iii) above implies that the characteristic function of a p-Sidon set is always the transform of a multiplier. The analogous statement for Sidon sets does not hold; indeed it is known that a Sidon set whose characteristic function is the Fourier-Stieltjes transform of a measure is necessarily finite.

REFERENCES

1. N. Dunford and J. Schwarz, *Linear operators*, Vol. I, Interscience, New York, 1958.
2. L. Hörmander, *Estimates for translation invariant operators in L^p spaces*, Acta Math. **104** (1960), 93–140.
3. ———, *Linear partial differential operators*, Academic Press, New York, 1963.
4. S. Kaczmarz and H. Steinhaus, *Theorie der Orthogonalreihen*, Monografje Matematyczne, Warsaw, 1935.
5. W. Rudin, *Fourier analysis on groups*, Interscience, New York, 1962.
6. ———, *Trigonometric series with gaps*, J. Math. Mech. **9** (1962), 203–227.
7. I. Segal, *The two-sided regular representation of a unimodular locally compact group*, Ann. of Math. (2) **51** (1950), 293–298.

UNIVERSITY OF CALIFORNIA, LOS ANGELES

Comments on the Article

The preceding illustrates practice recommended in the "Manual" of the American Mathematical Society.

The *title,* as required by the "Manual," places the article in a recognizable section of the body of mathematics. This is done by the effective use of keywords drawn from the preestablished areas or sections of the discipline.

The *headnote* to the title serves two purposes. First, it gives credit to the Office of Naval Research which supported the work. The reprinting and proprietary rights (see p. 107) to the article are released by ONR "for any purpose of the United States Government." This does not constitute copyright release (see p. 126) for any nongovernmental use. As a result, a fee had to be paid to the *Bulletin of the American Mathematical Society* for permission to reprint the article here. Second, the note fixes the article as part of a PhD dissertation to indicate where a complete discussion of the topic may be found. The name of the dissertation director is given both as a professional courtesy and to meet a requirement for dissertation publication. The final sentence of the headnote promises future articles. This is fair usage; the statement indicates that the author is still involved with the topic, but it does not preclude other persons from publishing results of their own work on the topic.

The *by-lines* are required by the "Manual." The author credit follows the title as in almost all professional journals, but in semiprofessional and popular journals (house organs and specialized publications like *The New Yorker*) the author's name is placed at the end of the article. Beneath the author's name is the name of the member of the society who submitted the article to the journal. His sponsorship presumably guarantees the merit of the content while the date of submission guarantees its immediacy.

The *introduction,* in the words of the "Manual," presents a "rough statement of the principal results." The statement is conceptual rather than computational. It refers to computations in the body of the article, but even there the goal is maximum generality. As a result, the introduction to a mathematical article resembles the abstract of a chemical article.

The *body* of the article illustrates that in mathematical articles content determines form. Data are arranged according to the logic of the proof selected by the author. The sequential numbering of definitions, theorems, corollaries, and remarks reinforces this logic. The use of different families of type indicates a change in topic.

The *references* are specific; page numbers are included. Abbreviations of journals are those illustrated in the "Manual." No foreign titles are translated unless a different alphabet is involved.

G. MEDICAL JOURNALS

The eleven journals published by the American Medical Association follow practice recommended in the *Style Book and Editorial Manual* published by the Scientific Publications Division of the AMA in 1966. These journals are the following.

The Journal (JAMA)
Archives of Internal Medicine (Arch.Intern.Med.)
American Journal of Diseases of Children (AMER.J.DIS.Child.)
Archives of Otolaryngology (Arch.Otolaryng.)
Archives of Dermatology(Arch.Derm.)
Archives of Neurology (Arch.Neurol.)
Archives of General Psychiatry (Arch. Gen. Psychiat.)
Archives of Surgery (Arch. Surg.)
Archives of Ophthalmology (Arch. Ophthal.)
Archives of Pathology (Arch. Path.)
Archives of Environmental Health (Arch. Environ. Health)

Writers planning to submit articles to the journals listed above are urged to follow the bulletin *Advice to Authors* published by the AMA in 1964. Key sections of that document are printed below with the permission of the American Medical Association.

Preparation of Text

General Information—1.00

Manuscripts submitted to *JAMA* and the ten specialty journals are judged on the basis of subject matter, general excellence, manner of presentation, conciseness, and timeliness. Decisions regarding acceptance or rejection of a manuscript are the prerogative of the particular editorial board and are final, but they are based in part on opinions solicited from authorities who serve as consultants. When suggestions for revising the manuscript are made, the author is in no way obligated to make such revision, nor does making such revision ensure acceptance of the manuscript.

It is assumed that when a manuscript is submitted it is in a form suitable for publication from the standpoints of form and content. Therefore, it is the policy of the editorial department not to make extensive changes, except when absolutely necessary or when marking the copy for the printer. Although every reasonable effort is made to check the validity of data, the author bears the final responsibility for all statements in the published manuscript. For this reason, he is given an opportunity to see the edited copy on most material before publication.

Categories—2.00

2.01—General Information—Contributions to THE JOURNAL are usually in one of the following principal categories:

ORIGINAL CONTRIBUTIONS	CLINICAL NOTES
SPECIAL CONTRIBUTIONS and	CLINICAL SCIENCE
SPECIAL COMMUNICATIONS	NEGATIVE RESULTS
PRELIMINARY COMMUNICATIONS	LETTERS
CLINICAL MANAGEMENT	NARRATION and SATIRE

A prospective contributor to one of the specialty journals should consult an issue to determine the category into which his contribution might fall.

The editors determine under which heading a manuscript will be published, but it is suggested that prospective authors familiarize themselves with the criteria upon which these judgments are based (see under the specific category). Contributions to most other sections of THE JOURNAL are treated as special features and do not come under the scope of this guide.

2.02—Unacceptable Manuscripts—Literature reviews add no new information and therefore are unacceptable for publication in THE JOURNAL. The acceptance of literature reviews for the specialty journals is left to the editorial judgment of the chief editor and editorial board.

Descriptions of equipment, instruments, or procedures are usually unacceptable, except when they open a new field of clinical investigation. In such instances, the emphasis is on the application of the method, rather than on the instrumentation or procedure itself. THE JOURNAL will not accept any manuscript in which the author holds exclusive right to any procedure, therapeutic agent, instrument, or the like.

Material prepared for oral presentation is usually unsuitable for publication in the same form. It is expected, therefore, that all such manuscripts will be revised before they are submitted to THE JOURNAL. This includes text as well as illustrations and tables.

THE JOURNAL will not review any manuscript which is simultaneously being considered by any other publication. In such instances, the manuscript will be returned to the author as soon as the facts become known, even if it has already been accepted by THE JOURNAL and set in type (see also Rights and Responsibilities of Author and Journal, 10.00).

2.03—Original Contributions *(JAMA)*—Classified as ORIGINAL CONTRIBUTIONS are all manuscripts which present new and important clinical data, extend existing studies, or provide a fresh approach to a traditional subject. Such contributions generally are the result of studies of a relatively large number of cases.

The organization of an ORIGINAL CONTRIBUTION will be determined largely by the data to be presented. It is beyond the scope of this guide to mention all possible variations. An examination of the literature and some thought regarding the reader's needs will solve the problem of presenting the material effectively and logically. The only JOURNAL specification regarding an ORIGINAL CONTRIBUTION is that a synopsis-abstract (3.06) be submitted with the manuscript. The summary is omitted (3.07).

The length of an ORIGINAL CONTRIBUTION will also vary, but a general guide is that it should not exceed four or five JOURNAL pages, including figures, tables, and references. The number of figures, tables, and references should be limited to those needed to clarify, amplify, or document the text. In order to achieve balance, consideration must also be given to the amount of space required for illustrations and tables in relation to the space devoted to text material. For example, if there is more than one figure or one page of tabular material for every two pages of typewritten copy, a careful appraisal of the necessity for including the extras should be made. Experience has indicated that 18 references or less are usually sufficient. Limitation may be accomplished easily by avoiding multiple citations for a single fact. Failure of the author to eliminate unnecessary figures, tables, and references results in publication delays or perhaps in rejection of the manuscript.

2.04—Clinical Notes—A CLINICAL NOTE is a brief report of an observation which is interesting, new, or of otherwise sufficient importance to warrant attention. Most often the observations are based on only one or two case reports.

The preparation of a CLINICAL NOTE follows the same general pattern as that of an ORIGINAL CONTRIBUTION, with two major exceptions: The length should be not more than 1,000 to 1,500 words (three to five typewritten pages); the synopsis-abstract is omitted and a brief summary (3.07) is included at the end of the manuscript instead.

For many CLINICAL NOTES, a workable outline will consist of the following elements: (1) a brief introduction (no more than one or two paragraphs); (2) one or two case reports, summarized (3.04); (3) a method or plan of study (if applicable); (4) a brief comment, which will discuss the general background of the clinical condition under consideration as well as its specific relationship to the case(s) presented; and (5) a summary (3.07), consisting usually of three or four sentences.

An extensive review of the literature is unnecessary. Not more than eight references are allowed. Similarly, illustrations and tables should be limited to those necessary for understanding the text, and should not be selected for mere embellishment. Many CLINICAL NOTES require no illustrations. Generally, one or two illustrations and/or tables will be sufficient, and each should be selected for its teaching value.

2.05—Clinical Science—Contributions which correlate the use of a recent instrument or technique with clinical data are classified as CLINICAL SCIENCE communications. They are primarily reports of experimental studies or of procedures not yet established as part of general medical practice. Recent excellent examples of this type of manuscript are "Determination of Red Blood Cell Life Span" by Berlin (*JAMA* 188:375-378 [April 27] 1964); and "Human Systemic-Pulmonary Arterial Collateral Circulation After Pulmonary Thrombo-Embolism" by Smith et al (*JAMA* 188:452-458 [May 4] 1964).

A CLINICAL SCIENCE contribution is prepared in a manner similar to an ORIGINAL CONTRIBUTION, except that a synopsis-abstract is not required. Instead, a summary is used at the end.

2.06—Special Contributions and Special Communications—The designations SPECIAL COMMUNICATIONS and SPECIAL CONTRIBUTIONS are reserved for reports of outstanding work, matters of extreme timeliness, new concepts of especial interest, and subjects of a nonclinical nature such as law, government, health legislation, or finances. The subject matter and length usually differ greatly from those of ORIGINAL CONTRIBUTIONS and CLINICAL NOTES. Exceptions are made regarding the number of illustrations, tables, and references. The decision as to what is appropriate for a SPECIAL CONTRIBUTION or a SPECIAL COMMUNICATION rests with the editorial staff. A summary is usually necessary, but the synopsis-abstract is omitted. After acceptance of the manuscript, the author may be requested to prepare a "deck." This is a brief phrase or sentence (5 to 50 words), which appears on the first page of the published manuscript, and which is designed to stimulate reader interest (see also Deck, 3.09).

2.07—Preliminary Communications—A PRELIMINARY COMMUNICATION presents results which, though not conclusive, are sufficiently tenable to merit serious consideration. It is assumed that these communications will be followed by a more complete study which will yield similar results. Generally, a PRELIMINARY COMMUNICATION consists of a brief introduction, a description of materials and experimental method, results obtained, tentative conclusions based on these results, and a summary. The allowable length is 1,000 to 1,500 words of text. Illustrations, tables, and references should be kept to a minimum and in most instances there should not be more than three illustrations, one table, and ten references. Since a PRELIMINARY COMMUNICATION should not be longer than three printed pages in THE JOURNAL, text must frequently be cut to accommodate the desired ancillary information. Suitable PRELIMINARY COMMUNICATIONS recently appearing in *JAMA* are "Renal Cadmium and Essential Hypertension" by Schroeder (187:358 [Feb 1] 1964); and "Regional Pulmonary Blood Flow in Man by Radioisotope Scanning" by Wagner et al (187:601-603 [Feb 22] 1964).

2.08—Negative Results—When clinical or laboratory studies fail to yield "positive results," further investigation is frequently abandoned and the study is not reported. The purpose of the NEGATIVE RESULTS section is to publish such studies in summary form in order to inform other investigators who may plan to undertake similar studies, and to provide clinicians with experimental data upon which they may base their own conclusions concerning the efficacy of certain therapeutic agents, diagnostic signs, and laboratory determinations.

Contributions to the NEGATIVE RESULTS section usually consist of a brief explanation of procedure, description of materials and methods, brief comment, and a conclusion or summary. These contributions are limited to a single page in THE JOURNAL and therefore should be restricted to 500 to 600 words of text. Generally, one or two short tables and four to six references may be needed to support the conclusions. Illustrations are usually unnecessary. However, when such ancillary materials are used, the text must be correspondingly condensed to meet the space limitations. Recent examples of this type of communication are "Diseases of Hypersensitivity and Childhood Leukemia" by Fraumeni and associates (*JAMA* 188:459 [May 4] 1964); and "Pesticide Storage in Human Fat Tissue" by Hoffman et al (*JAMA* 188:819 [June 1] 1964).

2.09—Clinical Management—The CLINICAL MANAGEMENT section, begun in July 1964, provides the practicing physician with brief reviews of therapy and diagnostic methods. It is designed for immediate application in the practice of bedside medicine. The term CLINICAL MANAGEMENT is used in its broad sense to include laboratory aids, prognosis, differential diagnosis, and etiologic considerations. In order to do justice to the subject, however, not more than one of these aspects should be detailed in the contribution.

In published form the contributions should not be longer than two JOURNAL pages; if no illustrations are used, this is approximately 1,700 words. Illustrations will decrease the amount of copy that can be accommodated correspondingly. References may not exceed five.

2.10—Letters—LETTERS TO THE JOURNAL are short communications which report new or unusual clinical observations or experimental findings, or which express personal opinions on a subject of interest to physicians in general. LETTERS should not exceed 250 words in length; they may be further condensed or edited at the discretion of the editors. LETTERS are judged on the same selective criteria as are other manuscripts: clarity, general excellence of composition, conciseness, timeliness, and pertinence. When many letters are received on a single topic, the editors usually select for publication one or two which are most representative; the names of other writers are acknowledged in the column. LETTERS which express opinions contrary to those of the editors are judged solely on the above-mentioned attributes. References are not essential but, when used, must be limited to two.

Since the LETTERS column is one of the most rapid means for dissemination of current information, physicians are urged to contribute to this section, particularly as a substitute for CLINICAL NOTES. The column lends itself particularly well to the presentation of isolated clinical observations that do not require lengthy comment or excessive documentation. This, in turn, permits an exchange of opinion with other physicians while a topic is still timely. The time between receipt of a letter and its appearance in THE JOURNAL can be considerably shortened if the writer will indicate "for publication" in his letter.

LETTERS are indexed in *Index Medicus.*

2.11—Narration and Satire—Worthy contributions are not always restricted to science. The physician's intimate touch with life frequently inspires him to relate stories suitable for the NARRATION section of THE JOURNAL. Items for NARRATION may tell of a young physician's experience with the Peace Corps, irritation at adjusting to daylight saving time, the makings of a medical secretary, the varieties of primitive childbirth in South America, or the means for providing medical care for the Lapps of Lapland. "Slocum Heath—The Prosthetic Wonder" *(JAMA* 188, May 25, 1964, adv p 198) is an interesting first-person account which began with a physician's inquiry into an unusual tombstone epitaph.

"Venting the spleen" might some day be a laboratory procedure, but in the present context it means a literary effort bordering on the "pet-peeve" fringe and is commonly called SATIRE. There is much in medicine that physicians want to poke fun at or even deplore in an effective manner. The ivory tower of research is a favorite topic, but such earthly matters as baldness (the poor man's alopecia) and the growth of "anonymous" clubs (Fatties-Anonymous, Sex-Anonymous, etc) have also had their airing on JOURNAL pages. "Study of a Remarkable Anti-Inflammatory Agent (Cucumase)" *(JAMA* 187, Feb 22, 1964, adv p 198) proved that the cucumber is not free from medical peregrinations; and "The Rat Index: Rapid Method of Evaluating Medical Journals" *(JAMA* 185, Aug 24, 1963, adv p 168) came very close to home.

Specific Parts of Text—3.00

3.01—General Organization—A single outline cannot possibly serve for all scientific manuscripts. Thoughtful consideration of the subject itself and anticipation of the reader's needs and questions will usually indicate to the writer the type of organization required. A logical, rather than a chronological, plan should be followed in most instances.

Certain pieces of copy, however, are standard parts of all manuscripts or are used so frequently as to merit special instructions. These parts are discussed below.

3.02—Title—The title of the manuscript should be short, specific, and clear. Omit phrases such as "The Use of," "Treatment of," and "The Report of a Case of." Select titles which are specific rather than general. This enables the reader to invest his time with maximum return and also ensures the writer that his report will be indexed accurately. For example, "Test for Circulatory Integrity" or "Response to Valsalva's Maneuver" says little, whereas "Heart Rate and Valsalva's Maneuver as Test of Circulatory Integrity" indicates more accurately the actual subject of the report.

Subtitles may occasionally be used to advantage, but they should be used to amplify and not as a substitute for or duplication of the main title. They are retained or deleted during editing, depending upon makeup requirements, and should therefore not contain essential material.

The title must be grammatically correct. In addition, particular attention must be given to placement of modifiers. "Recurrence After Curative Excision of Carcinoma of the Large Bowel" reads better as "Recurrence of Carcinoma of the Large Bowel After Curative Excision." Although no misunderstanding is likely in the first version, it requires more than a glance to know what is meant.

The use of "catchy" titles is generally discouraged in scientific contributions. On the other hand, they are admissible, and sometimes even desirable for some contributions, eg, LETTERS.

3.03—By-Line and Supplementary Information—Every manuscript must have a by-line, which will contain the following information: (1) author's first name, middle initial (or first initial and middle name if the writer habitually signs his name that way), and surname; (2) highest degree held by the author; (3) city in which the study was conducted. If the writer holds two degrees (MD and PhD or MD and DDS, for example), he may use either or both, depending on the nature of the work he reports. It is rarely acceptable, however, to use the combination MS, MD. If the author is on active duty in the armed services, he should use only his service designation.

```
Henry J. Smith, MD, PhD, and Capt E.
    William Jones, MC, USA, Detroit
```

Supplementary information, which appears in an affiliation footnote on the title page (see Footnotes, 3.10), *must include* the hospital, university, or other facility at which the study was conducted or the report was written and the present affiliation and title of the author if different from that held when the work was done. It must also include a street address for reader correspondence (see Reprint Address, 9.02).

```
From the Department of Microbiology, University of Chicago. Dr. Jones is now at the Mayo Clinic, Rochester, Minn.
Reprint requests to 5307 Blackstone, Chicago 60615 (Dr. Pell).
```

If the manuscript was read at a scientific meeting, the name of the organization, the city, and the date (month, day, year) should also be given.

```
Read before the Surgical Society of America, St. Louis, June 15, 1964.
```

Few scientific investigations are the work of only one person, and thus most reports of such work will bear more than one author's name, even if the manuscript itself was written by one person. However, only the names of those who have contributed essential portions of the work should be listed as authors. Generally, the editors will accept not more than six authors. Persons who have given substantial but nonetheless secondary aid, such as the statistical analysis of data or the interpretation of histologic specimens, may be listed at the end of the manuscript (see Acknowledgment, 3.11). Inclusion in the by-line of individuals who have simply approved publication of the manuscript is not acceptable.

The first-named author is considered the senior author, and it is to him that all correspondence will be directed, unless otherwise requested. Complete bibliographic data in the *Index Medicus* is generally given only under the name of the senior author.

3.04—Report of Cases—Case histories should be concise and devoid of all details which, though of interest to the writer, are not essential for the reader. Negative and normal results should not be given unless the data are contributory. Operative reports, autopsy protocols, and similar hospital records should always be condensed. Pertinent laboratory data must be presented in sentence form. Omit such details about the patient as his initials and name of referring physician or hospital. Avoid such phrases as "local physician" and "local hospital." Use no abbreviations or jargon such as "W/D, W/N, 75-y.o. W/M." Specify time relationships clearly, using a single base-time whenever possible (such as *x* number of days after admission to hospital or *x* number of days after surgery). Use the past tense throughout.

3.05—Data—The investigator is responsible for the proper collection, analysis, and reporting of data used in his study; yet few physicians have more than rudimentary knowledge in the highly specialized field of statistics. Unless the ingestigator is also an experienced statistician, therefore, he is urged to utilize the services of a biostatistician.

Consultation should be initiated not when the manuscript is about to be written, but before the study is begun, since the validity of the inferences depends not only upon proper classification and analysis of data but also upon collection of appropriate facts. To the nonstatistician, the necessity for these facts frequently becomes evident only in retrospect, when they are no longer available.

The classification of data, and inferences drawn therefrom, also depends on highly specialized techniques. Detailed descriptions of these methods should be omitted from the manuscript, however, except when the primary purpose of the report is statistical.

For additional information on the collection, classification, and analysis of data, see "Statistics, Science, and Sense" by Sidney Shindell, MD (*JAMA* **186**:499, 570, 637, 780, 849 [Nov 2, 9, 16, 23, and 30] 1963).

3.06—Synopsis-Abstract—A synopsis-abstract must accompany each manuscript submitted as an ORIGINAL CONTRIBUTION. It is similar to and replaces the summary. The principal differences are these: (1) It may not exceed 135 words in length; (2) it is handled as a separate piece of copy, ie, typed on a separate sheet of paper; (3) it is placed at the beginning, rather than at the end, of the contribution. Like a summary, the synopsis-abstract must contain only the essential features of the report. It should be complete enough, however, for accurate indexing with no need to refer to the remainder of the report. It should not repeat the title.

3.07—Summary–A summary should accompany most scientific manuscripts. Exceptions are ORIGINAL CONTRIBUTIONS, in which it is replaced with a synopsis-abstract, and LETTERS TO THE JOURNAL. Even in a letter, however, literary and scientific value is enhanced by a single concluding or summary sentence at the end.

A summary must be factual, not descriptive. It should review only the highlights of the report: statement of the problem, method of study, results, and conclusions, if any. Rarely should a summary be longer than 150 to 200 words.

Introduce no new material into the summary, ie, state no facts which are not in the body of the report. References to tables, illustrations, or bibliography are improper.

The rules of formal writing apply to summaries. Use of the third person is preferable, although not mandatory. Verb tenses must be consistent. Avoid phrases such as "a study is reported," "a case is described," "it was found that," "on the basis of these results it may be concluded that."

3.08—Headings–The use of headings within the text of the manuscript facilitates understanding, relieves tedium, and creates a more attractive page. They should be used generously.

Headings consist of one or several terms which indicate the nature of the material immediately following. Subheadings (centered) should usually be general terms, such as "Report of Cases," "Comment," and the like. Introductory material bears no heading. Side headings are more specific: "Isolation of Virus," "Distribution of Virus," "Other Viruses."

All material following each heading must be pertinent to its heading. If such is not the case, either an additional heading is indicated, or the present heading should be made more general.

Subheadings are typed with initial capitals and are centered on the page; side headings are indented and typed with initial capitals, followed by a period and two hyphen signs.

Comment (Subheading)

Physical Signs.--(Side heading)

3.09—Deck–For certain contributions, especially SPECIAL COMMUNICATIONS, SPECIAL CONTRIBUTIONS, SYMPOSIA, and CLINICAL SCIENCE communications, THE JOURNAL uses a "deck." Its purpose is to stimulate reader interest. Patterns for decks vary. They may consist of a short phrase, a question, or even one or two complete sentences. The deck is not a summary, and may therefore include data not given in the report. The maximum length of a deck is 50 words, the minimum as few as 4 or 5 words. Generally the deck is written by a member of the editorial staff, but occasionally the author of the report is requested to supply it.

3.10—Unnumbered Footnotes–The following information appears as an unnumbered footnote: author's affiliation at the time the work was performed; his present affiliation if different; name, place, and exact date of meeting if manuscript has been prepared for that purpose (see By-Line and Supplementary Information, 3.03); and sponsoring agency if the work was supported by a financial grant (see Acknowledgment, 3.11).

An unnumbered footnote is also used to indicate the death of the author between the time the study was completed and the manuscript published. The death dagger (†) is used. This footnote appears immediately after the affiliation footnote.

†Dr. Doe died June 2, 1964.

Unnumbered footnotes are also used in tables, with reference symbols given in the following order for each table: * (asterisk), † (dagger), ‡ (double dagger), § (section mark), || (parallels), ¶ (paragraph mark), # (number sign), **, etc. Within the columns, the symbols are placed from left to right. The footnote itself, with identifying symbol, is typed double-spaced immediately below the lower rule of the table. Each footnote begins on a new line and is paragraphed.

3.11—Acknowledgment–An acknowledgment to persons or agencies who have contributed substantially to the author's work may be published at his discretion. The following types of assistance may be acknowledged: compilation and statistical analysis of data; preparation of pathology reports and illustrations; performance of special examinations, research, or operative procedures; provision of financial support in the form of grants, supplies, or equipment; and permission to use portions of material published elsewhere.

Acknowledgment should be omitted when such assistance is part of the regular duties of the persons or agencies involved, eg, technicians, nursing and house staff, typists, and drug suppliers or equipment manufacturers, unless such material was supplied free of charge. Omit acknowledgment also for reviewers and referring physicians unless a substantial contribution to the work has been made. When acknowledgment is made to a sponsoring agency, grant numbers, contract numbers, and other similar information should be supplied.

To preserve the dignity of an acknowledgment, the recognition of assistance should be stated as simply as possible, without effusiveness or superlatives. Omit phrases such as "express our gratitude and appreciation," "who greatly aided," "through the courtesy of." Examples of acceptable acknowledgments are:

J. H. Johnson, MD, supplied the electrocardiographic interpretations.

This study was supported in part by Public ealth Service research grant OH 00064.

The oxytetracycline used in this study was ıpplied as Terramycin through J. M. Albert, D, of Chas. Pfizer & Co., Inc., New York.

Figures 2, 3, and 5 are reproduced with perission from Surgery, Gynecology & Obstetrics 98:546-552 [May] 1954.)

The acknowledgment is typed double-spaced on separate sheet of paper and appears at the end of he manuscript as an unnumbered footnote.

3.12—Addendum—The use of an addendum to ny manuscript is discouraged. If additional, esential data should become available after compleion of the manuscript, such data should be written nto the body of the manuscript. In rare instances an ddendum may be permitted, but only after consulation with the editor. In no case should data resented in the addendum be included in the ynopsis-abstract or in the summary, since such lata are *not* part of the body of the manuscript.

3.13—References—A good index to the reliability nd importance of an author's work is frequently ound in the type and number of references he seects. Use of a great many references may indicate lack of critical thinking; too few references, the ossibility of unwarranted speculation; many refrences to one's own published work, a lack of obectivity; many older references, a failure to keep ıp to date; and many textbook references, a failure o verify original work.

References have two major purposes: documenation and acknowledgment. Even with this in nind, however, it is not expected that each conribution will carry an exhaustive review of the iterature. It is unnecessary, for example, in a CLINICAL NOTE to document each published case report. Acknowledgment references should be confined only to those reports which have contributed substantially to the author's own current work.

It has been found from experience that for the various types of contributions, the following *maximum* range of references is generally adequate: ORIGINAL CONTRIBUTIONS, 18 or less; CLINICAL NOTES, 6 to 8; NEGATIVE RESULTS, 4 to 6; PRELIMINARY COMMUNICATIONS, 8 to 10; LETTERS, 0 to 2. For manuscripts in other categories, the number permitted is determined at the time of review.

All references accepted with the manuscript are published in THE JOURNAL. The footnote, "Additional references may be obtained from the author," is rarely acceptable.

The author is responsible for the accuracy and completeness of all references. He is urged therefore to make a full record of the necessary bibliographic data (listed below) at the time he examines the literature. When information is missing, the manuscript will be returned to the author for completion.

The list of references appears at the end of the manuscript, beginning on a new page headed "References." It should be typed double-spaced, each item with a separate number corresponding to the text reference, and each number paragraphed.

Within the text the reference numbers appear consecutively as superscript Arabic numerals, beginning with 1. Do not assign a new number to the same reference, and do not include more than one reference for each number (subreferences such as 1a, 1b are not acceptable). The superscript number is placed adjacent to author, article, or publication in that order of preference:

Jones and his group[1] report similar results.

Several recent reports[1,2,3,4] note similar findings.

A recent report in the Archives of Surgery[7] has similar results.

When more than one reference is cited at a given place, all numbers should be listed in the copy in series, separated by commas, even when the numbers are consecutive.

Below is a list of types of publications most frequently cited. For each, the information which must be supplied and the form in which it must be supplied are given, with examples. *When in doubt, supply too much rather than too little information.*

I. *References to articles appearing in journals*

A. Last name(s) of author(s) and two initials. If there are more than three authors, only the first is listed followed by *et al.* Include also *Jr., Sr., III,* if applicable.

Betcher, A.M.; Wood, P.; and Wright, P.H., Jr.: . . .

B. Complete title of article, including subtitle, in capital and lower-case letters, without quotation marks.

Blue-Domed Cysts and Cancer of Breast: Report of 20 Cases, . . .

C. Name of journal, abbreviated according to the list in *Index Medicus,* and underlined. If in doubt, spell out the name of the journal.

D. Volume number, *not* underlined.

E. First and last page numbers of article, separated by a hyphen.

F. Month of issue, enclosed in parentheses, or if issued weekly, the month and date of month.

G. Year of publication.

1. Betcher, A.M.; Wood, P.; and Wright, P.H., Jr.: Blue-Domed Cysts and Cancer of Breast: Report of 20 Cases, JAMA 159:766-770 (Oct 22) 1955.

The above information will cover most instances of references to communications in journals. Handling of special cases, such as journals in more than one part, supplements to journals, special issues, etc, is covered in the AMA *Style Book and Editorial Manual*. If the author does not have access to a copy of this book, he should supply the additional information in a form complete enough for the copy editor to mark.

II. *References to books*

A. Name(s) of author(s) in form specified for journal articles.

B. Complete title of book, including subtitle (both underlined), in form specified for journal articles.

C. Number of edition, after the first.

D. Name(s) of translator, editor, or both, whenever applicable. (See example 3 below for instances in which reference is not to author or editor of a whole book, but only to a contributor to a part of it.)

E. City of publication.

F. Name of publisher in full.

G. Year of publication of edition cited.

H. Volume number.

I. Page or pages cited.

2. Doe, J.T.; Roe, H.L.; and Jones, G.E.: Introduction to Medical Parasitology, ed 4, W. A. Green (ed.), F. S. Scott (trans.), Baltimore: Williams & Wilkins Co., 1962, vol 3, pp 890-894.

In certain instances, the reference will be not to the author or editor of a book as a whole but only to the work of one contributor or to a single chapter. In such instances, the chapter title should be included, in quotations, and the reference will be written as follows:

3. Gotch, F.A., and Edelman, I.S.: "Fluid and Electrolyte Disorders," in Brainerd, H.; Margen, S.; and Chatton, M.J.: Current Diagnosis and Treatment, ed 3, Los Altos, Calif: Lange Medical Publications, 1964, chap 2, pp 26-44.

III. *References to government bulletins*

A. Name(s) of author(s), if given.

B. Title of bulletin, underlined.

C. Number of bulletin or other similar identification.

D. City of publication, except if Washington, DC.

E. Name of issuing bureau, agency, department, or other governmental division.

F. Date of publication.

G. Page or pages cited.

4. Hoffman, F.L.: Problem of Dust Phthisis in Granite-Stone Industry, Bull 293, US Dept of Labor, Bureau of Labor Statistics, 1964, pp 8-13.

5. Marinell, L.D., and Hill, R.F.: "Studies in Dosage in Cancer Therapy," Brookhaven Conference Report, in Symposium on Radioiodine, unclassified document AECU, BNL--C-5, Oak Ridge, Tenn, Atomic Energy Commission, Technical Information Branch, 1949, p 98.

IV. *References to monographs, reports, serials*

A. Include usual bibliographic data as given for journals, books, and government bulletins.

B. List, in addition, the number of the report, if it is one of a series, and the title of the series.

6. Pulaski, E.J.: Surgical Infections: Prophylaxis-Treatment-Antibiotic Therapy, publication 170, American Lecture Series, monograph in Bannerstone Division of American Lectures in Surgery, New York: McGraw-Hill Book Co., 1961.

V. *References to theses*

A. Include usual bibliographic data (see books, journals).

B. In addition give name of university (or other institution), its location, and name of publisher when available; for foreign theses, give city of publication.

7. Pollock, B.E.: Effect of Carbon Arc Radiation on Cardiac Output in Dogs, thesis, Tulane University Graduate School, New Orleans, 1953.

VI. *References to unpublished material*

A. For material read before a society or other organization, but not published, use the following form:

8. Spurling, R.G.: Notes on a Conservative Treatment of Third Ventricle Tumors, read before the Harvey Cushing Society, Rochester, Minn, May 15, 1964.

B. For material *accepted* for publication, but not yet published, use the following form:

9. Kornetsky, C.H.: Psychological Effects of Chronic Barbiturate Intoxication, Arch Neurol, to be published.

This form of reference should be used only for manuscripts actually accepted in writing. Date of publication and additional data should be added if they are available by the time the author receives the typescript.

C. For material *submitted* for publication, but not yet accepted, use the following form:

10. Kleinert, H.E.: Homograft Patch Repair of Bullet Wounds of Aorta, unpublished data.

If, when typescript is received by the author, final disposition (acceptance or rejection) is known, the reference should be revised as follows:

(1) If accepted:

11. Kleinert, H.E.: Homograft Patch Repair of Bullet Wounds of Aorta, Arch Surg, to be published.

(2) If rejected:

12. Kleinert, H.E.: Homograft Patch Repair of Bullet Wounds of Aorta, unpublished data.

(3) If rejected, but submitted to another journal, also treat as "unpublished data."

D. For "personal communications," the name of the writer, the name of the recipient, and the date must be given:

13. Alexander, E.G.: Personal communication to the author[s], Jan 15, 1964.

Personal communications are rarely acceptable as references. The document must be retrievable from the author's file. Verbal communications are never to be included in the list of references.

E. "Unpublished data" is not used as a reference except in the special circumstances noted. Data from correspondence or similar unpublished forms should be handled as indicated in item *D* above.

VII. *References to work of one investigator cited by another investigator (secondary references)*

The author must examine the original report whenever possible. When he is unable to do so, he must indicate that he is using a secondary source of information. He must also indicate whether the secondary author has actually quoted the original investigator, or whether he has merely cited his work. In the text, the citation carries the name of the original author, not the secondary author.

14. Lautenschlager, cited [quoted] by Eggston, A.A., and Wolff, D.: Histopathology of Ear, Nose and Throat, Philadelphia: W. B. Saunders Co., 1944, p 433.

3.14—Indexing and Keywords—All major scientific contributions to THE JOURNAL are indexed by author and subject in the *Index Medicus*. This also includes all correspondence published in THE JOURNAL which contains new or unusual facts or findings.

At the end of the synopsis-abstract of an ORIGINAL CONTRIBUTION, or after the summary for other contributions, from three to five indexing terms (keywords) should be listed. These are nouns taken from the contribution under which the author believes his work should be subject-indexed. Although this is merely a courtesy and not necessary for acceptance of the manuscript, compliance will permit faster preparation of medical indexes and will also ensure more accurate and more easily retrievable medical information.

Writers planning to submit articles to the journals of medical associations and groups other than the AMA should follow the style guide of the special journal.. Such a style guide was published as an article by *The New England Journal of Medicine* and is reprinted on pp. 138–142 by permission of that journal and of the author.

Nomenclature varies among journals. Because usage recommended for journals of the AMA is more strict than that practiced by the journals of other medical associations and groups, the recommendations of the American Medical Association's *Style Book and Editorial Manual* are reprinted by permission below.

NOMENCLATURE—14.00

DRUGS

14.01—The generic (nonproprietary) name of a drug should be used throughout the paper.

Formally adopted generic names are listed in *New Drugs, Pharmacopeia of the United States (USP), National Formulary (NF),* and *United States Adopted Names (USAN).* In addition the Council on Drugs reports in THE JOURNAL (NEW NAMES) drug names adopted by the USAN Council. These monographs include both the nonproprietary and trademark names for the newest drugs, usually prior to their publication elsewhere.

Also available to editors of the AMA scientific publications is the drug nomenclature printout, compiled by the Department of Drugs. This compendium provides an alphabetical, cross-indexed listing of US and foreign code names, trade names, and alternate nonproprietary names of drugs, as well as the formally approved generic names.

Where there is no nonproprietary name for a drug, give the chemical name or formula or description of the nonproprietary names of the active ingredients. If this terminology is unwieldy, the trade name may be used throughout the manuscript after the first mention.

If an author uses a chemical name or code number for a new unlisted drug, this should be called to the attention of the Council on Drugs and verified.

14.02—Use the complete name of a drug (eg, tetracycline hydrochloride, tetracycline phosphate complex; potassium penicillin G, sodium penicillin O) at first mention (in the text and synopsis-abstract or summary) and elsewhere in contexts involving dosage.

14.03—A trade (proprietary brand) name used by the author should be placed in parentheses immediately after the first use of the generic name, both in the text and in the synopsis-abstract or summary.

Trade names are capitalized, except for a few oddities (eg, pHisoHex). The following forms are correct:

tetracycline (Panmycin) hydrochloride

tetracycline phosphate complex (Panmycin Phosphate)

potassium penicillin G (Dramcillin)

sodium penicillin O (Cer-O-Cillin Sodium)

phencyclidine hydrochloride (Sernylan)

Trade names should not be used elsewhere except as specified in **14.01** and **14.06**. When the author uses only nonproprietary names, no trade names will be mentioned, except as specified in **14.06**.

14.04—Trade names should be avoided in titles or subtitles of communications other than monographs of the Council on Drugs. *All* drug names in titles and subtitles should be cleared with the Department of Drugs.

14.05—When more than one nonproprietary name of a drug is listed, the preferred name (first-listed; printout, 101) should be used. If the author used an alternate term (printout, 200), this may appear in parentheses after first mention of the preferred name in the text and synopsis-abstract or summary.

calcium leucovorin (citrovorum factor)

If the author used an alternate nonproprietary name *and* a trade name, both may appear in parentheses at first mention.

ascorbic acid (cevitamic acid; Cevalin)

Cyanocobalamin vs *vitamin* B_{12}: The substance used as a drug is cyanocobalamin; used as a food it is vitamin B_{12}.

Nitrogen mustard: Although nitrogen mustard is listed as an alternate name for mechlorethamine hydrochloride, the term refers to a class of chemicals and should usually be used in the plural. As identification or description of a specific drug such as mechlorethamine hydrochloride, say "a nitrogen mustard" or "one of the nitrogen mustards."

14.06—Names of drugs mentioned in major contributions in THE JOURNAL and the specialty journals are listed at the end of the paper under the heading "Generic and Trade Names of Drugs," in the following circumstances: (1) when a trade name is mentioned in the current *New Drugs,* either in a monograph or in brackets in an introductory section; or (2) when the drug is listed in the current *USAN;* or (3) when the drug is the principal subject of the contribution, eg, the drug name appears in the title or subtitle.

Each generic name is accompanied by all US trade names listed in the nomenclature printout (400), unless the number is excessive. If more than six or eight trade names appear in the printout, only those trade names mentioned in *New Drugs* are listed, together with trade names used by the author. Do not list a trade name which is identical with the generic name.

Complete drug names (both generic and proprietary) should be given. If in doubt regarding the form of a drug used, request more information from the author. (NOTE: *New Drugs* gives complete drug names only for those drugs that are monographed. Consult the nomenclature printout.)

Follow the style in examples given here. Note that trade names are listed alphabetically, that all parts of trade names are capitalized and italicized, and that a period is used at the end of each listing.

Furazolidone—*Furoxone.*
Neostigmine bromide—*Prostigmin Bromide.*
Hydrocortisone—*Cortef, Cortifan, Cortril, Hycortole, Hydrocortone.*

Do not list foreign trade names. If an author mentions a foreign trade name, this should be noted in parentheses at first mention in the text, together with the comparable US product.

phenprocoumon (Marcoumar [Britain]; Liquamar, comparable US product)

14.07—Brand names of manufactured items and names of manufacturers should be avoided when possible.

Most brand names can be replaced by generic terms or descriptive phrases: "cellophane tape" for "Scotch tape"; "petroleum jelly" for "Vaseline"; "rebound tumbling device" for "Trampoline." In some instances, however, the brand name must be retained as necessary information. In such cases it should appear in parentheses following a descriptive phrase. For example, a complex piece of equipment may be referred to as "a general-purpose digital computer (Bendix G-15)."

It is sometimes difficult to determine whether a name ascribed to equipment is a trade name or an acceptable nonproprietary designation. Questions should be referred to the Department of Medical Physics and Rehabilitation.

14.08—When two or more brands of the same product are compared, brand names and manufacturers' names can usually be avoided by providing code names (eg, clamp A vs clamp B). If necessary, to avoid confusion, the brand and manufacturers' names may be included in a footnote at the end of the paper.

14.09—Equipment or apparatus provided free of charge may be acknowledged in a footnote (**10.05**).

14.10—When a device is referred to as a "modified" type, the modification should be explained or an explanatory reference cited.

DISEASES

14.11—For correct names of diseases and syndromes, consult *Current Medical Terminology (CMT)* or *Dorland's Illustrated Medical Dictionary.*

14.12—Distinguish between a disease and a syndrome.

hyaline membrane disease	tegmental syndrome
Hodgkin's disease	Marfan syndrome

14.13—Eponyms should be avoided except when listed as preferred terminology in *CMT* or *Dorland's.* If an author uses an unpreferred eponymic term, this may appear in parentheses after the correct descriptive name at first mention in the text and first mention in the synopsis-abstract or summary.

exophthalmic goiter **(Graves' disease)**

diffuse interstitial pulmonary fibrosis (Hamman-Rich syndrome)

14.14—Capitalize an eponym but not the common noun associated with it (4.08). Words derived from eponymic names of diseases are usually not capitalized (4.07). Consult *Dorland's.*

Addison's disease	Parkinson's disease
addisonism	parkinsonism

14.15—A disease name derived from the name of an organism is not capitalized or italicized.

Infestation with *Schistosoma* results in schistosomiasis.

GENUS AND SPECIES

14.16—For the correct names of genera and species and rules for forming plurals, consult *Bergey's Manual of Determinative Bacteriology.*

14.17—After first mention of the singular form in the text, abbreviate genus name when used with species (1.08).

14.18—Italicize species, variety or subspecies, genus when used in the singular, and genus abbreviation. Do not italicize genus name when used in the plural, the name of a class, order, family, or tribe (12.05).

14.19—Capitalize genus when used in the singular, genus abbreviation, class, order, family, and tribe. In the text, do not capitalize species, variety or subspecies, or plural form of genus name. In titles, capitalize the plural of a genus name, but do not capitalize name of species or subspecies (4.16).

14.20—If correct plural can not be determined, add the word "organisms."

Salmonella enteritidis	*S typhimurium*	salmonellae
Treponema pallidum	*T genitalis*	treponemas
Proteus vulgaris	*P mirabilis*	*Proteus* organisms
Escherichia coli	*E aurescens*	*Escherichia* organisms

VIRUSES

14.21—Uniformity has not yet been clearly established for capitalization, spacing, and phrasing of virus designations. Some preferred forms are listed here. Note that Arabic numerals are used, not Roman. Alphabetical designations of virus groups or types are capitalized.

picornavirus myxovirus herpesvirus

adenovirus arbovirus

enterovirus rhinovirus poliovirus

influenza virus type A influenza virus

echovirus echovirus 30 echoviruses 29 and 32

coxsackievirus coxsackieviruses A and B

coxsackieviruses B3 and B5, or

coxsackievirus B types 3 and 5, or

group B coxsackieviruses 3 and 5

simian virus 40 (SV 40) Giles virus Frater virus

GLOBULINS

14.22—Use Greek letters (**23.00**), and subscript numerals where applicable, in designating globulins and globulin fractions.

α-globulin α_1-globulin α_2-globulin (Greek 25)

β-globulin (Greek 26) γ-globulin (Greek 27)

(*But:* agammaglobulinemia)

IMMUNOGLOBULINS

14.23—Nomenclature for human immunoglobulins has recently been revised.* At present, however, the older terminology is still used by some investigators. An author may use either terminology provided he does so consistently. Alternate forms may be indicated in parentheses if supplied by the author.

For heavy-chain immunoglobulins:

---

*Nomenclature for Human Immunoglobulins, *Bull WHO* **30**:447-449 (No. 3) 1964.

Older Usage	*Revised Usage*
γ, 7Sγ, 6.6Sγ, γ_2, γ_{SS}	γG or IgG
β_2A, γ_1A	γA or IgA
γ_1M, β_2M, 19Sγ, γ-macroglobulin	γM or IgM

For light-chain immunoglobulins:

Older Usage	*Revised Usage*
type I, 1, B	type K
type II, 2, A	type L

14.24—The two major groups of polypeptide chains should be designated by the terms "light" and "heavy." Do not designate as *L* and *H* or as *A* and *B*.

VERTEBRAE

14.25—For preferred terminology consult *Dorland's*.

14.26—In reports of clinical or technical data and in tables, specific designations of vertebrae and intervertebral spaces may be abbreviated without first being written out, if no confusion results. Abbreviate in accordance with examples.

vertebrae:	C-2	L-2	S-2	T-2
spaces:	C2-3	L2-3	S2-3	T2-3

ISOTOPES

14.27—Isotope numbers are used in AMA publications principally but not exclusively in connection with radioactive drugs. When the isotope number is being given for an element named by itself rather than as part of the name of a chemical compound, follow **14.28** if the element's name is spelled out and **14.29** if the chemical symbol is used instead. When the isotope number appears as part of the name of a compound, follow **14.30** if the compound has an approved nonproprietary drug name and **14.31** if it does not.

14.28—When the isotope number is given for an element named by itself rather than as part of the name of a chemical compound, it follows the element name in the same type face. It is not preceded by a hyphen even in adjectival use.

Why did God create uranium 235?

Substances labeled with iodine 125 generally have the same uses as those containing iodine 131.

He was studying phosphate 32 beta-particle emission.

As a solid therapy source, cobalt 60 (for which no nonproprietary name has been established) is marketed as cobium.
Exception: In its use as a drug (in the form of a colloid), radioactive gold (^{198}Au) should be referred to by its approved nonproprietary name: *gold Au 198.*

14.29—When the chemical symbol is used for an element referred to by itself rather than as part of a compound, the isotope number becomes a superscript and is placed to the left of the symbol.

Of the 13 known isotopes of iodine, ^{128}I is the only one that is not radioactive.

Use of the symbol as an abbreviation after the name of the element has been spelled out once should be handled as follows:

Radioactive chromium (^{51}Cr) has a half-life of 27.8 days. . . . The transmutation of ^{51}Cr to vanadium is accompanied by the emission of x-rays.

Do not use the symbol representing a single element (eg, ^{74}As) as an abbreviation for a compound (eg, sodium arsenate As 74).

14.30—Approved nonproprietary names of drugs containing radioactive isotopes consist of three parts: (1) the name that would be used for the drug if it contained only stable isotopes, followed by (2) the chemical symbol of the element that is radioactive, followed by (3) the isotope number of the radioactive isotope. The isotope number appears in the same type as the rest of the drug name (ie, not in superscript), and it is not preceded by a hyphen. When a drug has an approved nonproprietary name, that name is to be used in preference to the other ways of referring to the drug.

Sodium iodohippurate I 131 (*not* radioactive sodium idiohippurate, ^{131}I-labeled sodium iodohippurate, etc)

Cyanocobalamin Co 57 (*not* cyanocobalamin tagged with radioactive cobalt, etc)

Exception: The approved nonproprietary name *tritiated water* omits the chemical symbol and isotope number (**14.32**).

14.31—A compound that has not been given an approved nonproprietary drug name may be referred to in any of several descriptive ways.

Glucose labeled with radioactive carbon (^{14}C)

Glucose tagged with carbon 14

When the name of such a substance is abbreviated, use the superscript form of the isotope number, placed to the left of the chemical symbol, not the form prescribed in **14.30** for approved nonproprietary names.

Glucose ^{14}C (*not* glucose C 14)

14.32—In trade names of drugs, isotope numbers usually appear in the same position as they do in approved nonproprietary names (**14.30**), but they are usually joined to the rest of the name by a hyphen, and they are not necessarily preceded by the chemical symbol. Follow *New Drugs,* NEW NAMES, and the usages of individual manufacturers.

Sodium-Radio-Chromate Cr-51

Radio-Renografin-I 131

Racobalamin-57

14.33—The abbreviation *ul* (for *uniformly labeled*) may be used in parentheses to differentiate a preparation in which all molecules of a compound contain radioactive atoms from those in which the radioactive material has merely been admixed with the compound to be labeled.

Glucose ^{14}C (ul)

This form should be used in place of such designations as *glucose-U-*^{14}C.

14.34—Two isotopes of hydrogen have their own specific names, *deuterium* and *tritium,* which should be used instead of "hydrogen 2" and "hydrogen 3." In text matter, these specific names should also be preferred to the symbols ^{2}H or D (for deuterium, which is stable) and ^{3}H (for tritium, which is radioactive). The two forms of heavy water (D_2O and $^{3}H_2O$) should be referred to by the approved nonproprietary names *deuterium oxide* and *tritiated water.*

14.35—Except in trade names, do not use the prefix *radio-* with the name of an element. Use the word *radioactive* instead.

The illustrative medical article below is reprinted by permission of the *New England Journal of Medicine* and of the author. The format is appropriate to that journal and is a first-rate example of the practice recommended by the AMA.

OCCURRENCE OF ACUTE WERNICKE'S ENCEPHALOPATHY DURING PROLONGED STARVATION FOR THE TREATMENT OF OBESITY*

ERNST J. DRENICK, M.D.,† CLARICIA B. JOVEN, PH.D.,‡ AND MARION E. SWENDSEID, PH.D.§

LOS ANGELES, CALIFORNIA

THE more liberal use of prolonged starvation in the treatment of obesity has posed a number of questions regarding the metabolic requirements of body tissues under these conditions. The need for vitamin supplementation has not been studied adequately as yet.

It has been held that total fasting does not result in vitamin B_1 deficiency because vitamin need and vitamin excretion diminish after a few days.[1-3] Ziporin et al.[4] have postulated that thiamine requirements decrease with a lowering of caloric intake. Similarly, it has been noted that thiamine requirements fall with diminishing physical activity.[5] In the starving obese subject endogenous fat and small quantities of amino acids, the equivalent of 4 to 6 gm. of nitrogen per day, are being metabolized by various pathways to provide calories. Therefore, one might expect the need for thiamine to be minimal because this vitamin functions primarily as a coenzyme in the decarboxylation of alpha-ketoacids that arise in the metabolism of glucose derivatives and amino acids. Determinations of the urinary excretion of some of the components of the vitamin B complex have indicated a continued loss during starvation, with the possible depletion of tissue stores.[6,7] Gellene et al.[7] also found low serum levels of vitamins B_1, B_2 and B_6 and pantothenic acid. Nonspecific symptoms such as nausea, vomiting and hypotension were thought to have been ameliorated by vitamin supplementation.

Recently, an acute fulminating thiamine deficiency was observed in an obese man who had starved for a relatively short period. This unusual and unexpected development, to our knowledge, has not previously been described. The circumstances of this case are presented below.

CASE REPORT

M.W., a 35-year-old single truck driver, was admitted to the metabolic-balance ward of the Wadsworth Veterans Administration Hospital on February 16, 1965, complaining of marked overweight resulting in shortness of breath. He had started to gain excessively at the age of 20. Repeated and various reducing regimens in the past had proved ineffective. He claimed to be eating only enough food to satisfy his hunger, but calculations of his dietary record, obtained by the recall method, showed that he was consuming an average of 5600 to 6600 calories daily. His diet consisted of a normal variety of food, including ample meats, dairy products, bread and vegetables, with a calculated intake of thiamine ranging from 2.5 to 3.4 mg. per day. He denied the use of alcohol or drugs. There was no history of exposure to toxic agents, and he had had no significant illness in the past except for mastoid surgery in 1939.

On physical examination the patient was grossly obese, weighing 151.9 kg. (335 pounds) and being 180 cm. tall. The blood pressure was 130/86. No nystagmus was noted, and there was no double vision. The fundi were normal. The lungs and heart were normal. Examination of the abdomen was unsatisfactory because of obesity. The testicles were atrophic. Neurologic examination revealed no abnormalities.

A urinalysis, complete blood count, serum electrolytes and creatinine were within normal limits. The fasting blood sugar was 128 mg., the cholesterol 237 mg., and the uric acid 6.9 mg. per 100 ml. An electrocardiogram and x-ray film of the chest were normal.

The patient volunteered to be included in a group undergoing prolonged starvation as a treatment for "resistant obesity." During the starvation period a special study of vitamin-excretion rates was planned, and, therefore, no vitamin supplements were furnished. A 500-calorie diet was started on February 25. On this regimen, he lost 10 kg. (22 pounds) in 26 days. The starvation period lasted for 30 days (March 21 to April 20) and resulted in an additional weight loss of 17.2 kg. (38 pounds). Only water was permitted. The routine medication consisted of probenecid, 1 gm., and potassium chloride, 2 gm. daily. With starvation ketonemia developed (maximum level of 17.3 mg. per 100 ml.). The serum uric acid rose from 5.6 to 10.2 mg., the blood sugar fell to a low of 69 mg., and the cholesterol decreased to 115 mg. per 100 ml.; the serum sodium, potassium, carbon dioxide and chloride remained unchanged.

The patient did well until April 13 (23 days from the onset of starvation). From April 14 on, he had occasional mild nausea, with emesis, but remained ambulatory and active. On April 20, because of nausea, the fast was terminated, and for the following week, as is done routinely with refeeding, 90 gm. of glucose was administered daily in divided doses by mouth. On the following day, the 1st after termination of the fast, he complained of dizziness, which increased over a 5-day period. Nausea gradually subsided, and on April 28 orange juice was started (6 ounces 3 times daily) in addition to glucose. On May 2, 12 days after refeeding was begun, the patient complained of double vision. He had become inactive, remained in bed and refused to bathe. On May 3 he became highly irritable and unco-operative and complained of double vision and of ringing in the ears. Examination of the eyes on this day for the 1st time revealed a bilateral partial 6th-nerve paralysis. One day later he became incontinent, lethargic and confused; he did not know his age or where he was or the date. On May 5 his memory was more seriously impaired. He was unable to perform the simplest calculations or abstract simple proverbs. He appeared to be extremely dull mentally. The sense of smell was intact on the left but impaired on the right side. The visual fields were intact; the pupils were equal and regular, and the disk margins appeared sharp. There was a clockwise rotatory nystagmus, more severe on the left. Only a 2-mm. horizontal medial eye movement, with no excursion lateral to the midline,

*Supported in part by a grant from the National Vitamin Foundation.

†Section chief, Wadsworth Veterans Administration Center; assistant clinical professor of medicine, University of California, Los Angeles, School of Medicine.

‡Postdoctoral fellow, University of California, Los Angeles, School of Public Health.

§Professor of nutrition and biologic chemistry, University of California, Los Angeles, School of Public Health.

was discernible. Downward gaze was normal; upward gaze was limited. The patient manifested continuous facial grimacing and teeth gnashing. A partial weakness of the nerves of the left lower part of the face was noted. No abnormalities of the 5th, 8th, 9th, 10th, 11th and 12th cranial nerves were apparent. Motor and sensory perception was normal, as were superficial skin reflexes. The left ankle jerk was absent; Romberg's sign and gait could not be tested. There was no nuchal rigidity and no abnormal toe reflexes.

X-ray films of the skull, an electroencephalogram and a cerebral photoscan using radioactive mercury (Hg^{203}) were within normal limits, as were a complete blood count, urine, serum electrolytes, creatinine, uric acid and cholesterol on the same day. No ketone bodies were present in the serum or urine. No lumbar puncture was carried out, but on the basis of available evidence, a mass lesion or demyelinating disease was considered unlikely. At 5 p.m. on May 5 an intravenous infusion of 1000 ml. of 5 per cent dextrose in water containing 400 mg. of thiamine hydrochloride was started. Improvement was dramatic. By 7 p.m. the patient was sitting up to eat dinner. Another 400 mg. of thiamine hydrochloride was administered during the night. By 3:30 a.m. on May 6 he was completely alert and oriented. Two days later eye movements were fully restored, the mental state was normal, and he was up and about. A 500-calorie diet containing 50 gm. of protein was given, with 1 therapeutic vitamin capsule daily. Some rotatory nystagmus remained, and some diplopia on occasion. He continued to complain of this and a slight unsteadiness when walking. Within 4 weeks after the 1st administration of thiamine chloride all subjective abnormalities had subsided. A trace of rotatory nystagmus could still be elicited 3 months later but all other objective neurologic abnormalities had reverted to normal.

Complete urine collections were carried out before and throughout the starvation period. Thiamine chloride was determined daily.[8] Table 1 demonstrates total B_1 excretion for this subject and 4 other starving patients who did not receive any vitamin supplements. Cases 1 and 2 had been taking a 500-calorie diet before starvation whereas the other 3 had been eating normally and started starvation abruptly. The lower urinary thiamine content during the initial days of starvation in Cases 1 and 2 seemed to reflect the preliminary low-calorie intake.

Discussion

Wernicke's syndrome is characterized by paralysis of extraocular muscles, nystagmus, ataxia and mental changes. Petechial bleeding in the brainstem is found at autopsy. The syndrome and pathological findings have been reproduced experimentally in animals given vitamin-deficient diets.[9] This condition has been noted clinically in patients with chronic alcoholism or malnourishment.[10] In such patients, the paralysis of the eye muscles, the ataxia and the nystagmus were found to clear with administration of thiamine chloride, but little improvement was noted in the mental abnormalities such as impairment of memory and confusion.[11] Therefore, no reliable proof existed for the theory that the thiamine deficiency was responsible for part or all of these mental disturbances. In the case described, the rapid improvement of the mental picture that followed infusion of thiamine seems to furnish evidence that the severe progressive mental deterioration associated with the other typical changes of Wernicke's syndrome can be a sequel of the thiamine deficiency. The failure to find improvement of the mental changes in cases of long standing thiamine deficiency may be due to permanent damage to the structure of the cortex. The cerebral cortex may well be more vulnerable to the harmful effects of the vitamin deficiency than the lower centers in the brainstem. With extensive damage of long duration, specific therapy may therefore be only partially effective. In this case the syndrome developed rapidly, and treatment was begun before permanent cortical changes occurred.

Table 1. *Urinary Thiamine Excretion during Starvation.*

Calorie Intake*	Thiamine Excretion†							
	5 days before starvation	1-5 days of starvation	6-10 days of starvation	11-15 days of starvation	16-20 days of starvation	21-25 days of starvation	26-30 days of starvation	30 days of starvation
	micro-gm./24 hr.	*micro-gm./24 hr.*	*micro-gm./24 hr.*	*micro-gm./24 hr.*	*micro-gm./24 hr.*	*micro-gm./24 hr.*	*micro-gm./24 hr.*	*micro-gm./24 hr.*
500:								
Case 1	122	54	37	27	20	0	0	0
Case 2	45	22	15	13	5	0	0	0
Ad libitum:								
Case 3	204	156	100	53	43	40	27	0
Case 4	–	175	60	33	31	15	–	–
Case 5	–	180	56	36	28	–	–	–

*Case 1 on this diet for 30, & Case 2 for 20 days; values represent last 5-day period on diet.

†5-day pools.

In this patient characteristic symptoms and signs of Wernicke's syndrome did not appear during the starvation period proper, but only after several days of carbohydrate feeding. This is in agreement with the observation of Phillips et al.,[11] who noted accentuation of ocular-muscle paralyses in 6 patients with Wernicke's syndrome after a regimen of glucose and saline solution had been instituted. They suggested that a further depletion of thiamine stores occurred as a result of the added caloric load.

Keys and his associates[12] did not note any vitamin deficiency symptoms in their volunteers on prolonged semistarvation. Gellene's[7] findings, in starving subjects, of nausea, weakness and hypotension were ascribed to thiamine deficiency. However, other metabolic abnormalities known to occur in fasting, such as ketosis, mineral and protein depletion and hyperuricemia, may have been responsible for these symptoms.

In the other 4 starving patients whose vitamin-excretion data were described above, as well as several others in whom vitamin excretion was not measured, no clinical evidence of a thiamine-deficiency syndrome developed during one or more months of fasting or during subsequent refeeding with glucose. This was true regardless of whether starvation was begun abruptly or after a preliminary period on 500 calories. Previously, we had encountered only 1 case in which multiple vitamin deficiencies developed after more than two months of starvation.[13] However, a history had been obtained suggestive of a poorly balanced diet, high in alcoholic beverages. This may have led to an asymptomatic vitamin depletion before starvation was instituted.

Brin[14] observed no objective characteristic clinical deficiency symptoms in patients given a virtually thiamine-free isocaloric diet for periods up to six weeks. In these subjects physical activity remained constant, and no weight loss was registered for the first three weeks on this regimen. Therefore, thiamine requirements presumably remained constant. Ziporin et al.[4] noted that thiamine excretion ranged around 14 microgm. in twenty-four hours and ceased within eighteen days in subjects fed a thiamine-restricted diet. He concluded that these phenomena indicated a depleted "thiamine status" of a subject. He further calculated the minimum need of an average healthy person as about 270 microgm. of thiamine per 1000 calories. If one assumes that only about 25 to 40 gm. of carbohydrate derivatives are metabolized in the starving person, theoretically, only a fraction of the thiamine requirement of 270 microgm. should be necessary to prevent depletion. During the 500-calorie food-restriction period the daily vitamin intake was about 380 microgm. During the same period nitrogen-balance data indicated daily losses of approximately 100 gm. of protein tissue. Assuming that this tissue contains quantities of thiamine similar to muscle, one could infer that 50 microgm. of thiamine is released daily.[15] Although these combined amounts of dietary and released thiamine seem to be in the range of a theoretically adequate minimum requirement the thiamine-excretion data in the first five-day period of starvation do not seem to support this hypothesis. In comparison with the 3 subjects who ate normally before starvation, the 2 patients on a preliminary low-calorie intake excreted much smaller amounts of urinary thiamine in the beginning of, and throughout, starvation, and the urine was devoid of detectable thiamine after a much shorter period of starvation. By Ziporin's criteria these findings indicated a "depleted thiamine status."

It can be calculated from urinary nitrogen loss that Case 1 metabolized 4.4 kg. of protein tissue during starvation and thus released 2.2 mg. of thiamine, which would provide 73 microgm. of this vitamin per day. An average of 6.9 microgm. per day was measured in the urine, denoting further thiamine depletion.

These data support the conclusion that in the starving or semistarving subject, endogenous release of thiamine does not satisfy tissue needs for this vitamin despite a minimal carbohydrate metabolism. Clinical deficiency manifestations have been shown to develop in 1 patient within one month of starvation.

The variable and unforeseeable responses to starvation in different persons suggest that this treatment regimen should only be carried out in a hospital environment with strict supervision of the patient and that ample vitamin supplementation during starvation is mandatory.

References

1. Caster, W. O., Condiff, H., Mickelsen, O., and Keys, A. Excretion of pyrimidine and thiamine by man in different nutritional states. *Federation Proc.* **4**:85, 1945.
2. Perlzweig, W. A., Huff, J. W., and Gue, I. Excretion of thiamine, riboflavin and nicotinic acid by fasting men. *Federation Proc.* **3**:62, 1944.
3. Wollenberger, A., and Linton, M. A., Jr. Metabolism of glucose in starvation and water deprivation. *Am. J. Physiol.* **148**:597-609, 1947.
4. Ziporin, Z. Z., Nunes, W. T., Powell, R. C., Waring, P. P., and Sauberlich, H. E. Thiamine requirement in adult human as measured by urinary excretion of thiamine metabolites. *J. Nutrition* **85**:297-304, 1965.
5. Keys, A., Brožek, J., Henschel, A., Mickelsen, O., and Taylor, L. H. *The Biology of Human Starvation.* 2 vol. Vol. 1. 763 pp. Minneapolis: Univ. of Minnesota Press, 1950. P. 468.
6. Swendseid, M. E., Schick, G., Vinyard, E., and Drenick, E. J. Vitamin excretion studies in starving obese subjects: some possible interpretations for vitamin nutriture. *Am. J. Clin. Nutrition* **17**:272-276, 1965.
7. Gellene, R., Frank, O., Baker, H., and Leevy, C. M. B-complex vitamins in total food deprivation. *Federation Proc.* **242** (2):314, Part 1, 1965.
8. Deibel, R. H., Evans, J. B., and Niven, C. F., Jr. Microbiological assay for thiamin using *Lactobacillus viridescens. Bact. Proc.*, p. 28, 1957.
9. Alexander, L., Pijoan, M., and Meyerson, A. Beri-beri and scurvy: experimental study. *Tr. Am. Neurol. A.*, pp. 135-139, 1938.
10. Riggs, H. E., and Boles, R. S. Wernicke's disease: clinical and pathological study of 42 cases. *Quart. J. Stud. on Alcohol* **5**:361-370, 1944.
11. Phillips, G. B., Victor, M., Adams, R. D., and Davidson, C. S. Study of nutritional defect in Wernicke's syndrome: effect of purified diet, thiamine, and other vitamins on clinical manifestations. *J. Clin. Investigation* **31**:859-871, 1952.
12. Keys, A., Brožek, J., Henschel, A., Mickelsen, O., and Taylor, L. H. *The Biology of Human Starvation.* 2 vol. Vol. 1. 763 pp. Minneapolis: Univ. of Minnesota Press, 1950. P. 469.
13. Drenick, E. J., Swendseid, M. E., Blahd, W. H., and Tuttle, S. G. Prolonged starvation as treatment for severe obesity. *J.A.M.A.* **187**:100-105, 1964.
14. Brin, M. Erythrocyte transketolase in early thiamine deficiency. *Ann. New York Acad. Sc.* **98**:528-541, 1962.
15. Ferrebee, J. W., Weissman, N., Parker, D., and Owen, P. S. Tissue thiamin concentrations and urinary thiamin excretion. *J. Clin. Investigation* **21**:401-408, 1942.

Comments on the Article

The *title* illustrates the recommendations of the AMA document *Advice to Authors*. The first word, "Occurrence," is used because the article reports a phenomenon both unexpected and hitherto unreported. The next phrase gives the specific term for what occurred. The third phrase indicates when it occurred, and the last phrase gives the context for the condition. The title is succinct, yet no ambiguities or complications exist. Keywords are given, and a reference librarian would have no difficulty cross-referencing the article.

The *headnote* gives credit to the foundation that partially supported the work. (It is an unnumbered note as are all notes except those referring to other literature.) Had other formal support been given the work, that support would have been noted here. Had a specific laboratory been involved, it would have been cited in the headnote. Because a hospital was involved, its name is given in the first sentence of the Case Report.

The *by-line* and *author credits* are required in medical articles. The senior author is listed first, the rest in alphabetical order. Names are given first-name-first and are followed by the highest degree. Had either of the PhD's also been MD's, both degrees would have been listed. The credits give the names and addresses of the facilities to which each author is attached. Had this article appeared in one of the eleven journals of the AMA, an address for reprint requests would have been given.

The *Case Report* is written entirely in the past tense. Only essential data are given, without comment, and only standard abbreviations are used. Because four other patients had received similar treatment, they are mentioned in the last paragraph of the case report. Their data, however are contained in Table 1. The use of the table is effective in presenting and contrasting similar data. A more elaborate and inclusive table might have been prepared, but it would serve no additional purpose in the article.

The *discussion* reviews the literature. Where there is a gap in knowledge that the authors wish to comment upon, they do so only to the extent their data permit. Their opinions are carefully hedged: "The cerebral cortex may well be...," "specific therapy may therefore be only...," "other metabolic abnormalities may have been responsible...."

The *summary* is contained in the last two paragraphs. (A journal of the AMA would have required a separate subheading.) It is written in the third person, contains no new data, makes no cross-references, and is less than 200 words in length.

The fifteen references mention only two textbooks. This reassures the reader that the work is timely and informed. The arrangement of the references is that preferred by the American Medical Association.

H. SOCIAL SCIENCE JOURNALS

Within the social sciences, journals in the disciplines of psychology, sociology, economics, history, government, and law use dissimilar practices. The examples that follow have been selected to show variations in style.

In the field of psychology, both content and organization are fixed by the *Publications Manual* of the American Psychological Association *for the twelve journals published by the association.* Other journals of psychology follow their own style guides. Because these journals greatly outnumber those published by the APA and because the APA *Publications Manual* is readily available for $1.00 from the Association's headquarters on 1333 16th Street N. W., Washington, D. C., only a description of the type of article sought by each APA journal is reprinted by permission below.

The *American Psychologist* (monthly) is the Association's professional journal. A large part of its annual contents consists of the official papers of the APA: the program of the annual meeting, the reports and proceedings of the Council of Representatives, the reports of boards and committees, and news notes. It also publishes some articles submitted voluntarily. Appropriate topics for contributed articles include the following: the practice of psychology as a profession, methods and resources for graduate and professional education, the work, status, and earnings of psychologists, the relation of psychology to other professions, novel applications of psychology, and scientific articles of interest to psychologists in many fields. Preference is given to articles of broad interest which cut across the subdivisions of psychology. The section *Comment* publishes correspondence on controversies of current interest, as well as brief articles and notes.

Contemporary Psychology: A Journal of Reviews (monthly) began publication in 1956 as a journal of book reviews. It took over all book reviewing from the *Psychological Bulletin*, the *Journal of Abnormal and Social Psychology*, the *Journal of Applied Psychology*, and the *Journal of Consulting Psychology.* The last named journal continues to review tests, but *Contemporary Psychology* covers books about tests. Short abstracts are avoided as belonging to *Psychological Abstracts.* Evaluative criticism is favored, and reviews are long enough to permit this sort of discussion. While all reviews are written on invitation, suggestions as to important books and appropriate reviewers are welcomed. The most important books, especially when they deal with controversial topics, may have more than one review printed simultaneously. There is a department, *Films*, in which educational and research films are described and assessed. A page of editorial comment discusses news about books and other matters that concern both book writing and this journal. Brief letters from readers and authors about the published reviews are welcomed insofar as space permits their publication.

Because *Contemporary Psychology* prints only reviews, many of the rules cited in this *Manual* do not apply. The editor supplies certain instructions with respect to literary style and form to each person who is invited to write a review.

The *Journal of Abnormal and Social Psychology* (bimonthly, two volumes annually) is devoted to both abnormal and social psychology. Its emphasis is scientific as distinguished from practical, and it is concerned with basic research and theory rather than with techniques and arts of practice. Abnormal psychology is broadly defined to include papers contributing to fundamental knowledge of the pathology, dynamics, and development of personality or individual behavior, including deterioration with age and disease. Articles concerned with psychodiagnostic techniques are evaluated with respect to their contribution to an understanding of the psychological principles of diagnosis; those concerned with psychotherapy are judged in terms of their contribution to an understanding of the therapeutic process. Case reports which serve to pose or to clarify important theoretical problems, or which promise to be useful in teaching, are often published. From the social area, this journal gives preference to papers contributing to basic knowledge of interpersonal relations, and of group influences on the pathology, dynamics, and development of individual behavior.

The *Journal of Applied Psychology* (bimonthly) favors manuscripts reporting original investigations in any field of applied psychology except clinical psychology and personal counseling. A descriptive or theoretical article is occasionally accepted if it deals in a distinctive manner with a problem of applied psychology. The policy is, however, to favor papers dealing with quantitative investigations of direct value to psychologists working in the following fields: vocational diagnosis and occupational guidance; educational diagnosis, prediction, and guidance at the secondary school level and higher; personnel selection, training, placement, transfer, and promotion in business, industry, and government service including the armed forces; supervisory training in business, industry, and government; biomechanics or design of machines to fit the human operator; illumination, ventilation, and fatigue in industry; job analysis, description, classification, and evaluation; measurement of morale of executives, supervisors, or employees; surveys of opinion on social or political issues; and psychological problems in market research and in advertising.

The *Journal of Comparative and Physiological Psychology* (bimonthly) contains original experimental contributions to physiological, comparative, and sensory psychology. Experiments utilizing human and subhuman subjects are given equal consideration. Physiological psychology is regarded as including correlational studies of any aspects of behavior and of the neurological and/or biochemical mechanisms underlying behavior. Theoretical interpretations of specific experimental discoveries are encouraged.

The *Journal of Consulting Psychology* (bimonthly) is the clinical journal of the APA. It is devoted primarily to original research relevant to psychological diagnosis, psychotherapy and counseling, personality, and the dynamics of behavior. Although quantitative studies are given priority, relevant theoretical contributions, case studies, and descriptions of clinical techniques may also be acceptable. The journal publishes *Brief Reports*, which are one-page condensations of minor or specialized research studies, if the author will make a full report available in mimeographed form and deposit it with the ADI (see sec. 6.2). The section *Psychological Test Reviews* publishes descriptive and critical reviews of newly published tests, questionnaires, projective techniques, and other assessment methods of interest to clinical psychologists.

The *Journal of Educational Psychology* is devoted primarily to the scientific study of problems of learning, teaching, and measurement of the psychological development of the individual. It contains articles on the following subjects: the psychology of school subjects; experimental studies of learning; the development of interests, attitudes, and personality, particularly as related to school adjustment; emotion, motivation, and character; mental development; and methods, including tests, statistical techniques, and research techniques in cross-sectional and developmental studies.

The *Journal of Experimental Psychology* (monthly, two volumes annually) publishes articles intended to contribute toward the development of psychology as an experimental science. Experimental work with normal human subjects is favored over work with abnormal or animal subjects. Studies in applied experimental psychology or engineering psychology may be accepted if they have broad implications for experimental psychology or for behavior theory.

Psychological Abstracts (bimonthly) contains noncritical abstracts of the world's literature in psychology and related subjects. Unlike other APA journals, it contains no original articles and therefore solicits no contributions. Competent abstracters are almost always needed, however, especially to cover foreign-language journals, and books and periodicals in fields related to psychology. Abstracters are appointed by arrangement with the Executive Editor.

Psychological Bulletin (bimonthly) contains critical reviews of the literature in all fields of psychology, methodological articles, and discussions of controversial issues. Reports of original research or original theoretical articles are not accepted.

Psychological Monographs[2] (appearing at irregular intervals making an annual volume of about 500 pages) publishes reports of fundamental research. Preference is given to reports of comprehensive experimental investigations which do not lend themselves to adequate presentation in other journals. There is no fixed limit to the number of pages permitted in any monograph. An important programmatic study, for example, will usually require more pages than a less extensive research.

The *Psychological Review* (bimonthly) is devoted to articles in general and theoretical psychology. This area is obviously difficult to define, but preference is given to manuscripts which contribute broadly to fundamental concepts and theory. Papers that present surveys of literature, report experiments, or deal with applications are not ordinarily appropriate. The *Psychological Review* does not publish book reviews.

[2] A monograph in the *Psychological Monographs* series differs in some respects from a journal article. Its form is more like that of a book. Authors of monographs should consult the article by Conrad (8) and also note the special instructions given in sections 4.9 and 9.11 of this manual.

The *American Journal of Psychology* is typical of the independent journals of psychology. Its "Style Sheet" is reprinted in its entirety by permission below. Following the "Style Sheet" is an illustrative article reprinted by permission from the same journal.

STYLE SHEET OF THE AMERICAN JOURNAL OF PSYCHOLOGY

The Style Sheet, published here for the information and guidance of contributors, is a summary of present usage in this JOURNAL. It is the product over the years of many minds—the present editors' inheritance from their predecessors. Changes in style have been made in the past and they will unquestionably be made, in keeping with the advance of the times, in the future, but for the present this Style Sheet will govern the preparation of copy and the handling of proof of the JOURNAL. Authors submitting manuscripts are requested to follow it. It makes no pretense of completeness; for more general rules, authors should consult standard style books.[1]

GENERAL

(1) Spelling. The AMERICAN JOURNAL spells the American way, *e.g. analyze, behavior, counseling, disk, esthetics, gray, questionary, synesthesis,* instead of *analyse, behaviour, counselling, disc, aesthetics, grey questionnaire,* and *synaesthesia,* respectively. The first spelling given in *Webster's New International Dictionary,* 2nd unabridged edition, 1949, is used.

(2)Taboos. Barbarisms, the adjectival use of nouns, and the contraction 'and/or' are taboo. If the avoidance of the adjectival use of nouns leads to involved constructions, the usage is permitted but in that case the nouns should be hyphenated, *e.g. stimulus-object.* The hybrid symbol 'and/or' evidences a poverty of words; it often accompanies barbarisms and the use of nouns as modifiers.

(3) Sex. Animals used as subjects are referred to as *male* or *female;* human beings, as *men* or *women,* or *boys* or *girls.*

PRINTING

(1) Type faces. The JOURNAL uses large capitals (three underlines), small capitals (two underlines), italics (one underline), and italic large capitals (four underlines). Boldface type (wavy underline) is used as occasion demands but it is avoided insofar as possible because it makes the printed page look spotty.

(2) Type-sizes. Three sizes of type are used. The sectional headings (Apparatus, Notes and Discussions, and Book Reviews) are set in 14 point (pt.); the titles of articles in 10 pt. large capitals; the names of authors below the titles in 10 pt. large and small capitals; the introduction, statement of problem, results, discussion, and summary in 10 pt. on 12 pt. matrix (10/12). Historical reviews, descriptions of method and procedure, short articles of six or less pages, and book reviews are set in 8 pt. on 10 pt. matrix (8/10). Footnotes are set on 8 pt. solid, *i.e.* 8 pt. on 8 pt. matrix (8/8).

---

[1] See Publication Manual, *Psychol. Bull.,* 49, 1952, 389-445, that was prepared by the Council of Editors of the American Psychological Association.

The Text

(1) Divisions. The text is divided by centrally placed divisions and marginal stubs at the left. The central divisions are set in large and small capitals and if numbered, as in reporting of separate experiments, Roman numerals are used. *E.g.* EXPERIMENT I.

In short articles use marginal stubs throughout to mark the divisions of the paper. Use stubs also in long articles to indicate subdivisions. The stubs are set in italics (lower case) and they should be short. They are to indicate subdivisions—not to tell a story. They should not be assumed in the exposition. The stubs are not numbered in short articles but in long articles they may be. When numbered, Arabic numerals (italic), set in parentheses before the subtitle, are used for the first order of subordination. For the second and third orders of subordination use italic letters of the alphabet and lower case Roman numerals set in parentheses. For example:

EXPERIMENT I

Problem.
Method and procedure.
  (1) Apparatus.
    (a) Accessories.
      (i) Loudspeaker.
      (ii) Microphone.
Results.
    (1) With tones of 8000 cycles and less.
    (2) With 10,000 cycles.
Discussion.

EXPERIMENT II

(Repeat as much of the above as necessary.)

DISCUSSIONS AND CONCLUSIONS

SUMMARY

In articles in which there are no center heads, *Discussion, Conclusions,* and *Summary* are run as marginal stubs.

(2) Titles in text: (a) Articles. Titles of articles and essays, chapters and sections of books, and unpublished works are enclosed in double quotation marks. The first letter of the first word is capitalized, all others are set in lower case.

(b) Books. Titles of published books, plays, pamphlets, periodicals, classical works, and noted poems and operas are set in lower case italics with the principal words capitalized.

(3) Quotations. All quotations should faithfully duplicate the originals in wording, spelling, and interior punctuation. Any exceptions, such as italicizing certain words for emphasis, should be explained in an accompanying footnote. When the quotations are short they are run in the text and set in double quotes; when long (five or more lines) they are set in a separate paragraph in 8 pt. solid and without quotes.

For ellipses within quotations use three spaced periods (. . .) and leave a space before the first period and the last word. If an ellipsis is made between two successive sentences, use three spaced periods in addition to the period ending the first sentence.

Interpolations (comments or explanations) within quoted matter should be enclosed in square brackets, not parentheses. A common interpolation is [sic] used to invite the reader's attention to the fact that the quotation is accurate.

Quotations from copyrighted books and periodicals legally require permission from the author or publisher or both. This requirement is ignored for short quotations but for quotations of 300 and more words it should be observed. The task of obtaining the permission is the author's.

Single quotes are used within double quotes. In cases when the quotation is set in 8 pt. solid and the double quotes are omitted, the subquotes are then raised to double quotes.

(4) Punctuation. Do not over-punctuate but use the various punctuation marks as clarity demands.

(a) Commas. In a series of three or more coördinates, use a comma before 'and' and 'or.' The comma should not be used in conjunction with a dash but it should be used after a parenthesis (such as this), if the context requires it.

(b) Dashes. The length of a dash is indicated by the editor. If he writes 'en' above a dash in the text, the compositor will set an 'en' dash, that is a dash as long as the pica 'N.' If he places an arabic 1, 2, or 3 on the dash in the text, the compositor will set a dash that is as long as 1, 2, or 3 pica 'Ms' respectively. The dash below the title and the author's name in the heading of an article is 4 pica 'ems' and is indicated by writing '4' above the dash. One-em dashes should be used to indicate pauses or breaks that are not as complete as those indicated by semicolons.

(c) Exclamation marks. Exclamation marks are rarely used.

(d) Hyphens. Hyphens are indicated by two short parallel vertical lines with short dash midway between them. They ligate two nouns (*e.g.* stimulus-point) into one word to avoid the adjectival use of nouns.

(e) Periods. In addition to closing sentences in text and footnotes, periods end all footnote citations. They are not used after incomplete sentences such as center headings, table captions, and legends to the figures. They do, however, follow marginal stubs.

(f) Parentheses. Parentheses are used to enclose an explanation, authority, definition, reference, translation, or other matter not strictly belonging to the sentence.

(g) Brackets. The use of square brackets is restricted to interpolations within quotations; to editorial comments, corrections, notes, or explanations; and to the numbering of formulas and equations on the righthand margin.

(h) Apostrophe. The apostrophe marks the elision of a syllable (*I've* for *I have*), of the century in dates (*The spirit of '76*), and the possessive case. Elisions rarely appear in scientific exposition. The possessive is indicated by an apostrophe and the addition of the letter *s* for all nouns in the singular number, whether proper or not, and all nouns in the plural ending with any letter other than *s;* as *man's, men's, Charles's, witness's.* The only exceptions are established idioms; *e.g. for conscience' sake, for goodness' sake, for righteousness' sake.* All plural nouns ending in *s* form the possessive simply by the addition of the apostrophe; as, *boys', horses'.* For monosyllabic proper names ending in *s* or another sibilant, add an *s, e.g., Keats's poems, Marx's doctrines.* For proper names of more than one syllable, add only the apostrophe, *e.g. Ebbinghaus' experiments.*

(i) Quotation marks. For quotations within the text, use double quotation marks, then, for quotations within quotations, single marks. Double marks are reserved for actual quotations; they are never used for emphasis nor to designate a special use of a word or phrase—single quotes, or italics, or capitalization are used for those purposes.

Quotation marks are set outside of periods and commas, but inside of colons, semicolons, and question and exclamation marks unless the punctuation is actually part of the matter quoted, in which case place the quotation marks outside.

(j) Accents, diereses, and diacritical marks. In quoting foreign articles, duplicate faithfully all accents and diacritical marks. In German words use the umlaut except for initial capitals where *e* is used to indicate it, as *Ueber* instead of *Über*. The dieresis is always used in words with doubled vowels; *as zoölogy, coöperate, reëxamine;* and with *naïve.*

(5) Numerals. Cardinal numbers below 10 are usually spelled out except when they are used before abbreviations; as *two rooms, nine stimulus-objects.* When used before abbreviations, they, in common with all numbers, are printed in Arabic; as *4 cm., 7 ft.* The numbers 10 and above are set in Arabic except at the beginning of a sentence, *e.g. Ninety Ss were used in this study.*

Exceptions to the rule of spelling out 'nine' and numbers below occur when numerals 10 and above occur in the same or contiguous groups of sentences; *e.g. Our Os, 15 in number (7 women and 8 men), were highly trained.*

Ordinal and nominal numbers are usually spelled out. Exceptions occur in the Book Review Section where different editions are indicated, as, *2nd ed., 3rd rev. ed.,* and in the text where points or results are listed in sequence when arabic numbers within parentheses may be used, as (1), (2), (3), etc., in place of *First, Secondly. Thirdly,* etc.

Dates and page-numbers are not spelled out. For dates use month, day, and year, as *October 20, 1954.* References to centuries are spelled out, as *eighteenth century, twentieth century.*

(6) Capitalization: (a) Titles. In English titles of publications, capitalize the first and all principal words; do not capitalize articles, prepositions, and conjunctions. For foreign titles, follow the foreign style.

(b) Abbreviation of psychological terms. Standard abbreviations of psychological terms and others that an author wishes to make to avoid frequent repetition of phrases are capitalized and run in italics without periods. *E.g.* (standard) *O, S, E, SD, CFF, RL, DL, UDL, LDL, PSE;* (author's) *AL* (adaptation-level), *TOE* (time-order effect), *CAL* (comparative adaptation-level), *HMD* (half-meridional difference). For the author's abbreviations, see this JOURNAL, 66, 1953, 630 ff., and 67, 1954, 327 ff.

(c) For emphasis. Capitalize such words as *experiments, conditions, trials,* etc., for emphasis—to make them stand out from the text—when they are used in connection with specific experiments, conditions, or trials; as, *Experiment II, Condition 1, Trial 3.* (For examples see this JOURNAL, 67, 1954, 264 f.)

(7) Italics: (a) For emphasis. Use italics sparingly for emphasis that they may be used effectively when occasion warrants them. They lose their significance when frequently employed.

(b) Foreign words. Single foreign words or short phrases are italicized. Foreign sentences are not; they are set in roman double quotes.

(c) Books and periodicals. Titles of books are set in lower case italics with principal words capitalized; titles of periodicals are abbreviated in accordance with the principles adopted by the editors of *A World List of Scientific Periodicals: 1900-1905* and set in italics.[2]

[2] For a list of periodicals and their abbreviations, see *Psychol. Abstr.,* 21, 1953, 835-840; for a list of abbreviations, see A publication manual of the American Psychological Association. *Psychol. Bull.,* 49, 1952, 436-438.

(d) Botanical and zoölogical names. The names of botanical and zoölogical classes, families, genera, and species are italicized. The first letter of the class, family, and genus is capitalized; of the species it is not.

(e) Formulae. Letters used in formulae and equations are set in italics, upper or lower case as the occasion demands.

(f) Signs. If the letters used in plates or figures as signs are printed in italics, italics should be used in the text when reference is made to them.

Documentation

(1) Footnotes. Terminal bibliographies are not run in the JOURNAL. All references and digressions (explanations and discussions not of immediate concern to the points being developed in the text) are given in consecutively numbered footnotes.

(a) The logic of footnotes. Footnotes save space as they are set solid whereas terminal bibliographies are usually leaded. Moreover, only items actually referred to in the text are given in footnotes whereas terminal bibliographies are often padded, *i.e.* items not referred to in the text are frequently included.

References are of greater aid and convenience to the reader if they are placed on the pages of their citations. It is much easier to cast one's eyes to the bottom of a page when a citation is met in the text than to turn to a terminal bibliography.

Footnotes permit of specific citations, *i.e.* the actual pages are given upon which the points referred to in the text appear. Terminal bibliographies give inclusive pages. Reference to an article given in a terminal bibliography is consequently 'blind,' *i.e.* the reader must search through the entire article or book to find the point or points at issue. Such references are of little help to him.

To avoid blind references, journals running terminal bibliographies sometimes number the items in the bibliography and give the number and specific pages in parentheses in the text; as (**9**, pp. 45-47). If the item-number is printed in boldface, as in some journals, the page appears spotty and is a poor example of the art of printing. Such references, moreover, catch the reader's attention and interrupt the flow of eye movements to a far greater extent than footnote references which are given in small superscripts at natural pauses, *i.e.* at punctuation marks.

(b) References in the text. Starting with 1, footnote references in the text are numbered consecutively in superscript Arabic figures. Insofar as possible, these figures are placed at punctuation marks, preferably at periods. When clarity (which takes precedence over every other rule) demands it, footnote references may be placed elsewhere.

(c) Footnotes concerned with books and articles. When the footnotes refer to published books or articles, the following information should be given and in the order indicated below.

(i) Name of author. The author's surname preceded by his initials (or first name if he has only one given name) is run first. Initials or given name are used only the first time an author is cited unless other authors of the same surname are cited. In subsequent citations to the work of a given author, drop the initials or first name—give only the surname. For example:

[1] E. B. Titchener, Structural and functional psychology, *Phil. Rev.*, 8, 1899, 290-299.
[6] Titchener, The psychological concept of clearness, *Psychol. Rev.*, 24, 1917, 43-61.

Exceptions are made to this rule, however, when subsequent citations are to articles of plural authorship. If the initials of any of the co-authors of a joint article are given, and they should be if any of the co-authors is cited for the first time, then give the initials for all. For example:

[5] T. A. Ryan, Interrelations of sensory systems in perception, *Psychol. Bull.*, 36, 1940, 659-698.
[6] T. A. Ryan, C. L. Cottrell, and M. E. Bitterman, Muscular tension as an index of visual efficiency, *Illum. Engng.*, 43, 1948, 1074-1081.

If, however, the same or another joint article or book by the same co-authors is cited, then drop the initials of all the authors, as:

[8] Ryan, Cottrell, and Bitterman, *op. cit., Illum. Engng.*, 43, 1948, 1080.
[9] Ryan, Cottrell, and Bitterman, Relation of critical fusion frequency to fatigue in reasoning, *Illum. Engng.*, 48, 1953, 385-391.

(ii) Titles of books. If a book is cited, then continue, after giving the author's initials and surname, with the title. This is set in italics with the first and principal words capitalized. The title is followed by the edition, if other than the first; then the volume-number in Arabic numeral if more than one; the year of publication; and the specific pages. All of these items are separated by commas, for example:

[3] Madison Bentley, *The Field of Psychology*, 1924, 321-323.
[4] E. G. Boring, *A History of Experimental Psychology*, 2nd ed., 1950, 21-24.
[5] William James, *The Principles of Psychology*, 2, 1890, 245.

(iii) Titles of articles. If the article cited is published in a periodical, give, after the author's initials and surname, the title in lower case with the first word capitalized; then the name of the periodical in italics and abbreviated in accordance with the usage of the *World List of Scientific Periodicals;*[3] the volume number in Arabic numerals; the year; the serial number of the publication, if there is one, set in parentheses; and the specific page or page numbers; all items being separated by commas. For example:

[4] M. F. Washburn, The function of incipient motor processes, *Psychol. Rev.*, 21, 1914, 376-390.
[5] Madison Bentley, The psychological antecedents of phrenology, *Psychol. Monogr.*, 21, 1916 (No. 92), 102-105.

(iv) Titles of chapters in a book. If a chapter in a book is cited, give after the author's initials and surname; the title of the chapter in lower case with the first word capitalized; the initials and surname of the author of the book, preceded by 'in' and followed by 'ed.' set in parentheses; the title of the book in italics with principal words capitalized; the edition, if other than the first; the volume number, if more than one; the year; and the specific pages. For example:

[3] Leonard Carmichael, Ontogenetic development, in S. S. Stevens (ed.), *Handbook of Experimental Psychology,* 1951, 304-329.

(v) References to this JOURNAL. If the citation is to an article published in the AMERICAN JOURNAL, use the phrase, 'this JOURNAL' in place of the standard abbreviation. The second word is set in capitals and small capitals; as:

[1] E. B. Newman, The patterns of vowels and consonants in various languages, this JOURNAL, 64, 1951, 369-379.

---

[3] See footnote 2.

(d) Digressions. Digressions, explanations, and discussions, that are not germane to the text, are placed in footnotes. When references are made in them to books or articles, they are given in parentheses in the form shown below.

[3] The hearing loss was surprisingly less than that obtained by a similar method of impairment. In the earlier study the loss was 65 db. at all frequency-levels and the *S*s. could not hear their footsteps nor understand speech unless shouted (Supa, Cotzin, Dallenbach, "Facial Vision": The perception of obstacles by the blind, this JOURNAL, 57, 1944, 169.

(e) Repetition of citations. At the first and subsequent repetition of a reference, the abbreviation '*op. cit.*' is substituted for the title. If only one article by a given author has been cited, *op. cit.* is followed by the page numbers. For example:

[1] S. W. Fernberger, Observations on taking peyote (Anhalonium Lewini), this JOURNAL 34, 1923, 267-270.
[6] Fernberger, *op. cit.*, 269.

If, however, two or more articles by the same author have been cited, the second reference to one of them is designated by '*op. cit.*' and, in the case of books, by the year of publication of the particular book referred to and the specific page numbers. for example:

[2] Madison Bentley, *The Field of Psychology: A Survey of Experience, Individual, Social, and Genetic,* 1924, 103-105.
[5] Bentley, *The New Field of Psychology: The Psychological Functions and Their Government,* 1934, 187-190.
[7] Bentley, *op. cit.*, 1924, 278-284.

In the case of multiple citations of articles by the same author, the second reference to one of them is designated by '*op. cit.*,' the abbreviated name in italics of the periodical of publication, volume, year, and specific page numbers. For example:

[1] K. M. Dallenbach, The temperature spots and end-organs, this JOURNAL, 39, 1927, 402-427.
[11] Dallenbach, The place of theory in science, *Psychol. Rev.*, 60, 1953, 33-39.
[12] Dallenbach, *op. cit.*, this JOURNAL, 39, 1927, 425.
[16] Dallenbach, *op. cit.*, *Psychol. Rev.*, 60, 1953, 35.

These rules apply only to citations in short articles and to articles with few footnotes. For the convenience of the reader they may be broken and the complete citation be repeated. Repetition may be desirable in long articles and in articles with numerous footnotes. The reader should be kept in mind; his ease and convenience are considerations that should not be overlooked. If he has to turn back many pages to find the first citation of a given reference, or if he has to search through numerous references to find it, the repetition of the title is advisable.

ABBREVIATIONS AND REFERENCE WORDS

Abbreviations and reference words may be divided into two classes: those reserved for the editor, and the standard ones that may be used by the author.

(1) Editor's abbreviations. The following reference words and abbreviations of Latin words should be left to the use of the editor.

cf., abbreviation for *confer,* meaning 'compare.' Do not use *cf.* when 'see' is intended. It is italicized.

ibid., for *ibidem,* meaning in the same place,' *i.e.* in the reference immediately preceding. It is italicized and should always be followed by specific page numbers.

idem, this is a reference word, not an abbreviation. It means 'the same,' *i.e.* the same person as cited in the reference immediately preceding. It is italicized. Though this reference word was for a time frequently used in the JOURNAL, it is now, under the rule described above (Section *1, c, 1*), rarely used. The author's name is instead repeated. This does not consume much more space and is clearer as the citation referred to may be on a preceding page and the reader, to discover it, may have to turn back. The repetition of the name avoids this inconvenience.

loc. cit., (plural *locc. citt.*) for *loco citato,* meaning 'in the same place cited,' *i.e.* in the same passage referred to in a recent note. It is italicized but is not followed by a page number. The page number or numbers given with the earlier reference are implied. If the same article but different pages are to be indicated, then *ibid.* or *op. cit.* should be used, depending upon which of the two is appropriate.

op. cit. (plural *opp. citt.*) for *opus citatum,* meaning work cited.

p., (plural pp.) for 'page.' Set in roman, not italicized.

q.v., for *quod vide,* meaning 'which see.' Rarely used in the JOURNAL.

vol., (plural vols.) for 'volume.' Set in roman and omitted when preceded by the name of the publication and followed by the year and inclusive pages.

The use of these tools of reference is restricted to the editor because he is the best judge as to which one of them, if any, should be used. Authors should not use them; they should complete their reference as it is much easier for an editor to cross out and interpolate the correct abbreviation than it is to cross out the incorrect abbreviation and substitute the correct one, or to repeat the reference as he may deem advisable.

(2) Standard abbreviations. Among the abbreviations frequently used in the text and footnotes are the following:

A.C., (alternating current), small capitals with periods but without spacing. Used only when preceded by a specified voltage; as 25 v. A.C.

A.D., (*anno Domini*), small capitals with periods but without spacing. Precedes numerals. Used only in contradistinction to B.C.

AD, (average deviation), italic capitals, without spacing.

A.M., (*ante meridian*), small capitals with periods but without spacing. Follow numerals denoting the hour.

anon., (anonymous).

AQ, (achievement quotient), italic capitals, without periods and spacing.

b., (born).

B.C., (before Christ), small capitals with periods but without spacing. Follows the numerals.

CA, (chronological age), italic capitals, without spacing.

ca., (*circa,* about). Precedes approximate date.

CFF, (critical fusion frequency), italic capitals, without spacing.

cycle, Spell out except when preceded by a numeral, then use sign (~).

d., (died).

db., (decibel). Spell out except when preceded by a numeral.

D.C., (direct current), small capitals with periods but without spacing. Used only when preceded by a specific voltage.

Dr., (doctor). Always use as a title before a name.

d.v., (double vibration), lower case with periods but without spacing. Used only when preceded by a numeral. Cycle sign is preferable.

E, Es, (experimenter, experimenters), italic capital. Possessive, *E*'s, *Es*'.

ed., eds., (editor, editors).

ed., (edition), preceded by the number of the edition other than the first; as, 2nd ed., 3rd ed.

e.g., (*exempli gratia,* for example), lower case italics without spacing and not followed by a comma.

etc. (et cetera, and so forth), this abbreviation should be used sparingly. It is of little help to the reader except as it denotes an orderly continuum.

f., ff., (and the following page or

pages). Preceded by a page number and a space. A single 'f.' means one page; double 'ff.' two or three pages. If more pages are intended, give inclusive pages.

Fig., Figs. (figure, figures). Always capitalized and followed by an Arabic number or numbers. Spell out when not.

g., (general factor).

GSR, (galvanic skin response), italic capitals, without spacing.

gm., (gram).

h, (index of precision).

i.e., (*id est*, that is), lower case italics, without spacing and not followed by a comma.

IQ, (intelligence quotient), italic capitals, without spacing.

Jr., (junior), follows name and is part of it. It takes the possessive, as "John Smith, Jr.'s study."

M, (arithmetic mean).

m., (meter).

MA, (mental age).

Messrs., (misters).

min., (minute).

mm., (millimeter).

Mr., (mister); never spelled out except with humorous intent.

Mrs., (mistress).

m.sec., (millisecond).

MS, MSS, (manuscript, manuscripts). Capitals without periods and spacing.

N, (number), mathematical symbol referring to number of cases.

N.B., (*nota bene*, mark well), capitals with periods but without spacing.

no., nos., (number, numbers).

O, *O*s, (observer, observers), italic capitals. Possessive, *O*'s, *Os*'. For difference between *O* and *S*, see Madison Bentley, 'Observer' and 'subject,' this JOURNAL, 41, 1929, 682 f.

ohms, Spell out except when preceded by a numeral, then use capital Greek omega (Ω).

PE, (probable error), italic capitals without periods and spacing.

percentage, Spell out except when preceded by a numeral, then use percentage sign (%).

P.M. (post meridian), small capitals with periods but without spacing. Follows numerals denoting the hour.

Professor, Always spell out.

PSE, (point of subjective equality), italic capitals without periods and spacing.

R, (Reiz = 'stimulus' in psychophysics, italic capital).

R, (response in behavorial psychology, roman capital).

rev., (revised), used before 'ed.'; as '2nd rev. ed.'

RL, (stimulus limen), italic capitals without periods and spacing.

r.p.m., (revolutions per minute).

r.p.sec., (revolutions per second).

S, *S*s, (subject, subjects), italic capitals. Possessive, *S*'s, *Ss*'. For difference between *S* and *O*, see Bentley, *op. cit.*, this JOURNAL, 41, 1929, 682 f.

S, (stimulus in behavioral psychology, roman capital).

SD (standard deviation), italic capitals, without spacing.

sec. (second).

sic (thus, so), between brackets in quoted matter when used as an interpolation; otherwise within parentheses.

TR (terminal stimulus).

trans. (translation).

viz. (*videlicet*, namely), roman, with period but not followed by a comma.

Abbreviations for times, weights, and other measures are the same for the plural as for the singular. Context determines which is intended; as 1 sec., 6 sec.; 1 gm., 10 gm.; 1 lb., 15 lb.; 1 ft. 7 ft.; 1 mm., 35 mm.

TABLES AND FIGURES

(1) Tables. The tables within every article are numbered consecutively in capital Roman numerals; as Table I, Table V. When the tables are referred to in the text by number, the first letter is capitalized, otherwise it is not. The word 'Table' and its number are set in capitals and centered on the page *above* the tabular matter. This is followed by an explanatory caption, set in capitals and small capitals, and centered upon the page. The caption should be brief but sufficiently detailed to permit the reader to discover what the table purports to show. The caption may have interior punctuation but it should not end with a period. Additional explanatory

material may be given in parentheses centered under the caption. This is set in lower case. Footnotes to the table, if required, should be referred to by the following printer's signs and in the order listed: asterisk (*); dagger (†); double dagger (‡); and sectional sign (§). For example:

* Significant at the 1% level.
† Significant at the 5% level.
‡ A paper-and-pencil version of the test used here.
§ Performed while standing and not on treadmill.

(2) Figures. The figures (line and half-tone blocks) within every article are numbered consecutively in Arabic numerals. The word 'Figure,' both in the legend and when referred to with its number in the text, is abbreviated; as Fig. 1, Fig. 2, Fig. 5. In the text, the first letter of the abbreviation is capitalized; under the figure, the abbreviation and explanatory legend are set in small capitals with principal words in large capitals. The legend does not end with a period; it is not a sentence. If further explanatory material is needed, give it in lower case, centered under the legend.

The figures should be drawn in India ink, with the size of the JOURNAL page in mind. It is 25 × 42 picas (approximately 4 × 7 in.). They should be drawn by a draftsman and lettered so large that they will easily be read after the necessary reduction. Amateurish drawings should not be submitted. Since reproductions can usually best be made from original drawings, these are preferred, though glossy prints of satisfactory drawings are acceptable.

The University of Texas **KARL M. DALLENBACH**

SKIN-CONDUCTANCE AND REACTION-TIME IN A CONTINUOUS AUDITORY MONITORING TASK

By JOHN L. ANDREASSI,
U. S. Naval Training Device Center, Port Washington, N.Y.

Activational theorists contend that level of performance rises with increases in physiological activity of an organism up to a point that is optimal for a given function and beyond this point further increases cause a drop in performance.[1] Thus, the basic proposition of this concept is that performance is best at some intermediate level of physiological activation and the relation between the two can be described by an inverted "U".

One study frequently cited in support of activation-theory is that of Freeman who studied the relation between palmar skin-conductance (*PSC*) and reaction-time (*RT*) with a single subject (*S*) under various states of alertness.[2] The results of 100 experimental sessions were recorded over a number of days. Freeman found an inverted "U" shaped relation between *PSC* and *RT*, in which *RT*s were slower at high and low *PSC* levels and fastest at the middle levels. Schlosberg reported that Freeman's result had been duplicated with another *S*, but a later study, using a greater number of *S*s (Schlosberg and Kling) failed to replicate these findings.[3] In none of these studies was a continuous measure of *PSC* obtained along with the *RT*s. The purpose of the present experiment was to explore the relationship between *PSC* and *RT* to an auditory stimulus when *PSC* was measured continuously.

Method. Sixteen students (9 women and 7 men) served as *S*s. *S*'s task was to sit quietly in a comfortable chair and to respond as quickly as possible to a 200-cps tone which was presented aperiodically over headphones, against a background of continuous white noise. The tone (70 db.) was generated by a Hewlett-Packard Wide Range Oscillator and a Scientific Prototype Audio Noise Generator was the noise source (60 db). Both tone and noise were presented to *S* via MSA Noisefoe Mark II headphones and the level of each was measured at the headphones with a General Radio Sound Level Meter. A tape programmer was used to actuate a three position relay. Holes were punched in a film strip at desired intervals and, when actuated, the relay performed three functions simultaneously: it started a Lafayette

---

* Received for publication November 2, 1965. This study was performed while the author was a USPHS Predoctoral Research Fellow at Western Reserve University. The writer's thanks are due Dr. R. C. Wilcott, for his assistance during the conduct of this experiment.

[1] Elizabeth Duffy, *Activation and Behavior,* 1962, 139-194; R. B. Malmo, Activation: A neuropsychological dimension, *Psychol. Rev.,* 66, 1959, 367-386.

[2] G. L. Freeman, The relationship between performance level and bodily activity level, *J. exp. Psychol.,* 26, 1940, 602-608.

[3] Harold Schlosberg, Three dimensions of emotion, *Psychol. Rev.,* 61, 1954, 81-88; Harold Schlosberg and J. W. Kling, The relationship between "tension" and efficiency, *Percept. mot. Skills,* 9, 1959, 395-397.

timer, it allowed presentation of the tone, and it caused an event marker to produce a vertical line on the ink writer used to record *PSC* continuously. Thus, *E,* who was situated in an adjacent room could record *RT* to the nearest hundredth of a second and make continuous readings of *PSC. S* responded by pressing a switch mounted on the right arm of the chair. None was given any knowledge of results.

The *PSC* electrodes consisted of 3-cm. zinc-plates, mounted in hard rubber cups,

TABLE I

DIFFERENCES BETWEEN MEAN *RT* FOR *PSC* TRIALS, AND LEVELS

H = 10 highest values, *M* = 12 intermediate values, and *L* = 10 lowest values

Comparison	Mean diff.	*t*-value	*df*	*p* (two-tailed)
RT				
H versus *M*	34.00	2.31	15	<.05
H versus *L*	31.00	2.80	15	<.05
M versus *L*	3.00	.25	15	
PSC				
H versus *M*	2.50	5.13	15	<.01
H versus *L*	4.05	6.10	15	<.01
M versus *L*	1.55	5.03	15	<.01

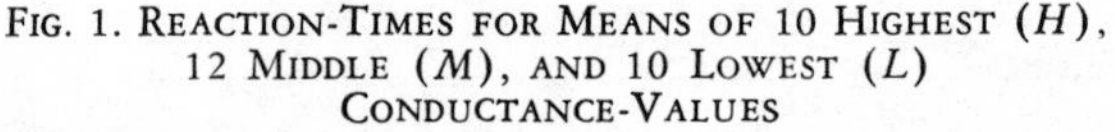

FIG. 1. REACTION-TIMES FOR MEANS OF 10 HIGHEST (*H*), 12 MIDDLE (*M*), AND 10 LOWEST (*L*) CONDUCTANCE-VALUES

with zinc sulfate paste as the conducting material. One electrode was placed on the palm of *S*'s left hand just below the base of the first finger, and the other electrode was placed below the base of the little finger of the same hand. Skin-resistance was recorded with a Wheatstone Bridge coupled into an Offner Type R Dynograph. The values of skin-resistance were later converted into conductance measures.

Results. The difference between the mean *RT*s for those trials with the 10 highest conductance-values (*H*), 12 middle conductance-values (*M*) and 10 lowest values (*L*) were analyzed by *t*-tests for correlated data (Table I). These results indicate that on those trials for which *S*s had the highest mean *PSC* values, *RT*s were significantly faster than those trials in which the *PSC* measures were of intermediate or low magnitude. The mean *RT*s for each third, by conductance, is shown in Fig. 1.

Mean differences for the four successive segments of the experiment were examined by *t*-tests for correlated data (Table II). Trials 1-8 are represented by (1); Trials 9-16 by (2); 17-24 by (3); and 25-32 by (4).

There was a sharp decrease in conductance between the first and second 10-min. segments of the experiment, and this was accompanied by a significant increase in *RT*. The differences between the first segment and segments three and four were also significant. For the second, third, and fourth segments decreases in *PSC* and increases in *RT* were consistent but not significant. These results are depicted in Fig. 2.

TABLE II

DIFFERENCES BETWEEN MEAN *RT* AND MEAN *PSC* FOR THE DIFFERENT SEGMENTS OF THE EXPERIMENT

(1), Trials 1–8; (2), Trials 9–16; (3), 17–24; (4), 25–32

Comparison	Mean diff.	*t*-value	*df*	*p* (two-tailed)
RT				
(1) versus (2)	36.00	2.98	15	<.01
(1) versus (3)	45.00	3.53	15	<.01
(1) versus (4)	51.00	2.82	15	<.05
(2) versus (3)	9.00	.69	15	
(2) versus (4)	15.00	1.51	15	
(3) versus (4)	6.00	.37	15	
PSC				
(1) versus (2)	2.33	3.91	15	<.01
(1) versus (3)	2.55	3.37	15	<.01
(1) versus (4)	2.98	3.29	15	<.01
(2) versus (3)	.23	.78	15	
(2) versus (4)	.65	1.61	15	
(3) versus (4)	.43	1.36	15	

To determine the association between percentage-change in conductance and that in reaction-time for the means of the 10 highest and 10 lowest conductance-values, a Spearman rank correlation coefficient was computed for the 16 *S*s. The Spearman method was used since the percentage-values obtained were not normally distributed. A rank correlation coefficient of 0.47 was obtained and, for 14 degrees of freedom (N-2) this value was significant at the 0.05 level, indicating that changes in *RT* were significantly related to changes in *PSC* level for the 16 *S*s as a group. This result indicates, for example, that *S*s who ranked high in terms of percentage-decrease in *PSC* also ranked high with respect to percentage-increase in *RT*.

Discussion. The results of the present study differ from those of Freeman in that an inverted "U" shaped relation between *RT* and *PSC* was not found, *i.e.* there was no indication of an upturn in *RT* at the highest values of *PSC*.[4] Freeman used one *S* and experimental sessions occurred at all times of the day, including early morning and late at night, with *S*'s condition ranging between half asleep and extremely tense. Perhaps Freeman's method is more similar to a situation where *S*'s state is more actively manipulated, as in the case of studies of Induced Muscle Tension (*IMT*). An

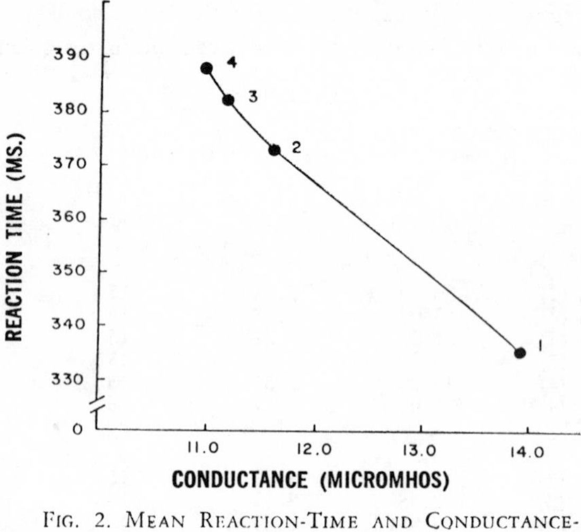

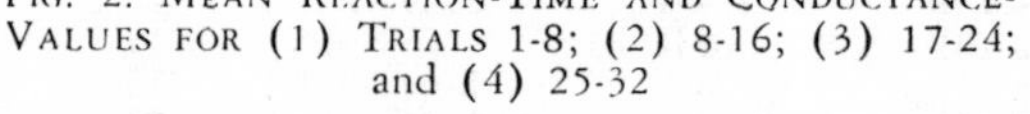

FIG. 2. MEAN REACTION-TIME AND CONDUCTANCE-VALUES FOR (1) TRIALS 1-8; (2) 8-16; (3) 17-24; and (4) 25-32

inverted "U" relation between level of effort and performance has been frequently reported in *IMT* studies.[5]

The findings of the present investigation are similar to those of Travis and Kennedy who investigated the relationship between frontalis muscular tension, reaction-time and level of performance in a continuous tracking task.[6] They reported that *RT*s became progressively slower with low levels of tension and faster when tension level was high. There was no evidence of an upturn in *RT* at the highest muscle tension levels, but there was no deliberate manipulation of tension-level by Travis and Kennedy, just as no

[4] Freeman, *op. cit.*, 602-608.

[5] F. A. Courts, Relations between muscle tension and performance, *Psychol. Bull.* 39, 1942, 347-367.

[6] J. L. Kennedy and R. C. Travis, Prediction and control of alertness: II. Continuous tracking, *J. comp. physiol. Psychol.*, 41, 1948, 203-210.

effort was made to change *S*'s state in the present investigation. *S*'s state, however, did change gradually in the present study, as indicated by the progressive decrease in *PSC*-values during the course of the experimental session, and this change was related to a change in *RT*s as shown in Fig. 2. Most *Ss* reported that the white noise presented continuously via the headphones produced drowsiness, making it difficult to stay awake. Thus, instead of showing an improvement in *RT* as the session progressed, *RT* became slower and the slowing was associated with decreased *PSC* values.

The fact that the results do not support the activational theory may be due to a limited range of arousal conditions. Sherwood made a similar suggestion as a result of two experiments which indicated that performance was high under high arousal and low under conditions of low arousal.[7] The results of the present study do not contradict activational theory. Rather, they suggest that a qualification be made to cover those situations in which *S*'s physiological state is not actively manipulated through physical work or task-induced stress, and in which high arousal conditions are not attained. It may be suggested that under such conditions *S*'s performance will improve with increases in physiological arousal, but that an inverted "U" shaped relation between level of bodily activity and performance will not be found.

SUMMARY

Continuous measures of palmar skin-conductance (*PSC*) were taken as *S* responded to aperiodic auditory signals presented against a white noise background. Thirty-two reaction-time (*RT*) trials were taken for each of 16 *Ss* over a 40-min. session. The results indicated that *Ss* had significantly faster *RT*s on the 10 trials in which *PSC* was highest as compared to the *RT*s for the 12 middle and 10 lowest *PSC* trials. There were decreases in *PSC* as the experiment progressed. The sharp decrease in *PSC* between the first and second 10-min. segments of the experiment was accompanied by a significant increase in *RT*.

There was no upturn in *RT* at the highest levels of *PSC* and it was suggested that in certain situations *S*'s level of arousal must be actively manipulated to achieve an inverted "U" relation between bodily activity level and performance.

[7] J. J. Sherwood, A relation between arousal and performance, this JOURNAL, 78, 1965, 461-465.

Comments on the Article

The *title* illustrates good use of keywords and phrases. "Skin-conductance," "Reaction-time," and "Continuous Auditory Monitoring Task" orient both the reader and the reference librarian. The title also limits itself and the article to "A...Task."

The *headnote,* which uses an asterisk so that Arabic numerals may be reserved for notes referring to literature, gives credit to the agency which supported the research and to the person who assisted in and advised upon its conduct.

The *by-line* and *author credit* are combined. This is standard practice in all the journals of psychology. It is dissimilar from the other fields of science—for example, medicine—which require that the author's affiliation be shown in a footnote.

The opening paragraph is a succinct review of the present state of knowledge. The second paragraph points out a conflict in the literature *without comment.* This paragraph closes with a statement as to the *purpose* of the present work. It falls logically into this paragraph because a partial purpose of the work is to resolve the conflict.

The *method* section is developed in some detail. Full development is essential in articles in experimental psychology, for findings may be justified only by sound methodology. Table 1 synthesizes the sequence and variables of the method.

The *results* section reports the findings without comment. Phrases such as "significantly faster," "sharp decrease," and "consistent but not significant" are fair usage because the data in Table 2 confirm them.

The *discussion* section places the results among the literature discussed in the two opening paragraphs of the article. The results "differ from those of Freeman"; they "are similar to those of Travis and Kennedy." The final paragraph evaluates the similarities and dissimilarities, but only to the extent that the data permit. Phrases such as "results...may be due to," "the results of the present study do not contradict...," "they suggest that a qualification be made...," and "it may be suggested..." are essential when the experiment is limited in size and when the writer is a professional.

The *summary* discusses only the experiment; it is not concerned with the state of knowledge or the contribution that the experiment makes to that knowledge. It is simply a declaration of what the experiment showed, and it is based entirely upon data that appear in the body of the article.

The journals of sociology loosely follow the University of Chicago *Style Manual,* a document that closely follows or duplicates the Modern Language Association *Style Sheet* (see p. 275). Some journals issue their own style guides such as the "Notice to Contributors" reprinted below by permission of the *American Sociological Review.*

Reprinted from AMERICAN SOCIOLOGICAL REVIEW, Vol. 27, No. 6, December, 1962

NOTICE TO CONTRIBUTORS

to the *American Sociological Review*

I. *Preparation of Articles, Research Reports, and Book Reviews sent to Editor*

Papers are evaluated by the editors and other referees and are judged without author's name or institutional identification. Therefore, contributors are asked to attach a cover page giving the title, author's name, and institutional affiliation; the manuscript should bear only the title as a means of identification. At least two, and preferably three, copies should be submitted to enable prompt evaluation, but the author should retain a copy in his own files.

Manuscripts are accepted subject to usual editing.

An abstract of about 100–125 words should accompany articles (but not research reports); it should present the principal substantive and methodological points.

Please prepare copy as follows:

1. *All* copy, including indented matter, should be typed *double spaced* on white standard paper. Lines should not exceed 5–6 inches.
2. A footnote to the title, author's name, or his affiliation should be starred (*). Other footnotes should be numbered serially, typed *double spaced,* and should be listed at the *end* of the article or research report. Sample footnote formats are presented below. Please note that full names are used, not initials.
3. Each table should be typed on a separate page. Insert a guide line, e.g., "Table 1 about here," at the appropriate place in the manuscript. See current issues of the *Review* for tabular style.
4. Figures should be drawn on white paper with India ink and the original tracings or drawings should be retained by the author for direct transmission to the printer. Copies should accompany the manuscript.
5. Mathematical notation should be provided both in symbols and words. Explanatory notes not intended for printing should be encircled in pencil.
6. If any symbols are used that might confuse the printer, please clarify in the margin of the manuscript.

Sample Footnote Formats

* Revision of a paper read at the annual meeting of the American Sociological Association, August, 1959.

1. Gordon W. Allport, *The Nature of Prejudice,* Cambridge: Addison-Wesley, 1954, p. 298.
2. *Ibid.,* pp. 299–300.
3. Seymour M. Lipset, "Democracy and Working-Class Authoritarianism," *American Sociological Review,* 24 (August, 1959), pp. 482–501.
4. Bruno Bettelheim and Morris Janowitz, *Dynamics of Prejudice,* New York: Harper, 1950.
5. Robert K. Merton, "Discrimination and the American Creed," in Robert M. MacIver, editor, *Discrimination and National Welfare,* New York: Harper, 1949, pp. 99–126.
6. Herbert Menzel, James C. Coleman, and Elihu Katz, "Dimensions of Being 'Modern' in Medical Practice," *Journal of Chronic Diseases,* 9 (January, 1959), pp. 20–40.
7. Committee on Nomenclature and Statistics of the American Psychiatric Association, *Diagnostic and Statistical Manual: Mental Disorders,* Washington, D.C.: American Psychiatric Association, 1952, pp. 12–13.

II. *Preparation of News and Announcements sent to Executive Office (New York City)*

These columns may include notices of academic appointments, promotions, resignations, visiting professorships, leaves of absence, special awards, appointments to governmental and private organizations, new training programs, and *major* curricular developments, special research projects and grants, special conferences and institutes, retirements, and deaths. *Do not include:* publications by department members (these will appear in "Publications Received" and many will be reviewed), appointments to graduate assistantships, the conferral of graduate degrees (which are reported annually in the *American Journal of Sociology*) or of graduate work in progress, public lectures, televised courses, papers delivered at conferences, and brochures. Notices should be *concise.* See current issues of the *Review* for editing style. It is suggested that each department assign one person the responsibility of assembling and transmitting news and announcements.

Notices of professional interest from governmental and other non-academic agencies are welcome.

Foreign sociologists planning to visit the United States who are interested in meeting American sociologists are invited to send their itineraries and other pertinent details to the Executive Office. These will be published under the heading, "Sociologists from Abroad." Please see *Time Schedule.*

The *Review* reserves the right to edit or exclude all items.

Time Schedule: To insure publication, announcements must be received no later than the beginning of the third month preceding the month of issue; for example, to be included in the October issue, material must be received by July 1.

The following illustrative article, reprinted by permission of The Sociological Quarterly, follows recommendations of both the University of Chicago *Style Manual* and the Modern Language Association *Style Sheet.*

Mate Selection in the Upper Class

LAWRENCE ROSEN *and* ROBERT R. BELL

DISCUSSIONS of social endogamy usually stress three social variables that influence mate selection. The three variables are race, religion, and social class. Race is the strongest endogamous force in American society today, with most estimates placing over 99 per cent of all marriages within racial limits. The estimates for marriages within religions are about 90 per cent, with Jews being the most endogamous, Catholics the least and Protestants in between.

Estimates of social class endogamy have been somewhat limited, and the rates presented are open to question because of the crude social-class measures used. The two most common measures of social class used in endogamy estimates have been occupation and residence. Five major studies, employing varying methodologies, modes of analysis and different populations show social class endogamy as ranging from 19 to 60 per cent, with most estimating about 50 per cent.[1] These studies also indicate that when contiguous social classes are considered, the endogamy rates increase so that the overall rate almost equals that of religious endogamy.

Because of the crudeness of social class measures, little can be inferred from the studies about endogamy in the very top social class, or what E. Digby Baltzell has defined as the "upper class":

> . . . a group of *families* whose members are descendants of successful individuals (elite members) of one, two, three or more generations ago. These families are at the top of the *social class* hierarchy;

[1] Residential studies are A. B. Hollingshead, "Cultural Factors in the Selection of Marriage Mates," *American Sociological Review,* 15:619–27 (1950); and Simon Dinitz *et al.,* "Mate Selection and Social Class: Changes during the Past Quarter Century," *Marriage and Family Living,* 22:348–51 (1960). Studies using occupation are T. C. Hunt, "Occupational Status and Marriage Selection," *American Sociological Review,* 5:495–504 (1940); Richard Centers, "Marital Selection and Occupational Strata," *American Journal of Sociology,* 54:530–35 (1949). The following study did not indicate its class measure but appeared to use an individualistic measure:

they are brought up together, they maintain a distinctive style of life and a kind of primary group solidarity which sets them apart from the rest of the population.[2]

The index of the upper class as defined by Baltzell is membership in the *Social Register.*

Studies of endogamy in the upper class (using Baltzell's definition) are nonexistent; yet, on the basis of upper-class values of lineage and the high degree of in-class association the assumption is often made that rates of intermarriage within the upper class are high, at least higher than for other social-class levels. Our interest in this paper is to test empirically the assumption of high upper-class endogamy. We will do so by examining the rates of intermarriage among the Philadelphia upper class (defined as being in the *Social Register*) for the years 1940 and 1961. These two points in time were selected to determine any possible changes over that period of twenty-one years.

The Social Register

The *Social Register* is published yearly (in November), with a summer edition (in June) published for the purpose of listing summer residences, by an organization called the Social Register Association (offices located in New York City). In addition to the two yearly editions, there is an erratum, called *Dilatory Domiciles,* issued in January and December. There are at present twelve separate social registers for twelve different geographical areas: New York, Chicago, Cincinnati-Dayton, Buffalo, Boston, Baltimore, Philadelphia, St. Louis, Pittsburgh, Cleveland, San Francisco, and Washington, D.C. In addition there is a central listing, the *Locator,* published yearly. The listings are by family units, with each nonmarried member listed under the head of the household (the head may be a female when the father is present). Children referred to in the *Social Register* as "juniors" appear to be automatically listed within certain age ranges (males, 14 to 20, and females, 12 to 17) and those beyond the upper age limits apparently have to apply separately for their own listing.[3]

Ernest W. Burgess and Paul Wallin, "Homogamy in Social Characteristics," *American Journal of Sociology,* 49:109–24 (1943).

[2] E. Digby Baltzell, *An American Business Aristocracy* (New York: Collier Books, 1962), p. 21.

[3] Because we were unable to acquire official interpretations of some procedures used in the Social Register we have carefully tried to make deductions on the basis of the limited available information.

Persons desiring to be listed must apply to the Social Register Association with references from *Social Register* members. The rest of the process is a well-kept secret, but there appears to be a review board that passes on the acceptability of applicants. What the specific criteria for approval are we were unable to determine. (We assume that young adults, those beyond "juniors", who are offspring of current members find it easy to obtain their own listing.) Continued listing in the *Social Register* is accomplished by simply paying the yearly subscription fee (usually ten dollars). It is unclear how one can be dropped, outside of someone's voluntary decision to do so.

The use of the *Social Register* has been criticized as an adequate measure of upper-class membership primarily because there are many persons in important functional social roles in the United States who are not included. For example, Bernard Barber states:

> Not all people in the *Social Registers* are of equally high social class position. Some people are included because their parents were, even though they themselves do not have very highly evaluated social roles. And of course the Social Registers leave out many people who are in the upper classes. For one thing, some who would be readily accepted for listing choose not to apply because of equalitarian beliefs. Also, on the whole upper-class Jews and Catholics are excluded. This is an example of ethnic prejudice, not of social class discrimination. And finally, many men who have just risen through distinguished occupational achievement into the upper class are not likely to be included.[4]

But Barber's criticism is not a great problem if we keep separate, as Baltzell does, the "upper class" and the "elite" (those who occupy the important functional roles). The very fact that the *Social Register* is an "indication of social intimacy" and not an objective measure (i.e., based on income, occupation, etc.) by disinterested third parties is what makes it a valid index of the upper class. Our position on this point is the same as Baltzell's:

> It is important to be clearly aware of the fact that any index used in the social sciences is only a convenient tool which is constructed to approximate any given concept one desires to use for purposes of abstract analysis . . . and the Social Register [is] nothing more nor less than the best available [index] of an upper class.[5]

---

[4] Bernard Barber, *Social Stratification: A Comparative Analysis of Structure and Process* (New York: Harcourt, Brace, 1957), pp. 141–42.

[5] Baltzell, *op. cit.*, p. 44.

Methodology

In the back of each November issue of the *Social Register* there is a listing of marriages made during the previous year, giving the last names of brides and grooms and the specific *Social Register* issue where information concerning the marriages can be found. The information usually includes the names of the fathers of both spouses and the time and place of the wedding. In cases where the bride was previously married, her maiden name is given as well as that of her father. In very few cases is less than this minimal body of information provided.

We decided to include in our sample only those marriages which were detailed in the November issue of the Philadelphia *Social Register,* primarily because of the difficulty in obtaining past issues of the *Dilatory Domiciles* and June editions of the *Social Register.* In the 1941 *Social Register* (listing marriages for 1940) there was a total of 246 marriages of which 130 were detailed in the November issue, representing 53 per cent of all marriages listed for that year. For 1961 approximately 63 per cent (140) of a total of 226 listed marriages were detailed in the November issue. We believe that the only reason for inclusion in the November issue is because of the time of year when the marriages occurred. A check of the dates of marriages show only three in 1940 and eight in 1961 that occurred before June. We decided not to include in our sample those marriages where the bride was being remarried (indicated by "Mrs."). It is probable that most of these women were widows, but we had no way of determining if there were any divorcees (male or female) or widowers remarrying. Table 1 shows that the sample sizes finally used were 103 marriages for 1940 and 104 marriages for 1961.

One problem in using marriage lists as we have is our lack of information on how the listings were compiled and therefore what the lists actually represent. However, each of the marriages listed included as least one spouse who was listed in the *Social Register,* either as a junior or under a separate listing, usually under the parents' name. In two cases there were individuals listed but their parents were not. But in no single case did we find an individual who was not a *Social Register member* but whose parents were. This means that we define a *Social Register* spouse as one who is himself listed.

The decision to list a marriage is probably a voluntary one and if this is true, there is no way of knowing how complete such a list is. We wondered if there were any *Social Register* marriages (those involving at least one spouse who is a *Social Register* member) which for some reason might not be included in the marriage lists. A systematic spot check of marriage announcements in the *Philadelphia Bulletin* for both years failed to turn up any such marriages. However, not all marriages may be listed in newspapers, and since there were a few marriages occurring in different years included in the *Social Registers* (see Table 1) for the time period

TABLE 1. SAMPLE DATA FOR SOCIAL REGISTER MARRIAGES, TAKEN FROM PHILADELPHIA SOCIAL REGISTERS, 1941 AND 1942

	1940		1961	
	No.	Percentage	No.	Percentage
Brides Listed as Mrs.	22	*17*	28	*20*
Marriages occurring in different years or no date given	5	*4*	8	*6*
Remaining (Sample Used)	103	*79*	104	*74*
Totals	130	*100*	140	*100*

used in this study, we expect that a few marriages did occur and were not listed. Despite the reservations, if the marriage listing is incomplete, we would expect it to underestimate the rate of intermarriage, since if both spouses were members of the *Social Register,* there probably would be a conscious attempt to publicize the marriage and make this information available to significant others; i.e., other *Social Register* members.

Both spouses' names were checked in the *Locator* for 1940 and 1961 to determine to what extent Philadelphia *Social Register* members were marrying into other metropolitan upper classes. This is in line with Baltzell's argument that membership in the *Social Register* represents a "national upper class with a similar way of life, institutional structure, and value system." [6]

Findings

Table 2 shows that the extent of endogamy is low and indicates some decreases from 1940 to 1961.[7] The majority of *Social Register*

[6] *Ibid.*, p. 24.

[7] A sophisticated analysis of intermarriage would involve the computation of

marriages in Philadelphia for both years were marriages which included only one *Social Register* spouse (69 per cent for 1940 and 79 per cent for 1961).

In order to determine the relative tendency for *Social Register*

TABLE 2. SAMPLE OF PHILADELPHIA SOCIAL REGISTER MARRIAGES, 1940 AND 1961

	1940		1961	
	No.	Percentage	No.	Percentage
Both Spouses in *Social Register* (any city)	32	*31*	22	*21*
Groom only in *Social Register*	30	*29*	41	*39*
Bride only in *Social Register*	41	*40*	41	*40*
Totals	103	*100*	104	*100*

males and females to marry outside, it is necessary to compute marriage rates on the basis of individuals in order to control for the unequal sex ratio in our sample.[8] Table 3 indicates that for

TABLE 3. PHILADELPHIA SOCIAL REGISTER INDIVIDUALS MARRYING NON*SOCIAL REGISTER* SPOUSES (ANY CITY) 1940 AND 1961

	1940			1961		
	N	No.	Percentage	N	No.	Percentage
Males	50	30	*60*	59	41	*70*
Females	68	41	*60*	56	41	*73*
Totals	118	71	*60*	115	82	*71*

both 1940 and 1961 there was close to an equal tendency for both males and females to marry outside the *Social Register*. These findings are somewhat surprising in light of the common belief that there is usually more pressure on the female than the male to marry within the upper class. Further evidence on this point is provided by the frequency of the married couples who continued to be listed in the *Social Register* after marriage. (The *Locator*

the ratio of actual endogamous marriages to that of "expected" endogamous marriages (if members of the group in question had married at random). In our case we are unable to perform such an analysis because one of the cells—non-*Social-Register* brides and grooms—is unknown. This would be similar to a study of intermarriages among Jews if the sample included only marriages involving at least one Jew. See Paul H. Besanceney, "On Reporting Rates of Intermarriage," *American Journal of Sociology*, 70:720 (1965).

[8] For a discussion of the differences between marriage rates based on marriages and individuals see Besanceney, *op. cit.*, p. 718.

was checked for the two-year period following the year of the marriage.) If a couple were listed after their marriage where one of the individuals was a nonregistrant, we assume that this marriage was probably acceptable to the *Social Register* family and the rest of the upper class. The figures in Table 4 indicate that

TABLE 4. FREQUENCY OF COUPLES WITH ONE NONSOCIAL REGISTER SPOUSE BECOMING LISTED IN SUBSEQUENT (TWO YEARS) SOCIAL REGISTERS

	1940			1961		
	N	No.	Percentage	N	No.	Percentage
Groom only in Philadelphia *Social Register*	30	25	*83*	41	35	*86*
Bride only in Philadelphia *Social Register*	41	30	*73*	41	31	*76*
Totals	71	55	*78*	82	66	*81*

there was a greater tendency for females than for males marrying outside the *Social Register* to drop their listing. Although this may lend support to the belief that it is more difficult for a female than a male to pull a spouse in, the relative tendency for males and females for both years to drop out of the *Social Register* was not very great. Certainly it is far from rare for women to pull husbands into the *Social Register*. The figures in Tables 3 and 4 also give a clue as to the number of individuals who leave the *Social Register* because of marriage (13 per cent in 1940 and 11 per cent in 1961 of Philadelphia social register individuals marrying).

We have suggested there are few empirical studies of endogamy in the upper class. There are no other studies using the *Social Register* as the index of the upper class with which we can compare our findings. However, there is one study with which we can make some limited comparisons. Hatch and Hatch analyzed the characteristics of a sample of marriage announcements in the *New York Times* for the period from 1932 to 1942.[9] One of the social characteristics studied by them was whether or not the individuals involved were listed in the New York *Social Register*. However, no attempt was made to determine if any of the spouses were listed in other social registers. They found a total of 188 marriages where

[9] David L. Hatch and Marya Hatch, "Criteria of Social Status as Derived from Marriage Announcements in the *New York Times*," American Sociological Review, 12:396–403 (1947).

at least one of the partners was listed in the *Social Register* in their sample of 225. Of the 188 marriages 40 per cent had both partners listed in the New York *Social Register.* Of the total of 263 social register individuals 43 per cent married persons not in the *Social Register.* We were unable to compute the male-female differential because of incomplete data.

In comparing our findings with those of the Hatch study several differences should be kept in mind. While one difference is the two cities studied, even more important are the methodological differences. The Hatch and Hatch study did not start with the known base of all New York *Social Register* marriages (those listed in the *Social Register*), but only those marriages which were listed in the *New York Times.* While we cannot be sure that the *Social Register's* list of marriages is complete it is probably more complete than the newspaper's list. If the situation in New York is similar to that of Philadelphia, where some marriages were not listed in the newspaper, then the intermarriage rate given by Hatch and Hatch would be biased upward. However, a second methodological difference would tend to underestimate the New York rate. That is, the Hatch study did not determine if any of the non–*Social-Register* spouses were members of *Social Registers* in other cities.

Conclusions

Our study has certain obvious limitations. It is only for one city in two specific years and it is possible that the years we studied are in some way atypical. However, the two different years studied allow us to look at the possible impact of World War II and the postwar period on upper-class mate selection. Our data offers no evidence of any significant mate selection changes in the Philadelphia upper class when the two different years are compared. Recognizing possible biases, our findings clearly indicate that members of the *Social Register* do not show a strong endogamous pattern, certainly no greater and even somewhat less than other social-class levels.[10]

A weak endogamous pattern does not necessarily mean that there is no pressure toward marriage within the social class. Certain factors, such as an unequal sex ratio, may limit the size of the potential marriage market among the upper class. But there may be strong pressure on those who do marry outside the *Social Reg-*

[10] See Centers, *op. cit.*, and Hollingshead, *op. cit.*

ister to marry individuals who possess characteristics acceptable to the upper-class members of the *Social Register.* Certainly the high percentage (78 per cent in 1940 and 81 per cent in 1961) of non–*Social-Register* spouses becoming members through their marriage would suggest their upper-class acceptability. We would hypothesize that a strong majority of non–*Social-Register* spouses come from contiguous social-class levels.

Baltzell has recently suggested that the upper class is becoming more castelike because of a decreasing tendency to recruit individuals successful in highly important functional roles when they are not White Anglo-Saxon Protestants.[11] Our findings clearly indicate that there is a recruitment into the upper classes through marriage. The question is, Who is being recruited? A check of *Who's Who* and *Who Was Who In America* indicates that the vast majority of *Social Register* recruits through marriage were not sons and daughters of Baltzell's "elite" class. In 1940 only three spouses (all males) were offspring of *Who's Who* members, and in 1961 only seven spouses (three males and four females) were offspring of *Who's Who* members, and two of those ten did not become listed in the *Social Register.*[12] Also, our subjective reading of the names of the *Social Register* recruits indicates the vast majority have WASP names. (This type of assessment is highly subjective and speculative because of the problems of name changes and the less than perfect correlation between names and ethnic membership).

While our data shows a high rate of marriage by upper-class members outside their social class, it appears that the non–*Social-Register* spouses come from contiguous and acceptable social-class groupings. But at the same time we have no evidence that mate selection is a means of incorporating into the upper class persons from functionally successful (elite) backgrounds. Therefore, our data does not refute Baltzell's thesis of the castelike characteristics of the upper class with regard to the nonwhite Anglo-Saxon Protestant population. We may think of the potential marriage market for upper class individuals as consisting of the *Social Register* upper

[11] E. Digby Baltzell, *The Protestant Establishment* (New York: Random House, 1964), pp. 70–75.

[12] Only those whose parents' names were given could be checked. (There were only four such cases for the two years covered.) Considering the small size of the *Social Register* population in Philadelphia, the number of non–*Social-Register* spouses whose fathers were listed in *Who's Who* could be somewhat large.

class as constituting a small group (A), a larger group (B) consisting of contiguous white Anglo-Saxon Protestant population, and (C), the great majority of the American population. Ideally, the upper class is probably encouraged to marry a partner from A or from B where the outside individual can then through his marriage be absorbed into the upper class. Whatever the forces, it appears uncommon for the upper-class individual to marry into C, and when one does the *Social Register* individual may often drop or be dropped from the upper class. Therefore, marriage may provide an important means of expanding the upper class with mate selection from a small acceptable outside group.

Comments on the Article

The *title* lacks the specificity of titles to articles in the basic and applied sciences. "Mate Selection" and "Upper Class" are terms that need both defining and limiting. Although they seem vague and general, they serve to orient readers trained in sociology to broad subject areas within which specificity is possible. The writer has chosen the title because he is aware that he must define his terms in the context of the article.

The *by-line* simply gives the authors' names without degrees or credits. This practice is consistent within the fine arts and humanities; it is dissimilar to practice within the sciences.

The first four paragraphs review the literature and the state of knowledge. The fifth paragraph points out a gap in knowledge: "Studies of endogamy in the upper class are nonexistent...." It then goes on to state the *purpose* of the paper: "to test empirically the assumption of high upper-class endogamy." The paragraph—and the introductory section—closes with a statement of method: "by examining the rates of intermarriage among the Philadelphia upper class (defined as being in the *Social Register)* for the years 1940 and 1961."

The next major section is subheaded "The Social Register" and it defines that document as a technique for defining "upper class" as used in the title. The definition is defended—"Our position...is the same as Baltzell's...."—in a manner that links the present article to other work in the cited literature. The extensive discussion of the *Social Register* is not a digression because the article can only be understood by readers who know the purpose of the Social Register Association and the twelve registers issued for different geographical areas.

The next section, "Methodology," not only shows what the authors did but *why* they decided to do it that way. The section is entirely candid; the authors admit limitations to their technique. Such candor is essential to an article whose methodology may be attacked upon statistical or other grounds. (An interesting psychological phenomenon exists here; by pointing out trivial but specific flaws, the authors build up the reader's confidence in the article as a whole.)

The section entitled "Findings" is comparable to the Results section in a scientific article. There is a difference in tone, however. Writers of sociological articles editorialize and comment; for example, "These findings are somewhat surprising in light of the common belief that...." Such comment is justified because there is no formal subsection for Discussion of Results.

The "Conclusions" section is dissimilar in content and tone from the factual and logical approach required in the same section of a scientific article. Instead of citing only those conclusions that may logically be drawn from data presented in a scientific article, a sociological article concludes with a prosy and argumentative discussion. New data are added ("Baltzell has recently suggested...."), and methodology is redefined. Conclusions are drawn in areas beyond the stated purpose of the article ("Therefore, marriage may provide an important means of expanding the upper class...."). In the hands of good writers such as Rosen and Bell, the seeming looseness communicates effectively and is highly readable. Less competent writers, however, might be more comfortable writing within a logical and rigidly structured framework.

The journals of economics follow no standard pattern. Each has its own style guide, and that of the *American Economic Review* is perhaps the most comprehensive. For that reason, it is reprinted by permission below.

AMERICAN ECONOMIC REVIEW

Style Instructions

"1. *Manuscripts* should be typed double-spaced throughout--quotations, citations and footnotes as well as text—on one side of 8½ x 11 white bond paper, with margins of at least an inch on all four sides. Submit only the original (ribbon) copy. Do not use scotch tape on the face of a manuscript.

"2. *Textual Divisions.* If a paper is divided into sections, the introductory one requires no heading or number; subsequent section headings should be centered and given Roman numerals. Headings of subsections (if any) should be flush with left margin. Headings within subsections (if necessary) should be indented, with text following on same line. Subsection headings should be given numbers (Arabic) only if necessary for clarity or reference purposes.

"3. *Footnotes* should be numbered consecutively and typed double-spaced in a separate section at the end of the manuscript. Exceptions: (1) The initial footnote, which identifies author by title and affiliation, carries an asterisk. (2) Footnotes to a table are indicated by lower-case Roman letters and typed double-spaced below the table.

"4. *Tables* should be numbered consecutively with *Arabic* numerals. Sources of data should be given below the table, either with a full citation or by the method of citation set forth in paragraph 11 below. The latter method should be used in all cases in which a source is referred to both in text and tables.

"5. *Figures* or diagrams should be numbered consecutively with Arabic numerals, and must be in black India ink on heavy white paper. Rough drawings may be submitted initially, and professional drawings prepared after acceptance of the article.

"6. *Mathematics.* Equations should be typed on separate lines and carry consecutive Arabic numbers (in parentheses) at the left margin. Mathematical expressions should be lined up accurately and superscripts and subscripts properly located. Use Greek letters only when necessary. Unless made unmistakably clear, ambiguities may arise between: capital and lower-case letters; zero and o; the letters eta and n, nu and v; the sign for partial derivative and lower-case delta. Use the typewriter to the full extent possible.

"7. *Quotations* should correspond exactly with the original in wording, spelling, and punctuation. Exceptions must be indicated: insertions, by brackets; omissions, by ...; omission of a paragraph, by a full line of periods. Quotations beginning or ending in the middle of a sentence should begin or end with....

"Quotations of over 50 words should be separated from the context, indented from both margins, quotation marks omitted (except a quotation within a quotation, which is set within double quotation marks), and typed double-spaced.

"8. *Spelling.* Authority for spelling, capitalization, hyphenation, and italicizing of foreign words is *Webster's International Dictionary.* Avoid over-capitalization and excessive use of italics for emphasis. Use quotes only for first occurrence of terms with special meaning. Such words as antitrust, underdeveloped, overproduction, neoclassical, macroeconomic, noncompetitive are not hyphenated. A priori, ex officio, per se, viz., and obiter dictum, for example, are not italicized; but *ceteris paribus, ex post,* and *laissez faire* are.

"9. *References to Individuals.* First reference should include initials or first name. Subsequent references should be by last name only. Omit titles except in referring to women, first reference to whom may be by first name with subsequent references by Miss or Mrs.

"10. *Abbreviation* of names of organizations or governmental agencies is desirable, provided first reference gives name in full. For example, second and succeeding references to Social Science Research Council should be abbreviated to SSRC.

"11. *Citations* of works should refer the reader to a list of *References,* typed double-spaced, placed after the text of the article and before the footnotes. See examples next page and note the following:

"a. Titles of books and articles should be listed alphabetically by authors and numbered consecutively. Titles without specified authors, such as government documents, reports, etc. should be at the end of the list, also arranged alphabetically.

"b. For books, give author's initials and name, title of book (underlined), place and date of publication (not publisher).

"c. For articles, give author's initials and name, title of article in quotes, title of periodical (abbreviated and underlined), date of issue, volume number (Arabic numerals, underlined), and inclusive page reference.

"d. *Capitalization of Foreign Titles.* Capitalize first word, lower-case all others except proper nouns in French, Italian, Spanish, Russian, Norwegian and Swedish. Capitalize all nouns in German and Danish; all nouns and adjectives derived from proper nouns in Dutch.

"e. *Abbreviations in Citations:* Vol., Bk., Pt., Ch., p. or pp. (use Arabic numerals for all of these). Vol. and pp. are omitted in case of articles. Abbreviations of some of the principal periodicals:

Am. Econ. Rev.	*Jour. Am. Stat. Assoc.*	*Oxford Econ. Papers*
Am. Econ. Rev., Proc.	*Jour. Econ. Hist.*	*Quart. Jour. Econ.*
Can. Jour. Econ.	*Jour. Farm Econ.*	*Quart. Jour. Econ.*
Can. Jour. Econ.	*Jour. Farm Econ.*	*Rev. Econ. Stat.*
Econ. Jour.	*Jour. Finance*	*Rev. Econ. Stud.*
Fed. Res. Bull.	*Jour. Marketing*	*So. Econ. Jour.*
Indus. Lab. Rel. Rev.	*Jour. Pol. Econ.*	*Surv. Curr. Bus.*

"*References in the text* of the article should be given by number of work cited, with page reference where appropriate, both enclosed in brackets. If surname of author is given in text before citation number, this reduces risk of error. A sample extract from an article followed by list of references is given on next page.

"12. *Suggested Sources for Style and Language Usage:* University of Chicago *Manual of Style,* Chicago 1949; Margaret Nicholson, *Dictionary of American-English Usage,* New York 1957, based on *Fowler's Modern English Usage.*"

EXAMPLE: EXTRACT FROM ARTICLE

There are three reasons for believing that firms gain through inflation. First, depreciations in the real value of money obligations, which are losses to creditors, are gains to firms [2, pp. 58-74] [5, p. 18]. Second, the lag of wage rates behind prices redistributes income from laborers to capitalists [4, p. 322] [6, p. 31]. Third, firms gain because they carry inventories, which are sold at prices that reflect mark-ups based on current prices rather than the lower prices at which they were purchased [5, pp. 18-19]. Evidence has been provided by investigators of particular inflations, such as Bresciani-Turroni [1], Graham [3], and others.

"With regard to the effect of inflation on the individual firm, consider first the case of banks. It can be shown that banks in general lose during inflation; Standard and Poor's index of New York bank shares increased by a mere 20 per cent between the end of 1942 and 1948 [7, p. 22] [8]."

REFERENCES

1. C. Bresciani-Turroni, *Economics of Inflation.* London 1937.
2. I. Fisher, *The Purchasing Power of Money.* New York 1920.
3. F. D. Graham, *Exchange, Prices and Production in Hyperinflation: Germany, 1920-23.* Princeton 1930.
4. E. J. Hamilton, "Profit Inflation and the Industrial Revolution, 1751-1800," *Quart. Jour. Econ.,* Feb. 1942, **56**, 256-73; reprinted in F. C. Lane and J. C. Riemersma, ed., *Enterprise and Secular Change,* Homewood 1953, pp. 322-36.
5. J. M. Keynes, *Tract on Monetary Reform.* London 1923.
6. E. M. Lerner, "Money, Prices and Wages in the Confederacy, 1861-65," *Jour. Pol. Econ.,* Feb. 1955, **63**, 20-40.
7. Standard and Poor's, *Industry Survey of Banks.* New York 1952.
8. *Wall Street Journal,* Nov. 5, 1951.

EXAMPLES OF OTHER TYPES OF REFERENCE

1. F. T. Bachmura, *Geographic Differentials in Returns to Corn Belt Farmers: 1869-1950*. Unpublished doctoral dissertation, Univ. Chicago, 1953.
2. R. Goldsmith, "The Growth of Reproducible Wealth in the U.S.A. since 1870," *Income and Wealth,* Ser. II, Cambridge 1952, pp. 247-328.
3. G. Haberler , *Prosperity and Depression,* 3d ed. New York, United Nations, 1946.
4. G. H. Moore, *Statistical Indicators of Cyclical Revivals and Recessions,* Nat. Bur. Econ. Research Occas. Paper 31. New York 1950.
5. Financing Highways, Tax Inst. symposium. Princeton 1957.
6. *Problems in the International Comparison of Economic Accounts,* Nat. Bur. Econ. Research Stud. in Income and Wealth Vol. 20. Princeton 1957.
7. U.S. Bureau of Labor Statistics, *Family Spending and Saving in Wartime,* Bull. 822. Washington 1945.
8. U.S. Congress, Joint Economic Committee, *Relationship of Prices to Economic Stability and Growth,* Hearings, 85th Cong., 2nd sess., Washington 1958.
9. *U.S. v. New York Great Atlantic and Pacific Tea Co., Inc.*, 67 F. Supp. 626 (1946).
10. ________________, Brief for the United States, filed Mar. 2, 1946, District Court of the United States for the Eastern District of Illinois.

The journals of history also follow the Modern Language Association *Style Sheet.* Many, however, introduce individual variations and issue their own supplemental guides. That of the *Hispanic American Historical Review* is reprinted by permission below.

INFORMATION FOR AUTHORS

GENERAL

"The Board of Editors has decided that the form of all material henceforth published in the *HAHR* will follow in general the *MLA Style Sheet* (revised edition), a copy of which can be obtained from the Modern Language Association, 6 Washington Square North, New York 3, N. Y. If the *Style Sheet* does not cover some point adequately, the latest edition of the University of Chicago Press *Manual of Style* or *Webster's New International Dictionary* is to be followed.

"Type manuscripts on one side of 8½ x 11 inch bond of medium weight, leaving ample margins, and double-space throughout, *including* footnotes and quotations to be set in reduced type. Footnotes should be typed with *double*-spacing on sheets separate from the text and placed after the last page of the article. DOUBLE-SPACE EVERYTHING.

"Book review and book notice authors should consult current issues for correct form of headings and signatures.

Rules Peculiar to the HAHR or Particularly Important Ones

"*HAHR* is to be used in text and footnotes when this *Review* is mentioned or cited.

"*Op. cit.* ordinarily is replaced by author's last name if only one work by that author is cited in the article, or by author's name and short title if several are cited. See section 24 of the *MLA Style Sheet.*

"Capitalize initials in titles of periodicals in foreign languages. Thus *Gaceta Ministerial de Chile and El Tiempo.*

"*HAHR* style on Dates: July 4, 1951, *and* seventeenth-century spelling persisted into the eighteenth century.

"Use no italics for the following words, which the *HAHR* considers as Anglicized: Armada, arroyo, audiencia, burro, cabildo, caudillo, cedula, conquista, conquistador, criollo, encomienda, gaucho, junta, legajo, padre, presidio, tomo, and in addition other foreign words considered Anglicized by Webster's. Italicize other foreign words *only* the first time they appear.

"*Do not* hyphenate Latin American or Hispanic American unless so used in the title of a book or an institution.

"Parenthetical documentation: References within sentences in notes or text are awkward and should be avoided if possible. Try to recast the sentence to make the reference come at the end.

CITATION OF ARCHIVAL MATERIAL

"In general, follow the principles used in the citation of printed material where applicable (but titles of unpublished works are set roman, not italics). First identify the specific document; then proceed to explain its location.

Identification of document. Avoid unnecessary detail. Usually suspensive points can be used to avoid lengthy titles in footnotes. After first citation give briefest possible form. In the case of a letter or dispatch, give name of writer (complete only on first reference), name of addressee, and date (but not place unless essential) of writing. In dealing with governmental documentation that is filed by administrative unit, it is more important to cite the office held by the correspondent than to give the officeholder's name.

Location of document. Proceed from the larger to the smaller and more specific indications. Thus, (1) name of archives in language of country and location of archives on first citation only, (2) section or subdivision of archive, (3) volume or other equivalent such as legajo, using Arabic, not Roman, numerals.

Use of abbreviations. After first citation use abbreviations as much as possible. In abbreviating name of archives use initial capitals without intervening space or periods. Thus AGI (for Archivo General de Indias) rather than A. G. I. Explain abbreviation used at the end of first citation in full (thus Archivo General de Indias, Sevilla, cited hereinafter as AGI). Abbreviate legajo to leg., volume or tomo to Vol., folio to fol. (not f. or ff). Usually it will not be necessary for sufficient identification to refer to recto and verso. If this is necessary these terms should be given after folio number in roman lower-case letters without intervening space or punctuation (fol. 495v). Abbreviate dates in notes. In referring to continuous foliation or pagination give number in full (thus pp. 445-487, not 445-87).

Punctuation. Avoid use of punctuation other than commas to set apart the elements in a single citation unless absolutely necessary. Separate successive citations of different documents in the same note by semicolons.

Capitalization. When identifying writers of documents by title rather than name, capitalize initials (thus Ministro de Relaciones Interiores). Abbreviate such titles after first use.

Examples

"Domingo Faustino Sarmiento to Juan Pujol, May 22, 1860, Archivo General de la Nación, Buenos Aires, Archivo del General Justo José de Urquiza (cited hereinafter as AGN, Archivo Urquiza), leg. 67.

[23]Lefebre de Bécourt to Minister of Foreign Affairs, Sept. 23, 1861, Archives du Department des Affaires Etrangeres, Paris (cited hereinafter as AAE), Correspondence Politique, Vol. 38, fol. 231.

Same material upon subsequent citation:

[54]Sarmiento to Pujol, May 22, 1860, AGN, Archivo Urquiza, leg. 67.

[38]Lefebre de Bécourt to Min. of For. Affairs, Sept. 23, 1861, AAE, CP, Vol. 38, fol. 231.

If various archives are cited frequently:

[1]This article is based on materials consulted in the following Spanish archives: Archivo General de Indias, Sevilla Archivo General de Simancas, and Archivo Historicó Nacional, Madrid (abbreviated hereinafter as AGI, AGS, and AHN, respectively).

The journals of political science and government loosely follow either the University of Chicago *Style Manual* or the Modern Language Association *Style Sheet*. A major attempt at standardization of bibliographies is being made by the United Nations. Specimen pages reprinted by permission below are from the United Nations *Bibliographical Style Manual* (United Nations Publication ST/LIB/SER. B/8, Sales No. 63.1.5, price $0.75).

ANNEX

SPECIMEN PAGES OF A BIBLIOGRAPHY

Adler, J. H. The underdeveloped areas: their industrialization. New Haven, Yale Institute of International Studies, 1949. 30 p. (Memoranda, 31)

Aleksandrov, B. Problemy sel'skogo khoziaistva slaborazvitykh stran. *Novoe vremia* (Moskva):9–15, 15 aprelia 1953.[a]

Arena, C. Il problema dell'industrializzazione. *Rivista di politica economica* (Roma) 37:577-588, maggio 1947.

Aten, A. Enige aantekeningen over de nijverheid in Indonesië.
Indonesië ('s-Gravenhage) 6:19-27, Juli 1952; 6:193-126, November 1952; 6:330–345, Januari 1953; 6:411–422, Maart 1953; 6:536–564, Mei 1953.

The two first articles are in Dutch, the others in English.

Benham, F. C. Reflexiones sobre los países insuficientamente desarrollados. *Trimestre economico* (México) 19:45–47, enero-marzo de 1952.

Blue print of Pakistan's industrial development. *Pakistan quarterly* (Karachi) Special number:85–90, 1949.

Buchanan, N. S. Deliberate industrialization for higher incomes. *Economic journal* (London) 56:533–553, December 1946.
For discussion see: T. Balogh. Note on the deliberate industrialization for higher income, *Económic journal* (London) 57:238–241, June 1947; H. Belshaw. Observations on industrialization for higher income, *Economic journal* (London) 57:379–387, September 1947.

Ceylon. Executive Committee for Labour, Industry and Commerce. Report on industrial development and policy. Colombo, 1946. 27 p. (Sessional paper XV-1946)

Corral, Enrique del. La industrialización de España. [Madrid, Publicaciones Españolas, 1952] 29 p. (Temas españoles, 19)

Federation of Greek Industries. Greek industries in 1945. Athens, 1946. 65 p.

France. Institut national de la statistique et des études économiques. La Grèce. Paris, Presses universitaires, 1952. 308 p.

"Industrie", p. 125–145.

[a]The rules for transliteration of Cyrillic alphabets used by the United States Library of Congress are followed for bibliographical purposes.

Institut international des civilisations différentes. L'attraction exerceé par les centres urbains et industriels dans les pays en voie d'-industrialisation. Bruxelles, 1952. 662 p. (Compte rendu de la XXVIIe session, Florence, juin 1952)
Bibliography.

Institute of Pacific Relations. 10th Conference, Stratford-on-Avon, 1947. Problems of economic reconstruction in the Far East; report. New York, 1949. 125 p.
"Industrial development", p. 71–80.

Inter-American Economic and Social Council. Secretariat report on economic conditions and problems of development in Latin America. Washington, Pan American Union, 1950. 156 p.
Published also in Spanish.

Itagaki, Y. Strategy and policy of economic development in underdeveloped countries; a significance of the stage theory. *Hitotsubashi journal of economics* (Tokyo) 2:1–12, September 1961.

League of Nations. Economic, Financial and Transit Department. Industrialization and foreign trade. Geneva, 1945. 171 p. (Ser. L.o.N.P.1945. II.A.10)

Liu, Ta-chün. China's economic stabilization and reconstruction. New Brunswick, Rutgers University Press, 1948. 159 p.
Published under the auspices of the Sino-International Economic Research Center, New York, and the China Institute of Pacific Relations, Shanghai.

Marette, Andre. Le problème de l'industrialisation des territoires français d'outre-mer. Paris, Librairie genéralé de droit et de jurisprudence, 1939. 238 p.
Bibliography.

Morruau, F.-X. L'industrialisation outre-mer. *In* L'economie de l'Union francaise d'outre-mer. Paris, Sirey, 1952. p. 233–253.

Philippines. Office of Economic Coordination. Reports of the Second Industrial Survey Group to Japan. Manila, Bureau of Printing, 1952. 43 p.

Rakowski, M. Ekonomiczne badanie i ocena inwestycji przemyslowych. Warszawa, Polskie Wydawnictwo Gospodarcze, 1952. 112 p.

Rosen, G. Patterns of Far Eastern industrial development; with special reference to their influence on foreign trade in the Far East. New York [1950] 32 p. (Institute of Pacific Relations. 11th Conference. Secretariat paper, 5)

Russenberg, H. Die Auswirkungen der Industrialisierung von Agrarländern auf Industrie-Exportstaaten. St. Gallen, Verlag der Fehr'schen Buchhandlung, 1949. 206 p. (Veröffentlichungen der Handelshochschule St. Gallen. Reihe A, Heft 20)

U.K. Board of Trade. Report of the United Kingdom Industrial Mission to Pakistan, 1950. London, H.M. Stationery Office, 1950. 79 p.

United Nations. Economic and Social Council. Bibliography on the processes and problems of industrialization in under-developed countries. 29 March 1954. 76 p. (E/2538)

U.S.A. Department of State. Division of Library and Reference Services. Point Four: Far East; a selected bibliography of studies on economically underdeveloped countries. Washington, 1951. 46 p. (Bibliography, 57)

____ Library of Congress. General Reference and Bibliography Division. China; a selected list of references on contemporary economic and industrial development, with special emphasis on post-war reconstruction. Comp. by Helen F. Conover. Rev. ed. Washington, 1946. 118 p.

____ Special Technical and Economic Mission to Thailand. Industrial activity in Thailand. Bangkok, 1953. 42 p.
Mimeographed.

The journals of law follow style guides issued by the particular graduate school of law that publishes the journal. All follow the forms of citation and abbreviation given in *A Uniform System of Citation,* tenth edition, published and distributed by the Harvard Law Review Association, Gannett House, Cambridge 38, Massachusetts. Nonacademic law journals are flexible, and they illustrate practice that is consistent only in citation techniques. The following article, reprinted by kind permission of the American Bar Association Journal, illustrates both the tone and the format common to nonacademic law journals.

Missouri Lawyers Evaluate the Merit Plan for Selection and Tenure of Judges

Twenty-five years ago Missouri became the first state to accept f some of its courts the American Bar Association's merit plan for t selection and tenure of judges. Since then some other states ha adopted all or parts of the plan, and campaigns are under way others. Thus far there have been only fragmentary and subjective a swers and opinions to the question of how well the plan has operat in Missouri. Now Professor Watson, a lawyer and a political scienti reports the objective findings of a survey made in the Missouri Bar.

by Richard A. Watson • ***Associate Professor of Political Science, University of Missouri***

THE AMERICAN Bar Association's merit selection and tenure plan for the judiciary celebrated its silver anniversary recently, since it was on November 5, 1940, that the voters of Missouri accepted the provisions developed by the Association and other bar groups for selecting judges in their state.[1] Since that time, the Appellate Judicial Commission, composed of lawyers, laymen and the Chief Justice of the Supreme Court,[2] have nominated candidates for appointments to the Supreme Court, as well as to the three intermediate courts of appeal of the state—St. Louis, Kansas City and Springfield. Similarly, circuit judicial commissions in St. Louis and Kansas City, with a parallel composition,[3] have nominated candidates for the circuit and probate courts of the two communities, as well as the St. Louis Court of Criminal Corrections.

Six governors of the state, who have served in that office in the last twenty-five years, have made some sixty appointments from among these nominees, and all judges so appointed have run subsequently "on their records" with the voters deciding whether or not they should be retained in office.[4]

In recent years this method of choosing judges has become the focus of a nationwide movement that has transformed Missouri from the "show-me" state to the "show-them" state. Since 1956 voters in four states, Alaska, Iowa, Kansas and Nebraska, have accepted the merit plan (also referred to as the Missouri Plan, as well as the nonpartisan court plan), and features of it apply to courts in six other states.[5] Moreover, in some areas where the plan has not been legally adopt its principles have been put into eff Governors William W. Scranton Pennsylvania and John A. Love Colorado, as well as Robert F. Wagr when he was mayor of New York C have voluntarily utilized nominat commissions for their appointmen Meanwhile, the plan is being deba in bar associations and state legis tures in other states in the Union, a it has been suggested recently for selection of federal judges.[7]

EDITOR'S NOTE: The material for this article is drawn from a general study conducted by the author and two of his colleagues, Rondal G. Downing and Frederick C. Spiegel, with the assistance of grants from the Social Science Research Council and the Research Center of the University of Missouri.

1. In addition to the American Bar Association, the American Judicature Society, the Missouri Bar Association, the Bar Association of St. Louis, the Lawyers' Association of Kansas City and the Missouri Institute for the Administration of Justice played important roles in developing the plan and working for its adoption. See PELTASON, THE MISSOURI PLAN FOR THE SELECTION OF JUDGES, Chapter II (Vol. XX, No. 2, the University of Missouri Studies, 1945).

2. The Appellate Nominating Commission includes, in addition to the Chief Justice, three lawyers elected by members of the Bar (one from each of the courts of appeals' jurisdictions) and three laymen appointed by the Governor from the same jurisdictions.

3. There is a separate nominating commission for Jackson County and St. Louis C County. Each consists of two lawyers ele by members of the Bar residing in the cir court jurisdiction, two laymen from the ju diction appointed by the Governor and presiding judge of the court of appeals of area, who serves as ex officio chairman.

4. In the twenty-five-year operation of plan, only one judge has been voted ou office. This occurred in 1942.

5. The plan is utilized to choose judges courts in Birmingham, Alabama; Dade Cou Florida; Denver, Colorado; and Tulsa, O homa. In addition, California and Illi utilize the tenure aspects of the plan, ur which incumbent judges run in noncomp tive elections.

6. 48 J. AM. JUD. SOC'Y 133 (1964). Ma Lindsay has made a pledge to continue practice in New York City and to work the permanent establishment of a nomina commission by constitutional amendment. Id. 124 (1965).

7. Savage, *Justice for a New Era*, 48 J. JUD. SOC'Y 47 (1965).

This concerted campaign has provoked a series of charges and countercharges among lawyers concerning how well the plan has operated in Missouri. Statements both pro and con that have appeared on the issue from time to time, while helpful and suggestive, have been based primarily on impressions of individuals or small groups of lawyers, rather than on reactions of a substantial number of attorneys of that state. As such, they are open to the criticism that they may not represent the views of the entire Bar. The findings reported here, however, are based on responses from a representative cross section of the Missouri Bar, numbering more than 1,200 lawyers, almost one fifth of those practicing in the state.[8]

The survey, part of a four-year study of the merit plan, sought information (provided anonymously by members of the Bar) on three major matters relating to the selection of judges. The first was the method lawyers prefer for choosing circuit and appellate judges—the merit plan, nonpartisan elections, partisan elections or straight gubernatorial appointment. The second was how they felt the plan had actually operated over the years in recruiting and selecting judges, as well as its effect on judges' behavior once in office. The third dealt with the improvements they felt should be made in the plan. The following discussion analyzes the responses of the Bar.

Preferences on Judicial Selection

The lawyers in the state clearly prefer the merit plan over other methods of selecting judges at the circuit court level. (These are the courts of original, general jurisdiction in Missouri.) Of the 1,233 respondents who replied to the questionnaire, 61 per cent favor the plan, 16 per cent prefer nonpartisan elections, 12 per cent partisan elections and 11 per cent have other suggestions or expressed no opinion. In the latter category, it is interesting to note that only about 1 per cent would like to see the governor on his own appoint judges (a practice that typically occurs in the actual operation of an elective system),[9] and an equally small number think that lawyers alone should choose them.

The strongest support for the nonpartisan plan comes from the state's urban areas. Lawyers practicing in Kansas City and St. Louis, where the circuit judges are selected under the plan, favor it by 79 and 70 per cent proportions. Those whose main practice is in St. Louis County, a suburban complex that surrounds the city, prefer the plan over the present system of choosing their judges in partisan elections by a 44-to-5 per cent ratio, with some 30 per cent supporting nonpartisan elections.

Lawyers with an equal amount of practice in the city and county in the St. Louis area, who have the opportunity to see both selection systems at work, support the merit plan even more: over half of them choose it, compared to 6 per cent who favor partisan elections and another 27 per cent who would like to see nonpartisan elections used to select circuit judges.

The major opposition that exists to the plan in the state is concentrated in the "outstate" areas, which at present choose their circuit judges in partisan contests. Lawyers practicing there generally prefer some kind of elective system, with 30 per cent choosing partisan and 27 per cent nonpartisan ballots. However, it is significant to note that a substantial group, the largest proportion favoring any one method, some 36 per cent, would like to see their jurisdictions adopt the merit plan system, a possibility that to date has never been authorized by the Missouri legislature.[10]

As far as choosing judges to the state's four appellate courts is concerned (the supreme court and three intermediate courts of appeal), lawyers in all areas of the state definitely want to retain the merit plan, which now governs these selections. Support for the plan in choosing appellate, as compared to circuit, judges increases in all geographical groups, but the contrast is most dramatic among outstate attorneys: some two thirds of them prefer the plan for upper court judges, almost twice the proportion that favors it to select those sitting on the lower courts.

Lawyers who prefer to elect their circuit judges, but who want to use the merit plan for appellate judges, were asked to give their reasons for choosing different selection methods for the two kinds of courts. A variety of explanations was given, but essentially they involved a few major arguments. The first is that the people, particularly in outstate areas, are in a position to know personally the lawyers practicing in their area, and thus they can make an intelligent and independent choice as to which of them should be judges. However, this is not the case with respect to the selection of appellate judges; the voter cannot be acquainted with their qualifications since they may come from distant areas of the state. The result is that the influence of urban political organizations, particularly in the primaries, would become the crucial factor in the selection of appellate judges. Therefore, these judges should be selected under the nonpartisan plan, which neutralizes the effect of political forces.

Other reasons advanced for preferring disparate systems at the two levels have to do with the different conceptions lawyers have of the functions performed by trial and appellate courts. Some attorneys feel that trial judges are "closer to the people" than appellate judges and that they should

---

8. Since The Missouri Bar is integrated, we used the membership of that association as our "population". We randomly selected 3,303 of the 6,606 Missouri lawyers who belonged to the association and mailed questionnaires to them in May of 1964. A comparison of the lawyers responding (1,233) with the entire Missouri Bar as of that same year analyzed in *The 1964 Lawyer Statistical Report* (American Bar Foundation, 1965) indicates that our sample is very representative on such characteristics as age, kind of practice arrangement (sole, firm, etc.), geographical location of practice and law school education.

9. Studies of the subject have shown that many state judges, as high as 50 per cent in some cases, come to the bench originally by appointment of the governor as a result of the retirement of a sitting judge. Once appointed, they run in subsequent elections as incumbents, which puts them at a decided advantage vis-à-vis other candidates. See VANDERBILT, MINIMUM STANDARDS OF JUDICIAL ADMINISTRATION 8 (1949) and Winters, *One-Man Judicial Selection*, 45 J. AM. JUD. SOC'Y 198 (1962).

10. The Constitution of the State of Missouri, 1945, Article V, Section 29(b), provides for the adoption of the plan by voters in other judicial circuits, but the Missouri Supreme Court has ruled that this section is not self-executing and requires legislation to make it effective. *State* ex rel. *Millar* v. *Toberman*, 360 Mo. 1101, 232 S.W. 2d 904 (1950). Several attempts to pass such legislation have failed.

be more responsive to public sentiment than higher court judges. As one respondent expressed it:

> I am of the opinion that the judges of the lower courts must feel the pulse of the people in order to accomplish justice according to current circumstances. Making the circuit judges more accountable to the people tends to insure this result. The higher courts should not be subject to the influence of circumstances of a particular locality.

Related to this same difference in function of the two kinds of courts is the opinion that the people are in a better position to evaluate judges for the lower courts, since (1) they observe these judges in action and (2) the conduct of trial proceedings does not involve as much legal *expertise* as work on an appellate tribunal. One lawyer suggested, "A circuit judge needs to be more understanding of human nature and observant of people, while an appellate judge applies more legal theory." Thus, the reasoning is that lay persons are capable of determining the qualities needed for a trial judge, while only lawyers and judges can properly evaluate those of candidates for the higher courts.

Preferences of Missouri lawyers concerning judicial selection for both the circuit and appellate courts are affected somewhat by the fields of legal practice in which they specialize. Defendants' and corporation lawyers most favor the plan; those representing plaintiffs, labor unions and persons accused of crime are least enthusiastic about it. However, it should be noted that in the metropolitan areas, all elements of the Bar, including lawyers with the last three specialties, prefer the merit plan over other selection methods, even though some of these groups opposed the plan when it was originally introduced in Missouri.[11] The same is true of sole practitioners, although their support for the plan falls considerably below that of attorneys working in law firms, as well as those employed by corporations or government agencies.

Political party loyalties also have some bearing, but their effect on the metropolitan Bar is not as great as one might expect. More Republicans than Democrats practicing in Kansas City and St. Louis favor the plan for the selection of both circuit and appellate courts. This is understandable since, given the Democratic majorities that exist statewide and in these areas, it is unlikely that as many Republicans would have been elected to these benches as have been appointed under the plan during the past twenty-five years.[12] However, the support of the Democrats for the plan in these two areas is nevertheless high (68 per cent for selecting circuit judges and 75 per cent for appellate), indicating that other factors besides partisan advantage are taken into account by urban Democratic lawyers in their preferences concerning judicial selection.[13]

Attitudes on the Operation of the Plan

In the personal interviews we conducted with over 200 lawyers in the state in connection with our general study of the merit plan, we asked lawyers to give their reasons for supporting or opposing the plan. From the comments advanced, we took six consequences lawyers mentioned most often as being associated with the operation of the plan (three favorable and three unfavorable) and asked the respondents in our sample survey to indicate whether they agreed or disagreed with each of them. We thus gained a view of which of the reasons advanced in support or opposition to the plan were most persuasive to the Bar as a whole.

The results show that the consequences of the plan about which most lawyers agree are that the plan recruits better judges than popular election does. The major reason advanced for this result is that able lawyers are more willing to seek a judgeship under a system that spares them the rigors, expense and uncertainties of election campaigns. Most lawyers also agree that the independence judges enjoy under the plan encourages them to make decisions based on the merits of cases, rather than subjecting them to pressures elective judges must often face. About two thirds of the attorneys see these as results of the plan, while only about one tenth of them do not.

Richard A. Watson received his law degree from the University of Michigan in 1951, and after private practice he returned there and earned his Ph. D. in political science in 1959. Since then he has taught political science at the University of Missouri, and during the 1965-1966 academic year he was on leave as a fellow at the Center for Advanced Study in the Behavioral Sciences at Stanford, California.

Also, by about a two-to-one ratio, lawyers in the state reject the comments often made by those opposing the plan that it tends to make judges arbitrary in their treatment of lawyers and laymen with business before the courts and favors the selection of defendants' and corporation lawyers as judges.

The allegations concerning the merit plan about which the Missouri Bar is most uncertain are those relating to the role that "politics" plays in choosing judges. The lawyers are almost evenly divided on the issue of whether the plan has succeeded for the most part in

11. Some elements of organized labor, together with some leading labor lawyers, opposed the adoption of the plan in Missouri. Also, The Kansas City Bar Association and the Lawyers' Association of St. Louis, both of which included in their leadership some prominent plaintiffs' lawyers, generally were opposed to the plan. PELTASON, *op. cit. supra*, note 1, Chapter III.

12. During that period, ten Republicans have been appointed to Kansas City and St. Louis circuit and probate courts, three to the courts of appeals of the two areas and three to the Missouri Supreme Court.

13. Among lawyers practicing in St. Louis County and outstate, the Democrats support the plan more than Republicans.

taking "politics" out of the selection of judges. Thirty-eight per cent say it has, while 41 per cent feel it has not. More than half the lawyers in the state agree, moreover, that the plan substitutes Bar politics and gubernatorial politics for the traditional politics of party leaders and machines, compared to about one fifth of the Bar that thinks this result has not occurred.

One of the difficulties with these statements, revealed by comments added to some questionnaires, is the fact that the respondents interpreted the word "politics" differently, some restricting it to party politics and others giving it a broader connotation, as evidenced by one lawyer's statement that, "Politics unavoidably exists whenever people are involved, since the personal element can never be discounted." Similar assumptions concerning the inevitability of politics led some lawyers to state that while they agreed that the plan substituted one brand of politics for another, they preferred that associated with the plan to the kind produced by the elective system of choosing judges. As one attorney put it, "Plan politics is the lesser of two evils inasmuch as party politics entails the additional demand of monetary expenditures." Thus, disagreements within the Bar on the part that politics plays in choosing judges under the plan are, at least in part, based on semantic differences rather than those of substance.

Suggested Improvements in the Plan

We followed the same procedure with respect to improvements in the plan as we did with attitudes on its operation. We asked lawyers we personally interviewed for suggestions as to how the plan might be improved, and we then requested respondents in our sample survey to indicate which of these suggestions they favored. They were also asked to specify any other improvements they would like to see made in the plan. Virtually all the lawyers had some ideas for improving the plan, including those who vigorously support it.

Clearly, the gravest weakness in the plan, as Missouri lawyers view it, is the problem of removing from the bench judges who are not capable of handling the duties of the office. Under present state statutes, appellate judges must retire at age 75, but there is no such requirement for circuit judges in Kansas City and St. Louis or, for that matter, for those outstate judges who are elected. About four fifths of all lawyers in Missouri favor extending this provision to the lower court judges, and some of them would like to see the age limit lowered to 65 or 70 for all judges. At the same time, about one half of the Bar would like to see the retirement benefits of judges increased beyond the present level of one third of the annual salary at the time of retirement.

Although the issue is most acute for older judges, there is also some concern among lawyers for developing a procedure for removing incompetent judges from the Bench regardless of their age. There is general agreement in the Bar that the present requirement that a judge be retained in office by an affirmative vote of the electorate is so easily met that in actuality it has meant life tenure for judges, and that some more effective method for dealing with the problem should be devised. One third of the lawyers favor raising the "yes" vote necessary to stay in office from a simple to a two-thirds majority. Other possibilities are also suggested, such as granting the nominating commission authority to remove judges who are not properly fulfilling the duties of their office, or giving the Bar a greater role in determining whether judges should be retained in office.[14]

Three other improvements in the plan drew the approval of at least one third of the Bar. One involves raising the compensation of judges to make it competitive with income received by leading members of the Bar engaged in private practice. Another is the selection of the commissioners of the appellate courts—who at present receive the same salary as the regular judges and share the duties of working on cases, but who have no vote in deciding them—through the provisions of the plan, rather than permitting the court members to choose them. (The fact that several supreme court commissioners have subsequently been chosen as judges is resented by some lawyers. They charge that judges in effect pick their future colleagues by first getting them in the court as commissioners and then using their influence on the appellate nominating commission[15] to get them selected as judges.) The third involves requiring the courts under the plan to be bipartisan, that is, to represent as equally as possible both major political parties. As might be anticipated, this latter improvement is much more favored by Republicans than by Democrats, since the latter party has had more of its members appointed to the bench under the plan than the former.[16]

How the Survey May Be Summarized

In summary, it may be said that after twenty-five years of experience with the merit plan, Missouri lawyers are for the most part satisfied with this system of selecting judges. It is particularly significant that the members of the Bar who have lived with it most closely—those practicing in Kansas City and St. Louis—are its strongest supporters. At the same time, they are willing to consider improvements that will strengthen the plan.[17] Their thinking is epitomized by the frank comment of one attorney: "Best system devised to date, but needless to say, far from perfect."

---

14. Since 1948 The Missouri Bar has conducted bar polls among lawyers practicing in merit plan jurisdictions to obtain their views as to whether judges up for re-election should be retained in office. The polls, however, are only advisory for the electorate.
15. It will be recalled that the Chief Justice of the Missouri Supreme Court serves as chairman of the Appellate Nominating Commission.
16. Of the sixty lawyers appointed to the bench under the provisions of the plan, 70 per cent have been Democrats and 30 per cent Republicans.
17. A committee was formed recently to look into the improvement of the plan.

I. EDUCATION JOURNALS

The journals of education for the past several years have tended to follow practice recommended by the National Education Association *Style Manual for Writers and Editors*. The 1966 edition of that document will quite likely become definitive in the field. One of its most significant sections, the discussion of bibliographical and footnote references, is reprinted by permission below. The discussion is unusual in that it recommends the same standard procedure for both bibliographical and footnote entries.

The illustrative article which follows the passage from the NEA *Style Manual* is reprinted by permission of the *NEA Journal*. A discussion of this article's effective use of graphs and tables appears on p. 144.

14. BIBLIOGRAPHICAL AND FOOTNOTE REFERENCES

In the interests of simplicity, consistency, and clarity, the following forms are intended for use in the composition of *both* bibliographical and footnote entries. The practice of referring to the bibliography rather than to a footnote by the use of one of a variety of numerical codes is becoming more and more common.

In footnote references, *ibid.* may be used to replace as much of the entry as is identical with the preceding entry. If an entry refers to the same book, chapter, or article as an earlier entry and no citation of another work by the same author intervenes, the author's name, *op. cit.*, and the specific location of the passage cited may replace a full citation. When an entry refers to a passage cited earlier and no citation of another passage by the same author intervenes, the author's name and *loc. cit.* may replace a full citation.

In a citation of a book, the desired information should be copied from the title page, not from the cover. In a reference to a periodical, data concerning author, title, and pages should be copied from the article itself, not from the table of contents of the periodical.

When information is secured from book and periodical indexes and not from the publications themselves, the author entry should be checked for the full name of the author and, in the case of a book, for the pages. *The Literary Market Place, Books in Print,* or the *Cumulative Book Index* should be referred to for the full name of the publisher, which should be given in the form current at the time of publication.

14.1. Abstract

Anderson, William F., Jr. *The Sociology of Teaching. I: A Study of Parental Attitudes Toward the Teaching Profession.* Doctor's thesis. Iowa City: State University of Iowa, 1952. 185 pp. Abstract: *Dissertation Abstracts* 12: 692; No. 5, 1952.

Lubin, Ardic. "A Rank Order Test for Trend in Correlated Means." (Abstract) *Annals of Mathematical Statistics* 28: 524; June 1957.

14.2. Alphabetizing

References should be alphabetized by author, last name first. When two or more references to one author give his name in different forms, the name should be spelled out to agree with the most nearly complete form if the references follow one another in the same bibliography.

After an author's name has once been given, a 3-em dash may be used in place of it in subsequent references.

Books and magazine articles by one author should be listed alphabetically by title, unless the logical sequence is by parts or dates.

Books and magazine articles written by an author should be listed before those edited by the same author. Any books or articles written and edited by the author in conjunction with others should follow.

In lists of NEA publications, general NEA publications should be first, arranged alphabetically by title; next, publications of departments, divisions, committees, and commissions, alphabetically arranged (see *NEA Handbook*); and, finally joint publications of the NEA and a department or other organization.

14.3. Articles

14.3.1. An article should be cited as follows:

Cumming, John R. "Educating Exceptional Children in Alaska." *Exceptional Children* 21: 82-83, 111; December 1954.

NOTE: When no author is given, an article should be listed as follows:

National Education Association, Department of Elementary School Principals. "Our New President." *National Elementary Principal* 37: 2; October 1957.

Scholastic. "Jam Session: Ideal Teacher." *Scholastic* 65: 34-35; September 22, 1954.

14.3.2. For a series of articles published in several issues of a periodical, only one reference should be made:

Eggert, Walter A. "Short-Term Financing." *American School Board Journal* 99: 40-42, 77; December 1939. 100: 29-30, February; 49-51, May; 30-31, June 1940. 101: 28, 88, September; 27-28, 94, November 1940.

Holmes, Warren S. "Economy, True and False in School Building." *American School Board Journal* 102: 21-22, 99, January; 45-47, February 1941.

14.4. Authors

14.4.1. When a publication has two or three authors, all of them should be named. A publication having more than three authors should be listed under the name of the first mentioned and *and others.* (When capitals and small capitals are used for printing authors' names, the *and* should be lower-cased, the *others* in small capitals:

Bagley, William C., and Keith, John A. H. *An Introduction to Teaching. . . .*

Reavis, William C.; Pierce, Paul R.; and Stulken, Edward H. *The Elementary. . . .*

BENSON, CHARLES E., and OTHERS. *Psychology for Teachers. . . .*

14.4.2. A publication that has been published by the author should be cited as follows:

Garber, Lee O. *The Yearbook of School Law, 1954.* Philadelphia: the Author (3812 Walnut St.), 1954. 119 pp.

14.5. Bibliography. A bibliography or other collection should be listed under the name of the compiler or editor when it is given; otherwise, under the name of the contributor mentioned first and *and others*:

Smith, John A., compiler. "Selected References on Secondary Education.". . . .

14.6. Books and Pamphlets. The essentials of the citation of a book or pamphlet are as follows:

(a) Author's name, as given on title page (inverted)

(b) Title, as given on title page (in italics)

(c) Publisher's name, with place of publication as given on title page

(d) Date of publication, as given on title page (if not given on title page, copyright date on reverse of title page should be used)

(e) Number of pages, inclusive, of chapter, section, or entire book (introductory pages numbered separately with Roman numerals should be ignored).

Wheat, Harry G. *How To Teach Arithmetic.* Evanston, Ill.: Row, Peterson and Co., 1951. 438 pp.

14.7. Census Report

U.S. Department of Commerce, Bureau of the Census. *Estimates of the Population of the United States and Components of Population Change: 1950-1955.* Current Population Reports, Series P-25, No. 111. Washington, D.C.: the Bureau, February 23, 1955. 4 pp.

14.8. Chapter. Arabic numbers should be used for chapter numbers; and the title of the chapter should be included, except when reference is made to more than one chapter (see 14.18., example 2, for chapters in edited books and 14.62. for chapters in yearbooks):

Havemann, Ernest, and West, Patricia S. *They Went to College.* New York: Harcourt, Brace and Co., 1952. Chapter 11, "What College Graduates Think About College," pp. 126-37.

14.9. Chart

United States Air World Map and Air Distance Time Chart. Chicago: A. J. Nystrom and Co.

America—A Nation of One People from Many Countries. Chart. New York: Council Against Intolerance in America (17 E. 42nd St.), 1940.

14.10. Commission, NEA

National Education Association and American Association of School Administrators, Educational Policies Commission. *Moral and Spiritual Values in the Public Schools.* Washington, D.C.: the Commission, 1951. 100 pp.

National Education Association, National Commission on Teacher Education and Professional Standards. *Milestones in the Professional Standards Movement.* Washington, D.C.: the Commission, 1954. 8 pp.

14.11. Committee, NEA

National Education Association, Educational Finance Committee. *Financial Status of the Public Schools.* Washington, D.C.: the Association, 1964. 35 pp.

14.12. Compiler

Martin, William E., and Stendler, Celia B., compilers. *Readings in Child Development.* New York: Harcourt, Brace and Co., 1954. 513 pp.

Davis, Clara. "Results of the Self-Selection of Diets by Young Children." *Readings in Child Development.* (Compiled by William E. Martin and Celia B. Stendler.) New York: Harcourt, Brace and Co., 1954. pp. 69-74.

14.13. Condensation

Weber, C. A. "Promising Techniques for Educating Teachers in Service." *Educational Administration and Supervision* 28: 691-95; December 1942. Condensed: *American School Board Journal* 106: 32; February 1943.

14.14. Congressional Bill

U.S. 83rd Congress, 2nd Session. *S. 2723: A Bill To Provide for a White House Conference on Education.* Washington, D.C.: Senate Committee on Labor and Public Welfare, 1950.

14.15. Department, NEA. If the name of the department does not contain the word *department,* the NEA should not be mentioned in the author entry, but the departmental affiliation should be indicated in the publisher entry. If the name of the department contains the word *department,* the NEA should be mentioned in the author entry but not the publisher entry:

American Educational Research Association. *Growing Points in Educational Research.* Official Report. Washington, D.C.: the Association, a department of the National Education Association, 1949. 340 pp.

National Education Association, Department of Classroom Teachers. *Official Report, 1953-54.* Washington, D.C.: the Department, 1954. 88 pp.

14.16. Discussion Pamphlet

National Education Association, Research Division and Department of Classroom Teachers. *Teacher Rating.* Discussion Pamphlet No. 10. Revised edition. Washington, D.C.: the Association, 1950. 24 pp.

14.17. Division, NEA

National Education Association, Research Division. *Economic Status of Teachers in 1960-61.* Research Report 1961-R4. Washington, D.C.: the Association, March 1961. 52 pp.

14.18. Editor

Taylor, Calvin W., editor. *Research Conference on the Identification of Creative Scientific Talent.* Salt Lake City: University of Utah Press, 1955. 268 pp.

Keesecker, Ward W. "The Rights of Pupils and Parents." *The Tenth Yearbook of School Law.* (Edited by M. M. Chambers.) Washington, D.C.: American Council on Education, 1942. Chapter 1, pp. 1-15.

14.19. Editorial

Nation's Schools. "Let the People Know." (Editorial) *Nation's Schools* 27: 17; March 1941.

Binnion, John E. "3-R's in the Teaching of Bookkeeping: Read, Reflect, and React." (Editorial) *Business Education Forum* 12: 4; December 1957.

14.20. Educational Research Service Circular

National Education Association, Research Division and American Association of School Administrators. *Education in Lay Magazines, Third Quarter, 1954.* Educational Research Service Circular No. 10, 1954. Washington, D.C.: the Association, October 1954. 25 pp.

14.21. Encyclopedia of Educational Research

Harris, Chester W., editor. *Encyclopedia of Educational Research.* Revised edition. New York: Macmillan Co., 1960.

Knight, Edgar W. "History of Education." *Encyclopedia of Educational Research.* Revised edition. (Edited by Walter S. Monroe.) New York: Macmillan Co., 1950. pp. 551-56.

14.22. Excerpt

National Education Association, Research Division. "The Status of Driver Education in Public High Schools, 1952-53." *Research Bulletin* 32: 52-99; April 1954. Excerpts: *Education Digest* 20: 26-28; October 1954.

14.23. Federal Law. There are two official citations to a federal law: to the U.S. Statutes, which correspond to session laws in the states, and to the U.S. Code. Page numbers should always be used in references to the Statutes, never in references to the Code:

U.S. Statutes. Vol. 53, Chapter 281, p. 1013. 76th Congress, 1st Session, 1939.

U.S. Code, Supp. IV (1941-1945). Title 42, Chapter 6A, sec. 282, subsec. (d).

U.S. Code. Title 36, Chapter 9, sec. 141.

14.24. Film

Summer Harvest. 29 min., 16mm, sound, color and b & w. National Education Association, Press, Radio, and Television Relations Division, 1201 Sixteenth St., N.W., Washington, D.C., 1962.

The Transplantation. 30 min., 16mm, sound, b & w. Indiana University, NET Film Service, Audio-Visual Center, Bloomington, Ind., 1956.

14.25. Filmstrip

Managing Your Money. 45 frames, color. McGraw-Hill Book Co., Textbook Film Department, 330 W. 42nd St., New York, N.Y., 1954.

14.26. Hearing

U.S. 84th Congress, 1st Session, Senate Committee on the Judiciary, Subcommittee To Investigate Juvenile Delinquency. *Hearings Pursuant to S. Res. 62, Investigation of Juvenile Delinquency in the United States.* Washington, D.C.: Government Printing Office, 1955. 141 pp.

14.27. In Press Publications. A notation should be made in parentheses following the reference: (In press)

14.28. Joint Committee. In references to publications produced by joint committees, the organization which published the work (holds the copyright) should be listed as publisher:

National Education Association and American Medical Association, Joint Committee on Health Problems in Education. *Health Education.* (Edited by Bernice R. Moss, Warren H. Southworth, and John Lester Reichert.) Fifth edition. Washington, D.C.: National Education Association, 1961. 429 pp.

National Education Association and American Medical Association, Joint Committee on Health Problems in Education. *A Story About You.* (Prepared by Marion O. Lerrigo and Helen Southard.) Chicago, Ill.: American Medical Association, 1962. 43 pp.

14.29. Judicial Decision. Although decisions of state and federal courts are published in several series, only the National Reporter System should be used. The citation of a decision includes the following elements: (a) the name of the case, which should be italicized unless used in a footnote; (b) the volume number of the report; (c) the abbreviated title of the reporter; (d) the number of the page on which the decision begins; and (e) the state and the year of the decision, in parentheses if a state court; the year only if a federal court. If a particular page is being referred to instead of a case as a whole, the particular page number should be given after the number of the first page of the case.

Abbreviations of the National Reporter System follow:

Atlantic—Atl. *or* A. (2d)
Northeastern—N.E. *or* N.E. (2d)
Northwestern—N.W. *or* N.W. (2d)
Southern—So. *or* So. (2d)
Southeastern—S.E. *or* S.E. (2d)
Southwestern—S.W. *or* S.W. (2d)
Pacific—Pac. *or* P. (2d)
New York Supplement—N.Y. Supp. *or* N.Y.S. (2d)
Federal Reporter—Fed. *or* F. (2d)
Federal Supplement—Fed. Supp.
Supreme Court of the United States—Sup. Ct.

Alston v. School Board of City of Norfolk, 112 F. (2d) 992 (1940)

Bopp v. Clark, 147 N.W. 172 (Iowa, 1914)

Missouri v. Canada, 59 Sup. Ct. 232,236 (1938)

14.30. Leaflet

National Education Association. *Learning Is Your Business.* (Leaflet) Washington, D.C.: the Association, n.d. 8 pp.

14.31. Mimeographed Publication. A notation should be made in parentheses following the reference: (Mimeo.) or (Offset):

National Education Association, Research Division. *Education in the United States.* Research Memo 1965-11. Washington, D.C.: the Association, May 1965. 8 pp. (Offset)

14.32. Monograph

National Association of Secondary-School Principals. *The American High School.* Bulletin No. 208. Washington, D.C.: the Association, a department of the National Education Association, February 1955. 308 pp.

14.33. Newspaper Article

New York Times. "The Mayor and Education." (Editorial) *New York Times,* December 20, 1954. p. 28.

Nettleton, Lewis L. "London Markets Offer Contrasts." *New York Times,* December 20, 1954. pp. 42-43.

14.34. Official Report of NEA Unit

American Association of School Administrators. *Building Americans in the Schools.* Official Report. Washington, D.C.: the Association, a department of the National Education Association, 1954. 254 pp.

14.35. Out-of-Print Publication. A notation should be made in parentheses following the reference: (Out of print).

14.36. Parts. Roman numerals should be used for parts of books:

National Society for the Study of Education. *The Community School.* Fifty-Second Yearbook, Part II. Chicago: University of Chicago Press, 1953. 292 pp.

14.37. Periodical (see 14.41. for treatment of publishers or periodicals)

14.37.1. The essentials of the citation of a periodical are as follows:

(a) Author or authors, as given with the article
(b) Title, as given in the article (As far as possible *a, and,* and *the,* which are omitted by *Education Index,* should be supplied.)
(c) Name of the periodical (in italics, omitting *The* from the title)
(d) Volume number (in Arabic numerals)
(e) Pages, inclusive numbers
(f) Date (with the month spelled out).

Smith, Constance C. "Sound Foundations in High School." *NEA Journal* 50: 21-23; October 1961.

14.37.2. When an entire issue of a periodical is devoted to one subject under a special title, it should be listed under the name of the editor, except for periodicals of NEA departments, which should be listed under the name of the department:

Belding, Anson W., editor. "World-at-Our-Doors." *Journal of Education* 130: 215-40; October 1947.

Mudgett, Helen Parker, issue editor. "Inter-Group Education: Practice and Principle." *Education* 68: 129-92; November 1947.

National Education Association, Department of Elementary School Principals. "Effective Written Communication." *National Elementary Principal* 33: 6-27; April 1954.

14.38. Prices. Generally, prices should be quoted only of NEA and U.S. Office of Education publications. (If the reference is to a chapter or section and not to the complete publication, the price should not be given.) The price, followed by a period, should be placed at the end of the reference:

15¢. $2. $1.25. Cloth, $3.25; paper, $1.25.

14.39. Proceedings. (a) Proceedings should be listed under the name of the association; (b) when an individual paper is referred to, it should be listed under the name of the individual:

(a) National Education Association. *Addresses and Proceedings, 1954.* Washington, D.C.: the Association, 1954. 413 pp.

Association for Higher Education. *Current Issues in Higher Education, 1954.* Proceedings of the Ninth Annual National Conference on Higher Education. Washington, D.C.: the Association, a department of the National Education Association, 1954. 321 pp.

(b) D'Evelyn, Katherine E. "How To Conduct a Parent Conference." *Forty-First Schoolmen's Week Proceedings.* University of Pennsylvania Bulletin, Vol. 55, No. 1. Philadelphia: University of Pennsylvania, August 1954. pp. 65-73.

14.40. Project. Publications of an NEA project should be treated in the same way as those of an NEA committee.

14.41. Publishers. The state should be given only for small cities or when there are several cities of the same name. When the title page mentions several cities, the first one mentioned should be used. *Inc.* and *Publishers* should be omitted; *Company* should be abbreviated:

Boston: Allyn and Bacon
New York: American Book Co.

In references to periodicals, the place of publication and the publisher should be included only (a) when the periodical is local, rather than national, in distribution; (b) when two periodicals are almost identical in name; or (c) when it seems desirable to indicate government and university periodicals:

(a) National Education Association and American Teachers Association, Joint Committee. "A Study of the Status of the Education of Negroes: Part 1, Legal Status of Segregated Schools." *Bulletin* 29: 6-19; May 1954. (Montgomery, Ala.: American Teachers Association.)

(b) Gilbaugh, John W. "The Superintendency and Boards of Education." *Bulletin of Education* (University of Kansas) 8: 65-71; May 1954.

(c) Pidgeon, Mary E. "Changes in Women's Occupations, 1940-50." *Monthly Labor Review* 77: 1205-1209; November 1954. (Washington, D.C.: U.S. Department of Labor, Bureau of Labor Statistics.)

14.42. Recording

Freedom's People. "Negroes' Contribution to Sports." 15 min., 33⅓ rpm. Educational Radio Script and Transcription Exchange, U.S. Department of Health, Education, and Welfare, Office of Education, Washington, D.C. (Available on loan)

14.43. Reprint. The original should be used if it can be obtained. If not, the reprint should be cited as a book or pamphlet; a note should be added in parentheses, indicating that a reprint is being cited and giving such information about the original as is available:

Fine, Benjamin. *Why Our Schools Are in Serious Trouble.* New York: New York Times, 1952. 10 pp. (Reprint of articles in the *New York Times,* January 14 to 19, 1952)

14.44. Revised Edition

Ross, Clay C. *Measurement in Today's Schools.* Third edition. New York: Prentice-Hall, 1954. 485 pp.

Douglass, Harl R. *Modern Administration of Secondary Schools.* Revision and extension of *Organization and Administration of Secondary Schools.* Boston: Ginn and Co., 1954. 601 pp.

14.45. School Publications. Authorship should be determined from the title page. When there is doubt, . . . *Public Schools* should be used as author, and *Board of Education,* as publisher. The name of the school superintendent should not be used:

Wellesley Public Schools. *Annual Report of the Wellesley Public Schools for the Year Ending December Thirty-First, 1953.* Wellesley, Mass.: Board of Education, 1954. 90 pp.

Warren Public Schools. *And Gladly Teach.* 1952 Report of the Superintendent of Schools. Warren, Ohio: Board of Education, 1952. 21 pp.

14.46. Series Notes

Brown, Francis J., editor. *University and World Understanding.* Report of a Conference of Fulbright Scholars on Education. Series I, Reports of Committees and Conferences, Vol. 8, No. 58. Washington, D.C.: American Council on Education, 1954. 97 pp.

Kaho, Elizabeth E. *Analysis of the Study of Music Literature in Selected American Colleges.* Contributions to Education, No. 971. New York: Teachers College, Columbia University, 1950. 74 pp.

President's Materials Policy Commission. *Foundations for Growth and Security.* Resources for Freedom, Vol. 1. Washington, D.C.: Government Printing Office, June 1952. 184 pp.

Eberle, August W. *A Brief History and Analysis of the Operation of the Educational Placement Service at Indiana University.* Bulletin of the School of Education, Vol. 31, No. 1. Bloomington: Indiana University, 1955. 30 pp.

14.47. Slides

Butterflies and Moths. 2 x 2, color. Coast Visual Education, 5620 Hollywood Blvd., Hollywood, Calif.

14.48. Speeches

Jones, Henry W. "Education in the Modern World." Address given at the tenth annual meeting of the Middletown Educational League, Middletown, Massachusetts, May 15, 1965.

Percy, Charles H. "Education in a Revolutionary Age." (Speech) *Addresses and Proceedings, 1962.* Washington, D.C.: National Education Association, 1962. pp. 56-61.

14.49. State Laws. Codes and session laws should be listed under the name of the state, and the states should be listed alphabetically. The title of a state code should include the date only if the date is part of the title. The session laws should always include the year as part of the title. Page numbers should never be used in either codes or session laws unless there is no better way of identifying the law. Usually chapter number or act number is sufficient for a session law, and if a particular part of it is being referred to, subsections, subdivisions, or paragraphs may be identified by number. In the state codes, the title and/or chapter number should be given unless the system of codification makes it unnecessary by use of identifying section numbers in sequence throughout the system. The words *Title* and *Chapter* should be capitalized whenever used. The word *section* should be written as *sec.,* not capitalized and always abbreviated in a reference. In a few states, the title of the code includes the compiler's name; otherwise the name of the compiler should be omitted from the reference.

In a reference to a school law, the state code or session law should be cited, rather than the publication of the state department of education. If possible, the code reference should be obtained, but if it is not available, a citation to the school law as issued by the department of education should, in general, follow the rules for citations to the code and session laws:

Remington's Revised Statutes of Washington, Annotated, 1932. Title 28, sec. 4884-92.

Wyoming Revised Statutes, 1932, Annotated. Chapter 99, sec. 913.

Iowa Code of 1946. Chapter 272.

Consolidated Laws of New York. Chapter 16, sec. 503, part 4.

14.50. State Publications

Montana State Department of Public Instruction. *Your Schools Today and Goals for Tomorrow.* Biennial Report of the Superintendent of Public Instruction of Montana, 1952-1954. Helena: the Department, 1954. 101 pp.

Nolan, William J. *Building a Community's Curriculum for the Mentally Handicapped.* Bulletin No. 58. Hartford: Connecticut State Department of Education, May 1952. 24 pp.

14.51. Subtitles. In general, subtitles should be omitted.

14.52. Supplements to Periodicals

Reeves, Floyd W. "Education and National Defense." *Ninth Educational Conference, 1940.* Educational Record, Vol. 22, Supplement No. 14. Washington, D.C.: American Council on Education, January 1941. pp. 12-22.

14.53. Surveys

Utah Public School Survey Commission. *A Survey of the Utah Public Elementary and Secondary School System.* An Interim Report to the Governor, Legislative Council, and Legislature. Salt Lake City: the Commission, February 1953. 288 pp.

Strayer, George D., and Yavner, Louis E., directors. *Administrative Management of the School System of New York City.* Report of Survey of the Board of Education and the Board of Higher Education. New York: Mayor's Committee on Management Survey of the City of New York, October 1951. 51 pp.

14.54. Symposium. A parenthetical notation should be used:

Potter, Gladys L., and others. "How Can We Help Emergency Teachers?" (Symposium) *Educational Leadership* 1: 9-12; October 1943.

14.55. Test

Spitzer, H. F., and others. *Iowa Every-Pupil Test of Basic Skills: Test A, Silent Reading Comprehension.* (Elementary Battery) Boston: Houghton Mifflin Co.

14.56. Thesis (see 14.1. for treatment of abstracts)

Kirkpatrick, Ervin E. *Problems in the Use of Instructional Films.*

Master's thesis. Pittsburg: Kansas State Teachers College, 1940. 146 pp. (Typewritten)

Davis, Hazel. *Personnel Administration in Three Non-Teaching Services of the Public Schools.* Contributions to Education, No. 784. Doctor's thesis. New York: Teachers College, Columbia University, 1939. 323 pp.

14.57. Translation

Az-Zarnuji, Burham ad-Din. *Instruction of the Student: The Method of Learning.* (Translated by G. E. Von Grunebaum and Theodora M. Abel.) New York: King's Crown Press, 1947. 78 pp.

14.58. U.S. Office of Education Publication. When an individual author's name is given on the title page, that should be used; otherwise, for publications after April 1953, *U.S. Department of Health, Education, and Welfare, Office of Education. U.S. Department of the Interior, Bureau of Education* should be used for publications issued before 1929; *U.S. Department of the Interior, Office of Education,* for publications issued from 1929 to July 1939; and *U.S. Office of Education, Federal Security Agency,* for publications issued from July 1939 to April 1953:

Hutchins, Clayton D.; Munse, Albert R.; and Booher, Edna D. *Federal Funds for Education, 1952-53 and 1953-54.* U.S. Department of Health, Education, and Welfare, Office of Education, Bulletin 1954, No. 14. Washington, D.C.: Government Printing Office, 1954. 130 pp.

Sanders, Jennings B. *General and Liberal Educational Content of Professional Curricula: Forestry.* U.S. Department of Health, Education, and Welfare, Office of Education, Pamphlet No. 115. Washington, D.C.: Government Printing Office, 1954. 12 pp.

U.S. Department of Health, Education, and Welfare, Office of Education. "Statistics of Special Education for Exceptional Children, 1952-53." *Biennial Survey of Education in the United States: 1952-54.* Washington, D.C.: Government Printing Office, 1954. Chapter 5, 78 pp.

U.S. Department of Health, Education, and Welfare, Office of Education. *Education Directory, 1954-55.* Washington, D.C.: Government Printing Office, 1954. Part 3, "Higher Education," 178 pp.

14.59. Unpublished Material. The title of the work should appear in Roman type with quotation marks rather than in italics, and a notation should be made in parentheses following the reference: (Unpublished)

14.60. Volume

14.60.1. Arabic numerals should be used for volume numbers:

Burke, Arvid J. "What Makes Good Teachers?" *NEA Journal* 43: 475-77; November 1954.

14.60.2. When the volume number of a magazine article is missing, the month may be used, as follows:

Viele, John A. "Let Us Go Forward Together." *Virginia Journal of Education,* April 1946. pp. 337-38, 410.

14.60.3. The paging of several volumes in one publication should be indicated as follows:

Chitty, Dennis, editor. *Control of Rats and Mice.* New York: Oxford University Press, 1954. Vol. 1, 338 pp. Vol. 2, 244 pp. Vol. 3, 240 pp.

President's Materials Policy Commission. *Resources for Freedom.* Washington, D.C.: Government Printing Office, 1952. Vol. 105, 819 pp.

Churchill, Sir Winston. *History of the English-Speaking Peoples: The Birth of Britain.* New York: Dodd, Mead & Co., 1956. Vol. 1, 521 pp.

14.61. "What Research Says to the Teacher" Series

Ojemann, Ralph H. *Personality Adjustment of Individual Children.* What Research Says to the Teacher, No. 5. Prepared by the American Educational Research Association in cooperation with the Department of Classroom Teachers. Washington, D.C.: National Education Association, October 1954. 32 pp.

14.62. Yearbook

National Education Association, Department of Elementary School Principals. *Guidance for Today's Children.* Thirty-Third Yearbook. Washington, D.C.: the Department, 1954. 278 pp.

Strang, Ruth. "Guidance in the Elementary School." *Guidance for Today's Children.* Thirty-Third Yearbook. Washington, D.C.: Department of Elementary School Principals, National Education Association, 1954. Chapter 1, pp. 2-10.

Stensland, Per G. "The Classroom and the Newspaper." *Mass Media and Education.* Fifty-Third Yearbook, Part II, National Society for the Study of Education. Chicago: University of Chicago Press, 1954. Chapter 10, pp. 217-42.

John Dewey Society. *The American Elementary School.* Thirteenth Yearbook. New York: Harper & Brothers, 1953. 434 pp.

Dale, Edgar. "Improved Teaching Materials Contribute to Better Learning." *The American Elementary School.* Thirteenth Yearbook, John Dewey Society. New York: Harper & Brothers, 1953. Chapter 10, pp. 233-50.

NOTE: When referring to a chapter in a yearbook that does not list the author of the chapter, the association or society name should be used as author, and the chapter should be referred to at the end of the entry (see also 14.8. for treatment of chapters):

Association for Supervision and Curriculum Development. *Creating a Good Environment for Learning.* 1954 Yearbook. Washington, D.C.: the Association, a department of the National Association for Supervision and Curriculum Development. High School Seniors," pp. 139-48.

Teacher Salary Trends

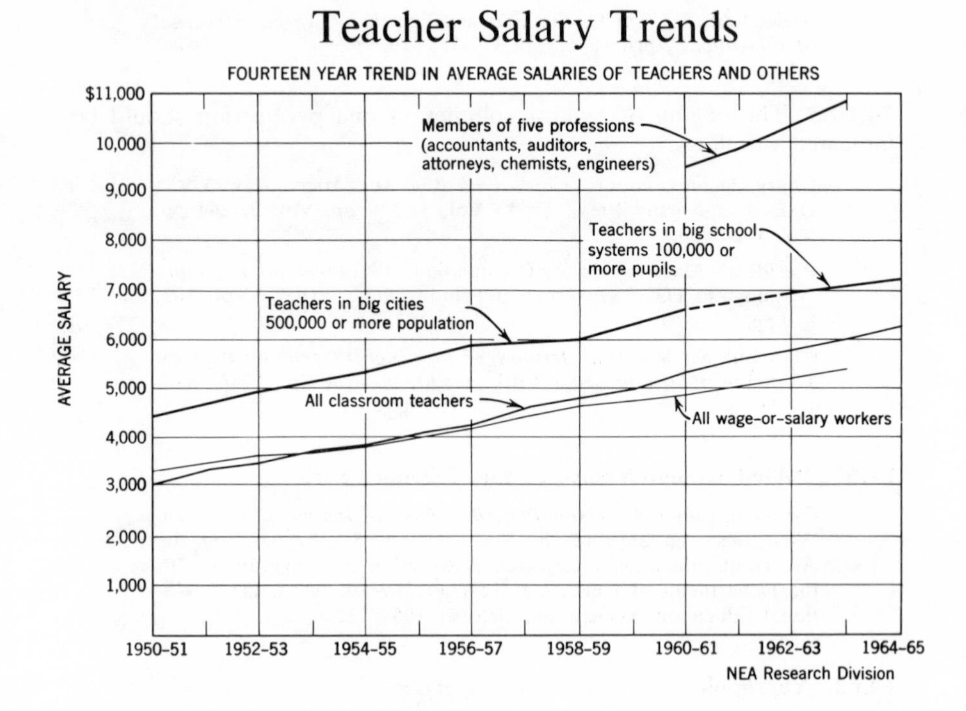

In 1950-51, the average salary of classroom teachers was about $200 lower than the average paid all wage-or-salary workers in the United States. By 1955 and 1956, the two averages were almost the same. Since 1956, the average for teachers has risen a little faster than the average for all workers, and by 1963-64, was about $600 (11 percent) higher than that for all workers.

Beginning in 1962-63, the basis for classification of school systems in the Research Division's biennial salary studies was changed from population of the district to pupil enrollment. Hence, the later figures are not directly comparable with those in earlier reports. However, urban school districts in centers having 500,000 or more *population* (there were 23 such districts in 1960-61) are nearly the same as the large school districts enrolling 100,000 or more *pupils* (there were 21 of these in 1964-65).

In 1950-51, the average salary of teachers in slightly more than a score of the largest school systems was about 47 percent above the all-teacher average. By 1960-61, the differential had shrunk to 25 percent. The figures for 1964-65 showed that the largest school systems paid an average salary only 16 percent higher than the average salary for all teachers.

Only in recent years have continuing series of salary information become available with regard to professional workers other than teachers. A recently established annual survey by the U.S. Bureau of Labor Statistics includes five professional groups, the salaries of which have been summarized by the NEA Research Division in a single average for the five groups—salaried accountants, auditors, attorneys, chemists, and engineers. These are neither the highest nor the lowest paid among professional workers, but they represent a substantial segment of the national total of persons employed in a professional capacity.

The average salary of the five professional groups in 1960, the first year the study was made, was about $4,200 above the average paid classroom teachers. In 1964, the most recent study reported, the gap had widened to $4,784, or 80 percent above the classroom teacher average.

—GERTRUDE N. STIEBER, *research associate, NEA Research Division.*

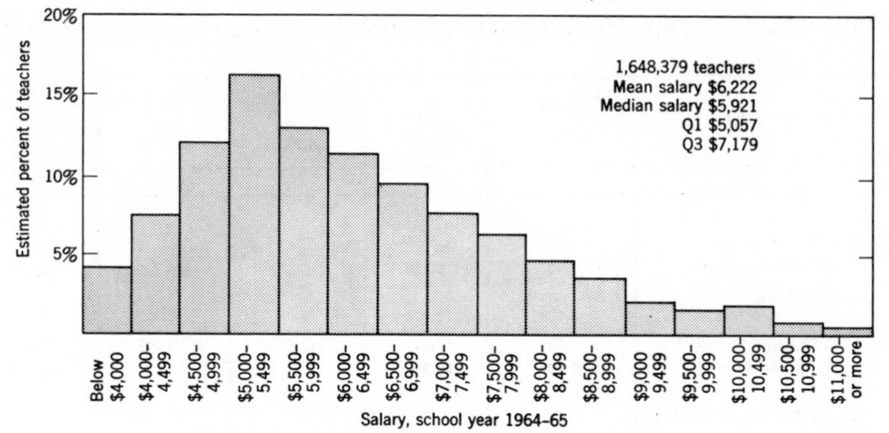

CURRENT SALARY REPORTS

The major salary reports issued in 1964-65 by the NEA Research Division are the following:

Salary Schedules for Classroom Teachers, 1964-65* Research Report 1964-R13. 195 pp. $3. Stock #435-13186.

Salary Schedules for Administrative Personnel. 1964-65* Research Report 1965-R2. 221 pp. $3.50. Stock #435-13200.

Economic Status of Teachers in 1964-65* Research Report 1965-R7. 44 pp. $1. Stock #435-13210.

State Minimum-Salary Laws and Goal Schedules for Teachers, 1964-65. Research Report 1964-R15. 47 pp. $1. Stock #435-13188.

Twenty-Second Biennial Salary Survey of Public-School Employees, 1964-65: Summary Data for All School Systems. Research Report 1965-R5. 41 pp. $1. Stock #435-13206.

Twenty-Second Biennial Salary Survey of Public-School Employees, 1964-65: Individual School Systems. Research Report 1965-R6. 225 pp. $3.50. Stock #435-13208.

*Will be repeated in 1965-66.

AVERAGE SALARIES PAID FOUR GROUPS OF PUBLIC SCHOOL PERSONNEL, 1964-65

Personnel group	Weighted national total	Enrollment grouping of school systems			
		25,000 or more pupils	3,000-24,999	300-2,999	Under 300 pupils
1	2	3	4	5	6
Mean salaries					
Classroom teachers.....	$ 6,222	$ 6,788	$ 6,276	$ 5,785	$4,909
Elementary school principals..........	8,903	10,973	9,046	7,792	6,210
High school principals...	9,457	12,817	10,023	8,569	7,111
Superintendents........	11,227	23,396	15,637	10,699	8,060
Estimated number of persons					
Classroom teachers.....	1,648,379	412,742	723,694	455,864	56,079.
Elementary school principals..........	49,036	9,789	22,580	13,574	3,093
High school principals...	13,925	1,611	4,713	6,703	898
Superintendents........	13,635	142	2,571	8,178	2,744
Estimated number of operating local school systems...............	25,656	142	2,646	9,397	13,471

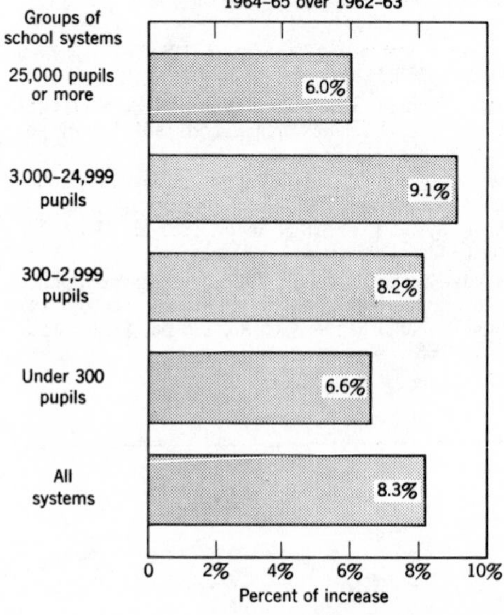

NEA Research Division

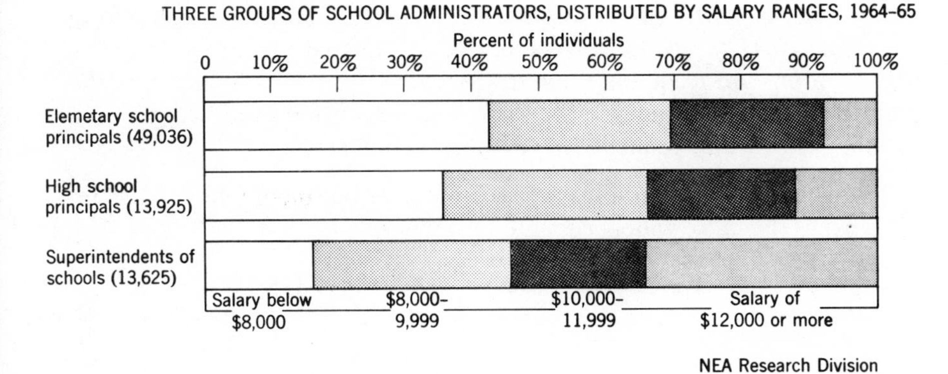

APPENDIX: ABBREVIATIONS

Universally accepted standards for abbreviations do not exist. In spite of international attempts to achieve standardization, only British journals are predictable in their usage. Even the English, however, have been unable to agree upon an interdisciplinary standard. As a result, their excellent document British Standard BS 1991 is divided into several parts to serve incompatible disciplines. Within disciplines, the English have achieved a high degree of standardization.

In the United States, there are no widely accepted standards. Those of the Department of Defense and of the American Standards Association are the most influential. They are not accepted by the various professional societies whose journals follow standards unique to a discipline or even to the current editor. (See, for example, the "Abbreviations and Symbols Used in Chemical Abstracts" on pp. 232–240, the "Signs and Symbols" approved by the American Mathematical Society on pp. 293–309, and "IEEE Recommended Practice for Units in Published Scientific and Technical Work" on pp. 157–177. Because of these variations, writers must follow the standards set by the style guides of either a specific discipline, a sponsoring agency, or a single journal.

The United States Navy Department *Standardization Manual* is pragmatic about abbreviation standards. Its requirements, printed in their entirety below, are that writers define their own terms.

"To assist the reader of a report, it may be desirable to include a sheet that will explain, clarify, or identify any symbols, letters, abbreviations, initials, etc. In some cases, this can be accomplished by reference to recognized—MIL or American—standards. Explanations of symbols and abbreviations are frequently important where the report includes mathematical discussions, because the same symbol is often used with different meanings in various branches of engineering and science."

The following bibliography is representative but incomplete. It is simply a list of the more widely accepted "standards."

ABBREVIATIONS STANDARDS

INTERNATIONAL

IEC Publication R27 (3rd Edition), *International Letter Symbols Used in Connection with Electricity—Quantity Symbols—Alphabets and Letter Type,* International Electrotechnical Commission, 1953.

ISO R31 Part I, *Fundamental Quantities and Units of the System and Quantities and Units of Space and Time,* International Organization for Standardization, 1956.

ISO R31 Part II, *Quantities and Units of Related Phenomena,* International Organization for Standardization, 1958.

UIP Document 6, Symbols and Units, International Union of Pure and Applied Physics, 1955.

BRITISH

BS 1991: Part 6, *Recommendations for Letter Symbols, Signs, and Abbreviations—Electrical Science and Engineering,* British Standards Institution, 1963.

BS 1409, *Letter Symbols for Electronic Valves,* British Standards Institution, 1959.

UNITED STATES AND CANADA

AIBS *Style Manual for Biological Journals,* American Institute of Biological Sciences, 1964.

AIP *Style Manual,* American Institute of Physics, 1959.

AMA *Style Book and Editorial Manual,* American Medical Association, 1966.

ASA Z10.1 *Abbreviations for Scientific and Engineering Terms,* American Standards Association, 1941.

ASA Z10.6 Letter Symbols for Physics, American Standards Association, 1948.

ASA Y10.9 *Letter Symbols for Radio,* American Standards Association, 1953.

Bigg, P. H., "The International System of Units (SI Units)," British Journal of Applied Physics, 1964, pp. 1243 ff.

OSA 285, *Standard Specification for Abbreviations for Scientific and Engineering Terms,* Canadian Standards Association, 1943.

Dictionary of Aeronautics and Aero Space Technology Abbreviations, Signs and Symbols, Odyssey Press, 1964.

Dictionary of Architectural Abbreviations, Signs, and Symbols, Odyssey Press, 1964.

Dictionary of Civil Engineering Abbreviations, Signs, and Symbols, Odyssey Press, 1965.

Dictionary of Computers and Controls Abbreviations, Signs, and Symbols, Odyssey Press, 1965.

Dictionary of Electrical Abbreviations, Signs, and Symbols, Odyssey Press, 1964.

Dictionary of Electronics Abbreviations, Signs, and Symbols, Odyssey Press, 1965.

Dictionary of Industrial Engineering Abbreviations, Signs, and Symbols, Odyssey Press, 1964.

Dictionary of Mechanics and Mechanical Engineering Abbreviations, Signs and Symbols, Odyssey Press, 1965.

Dictionary of Nuclear Abbreviations, Signs, and Symbols, Odyssey Press, 1965.

Dictionary of Physics and Mathematical Abbreviations, Signs, and Symbols, Odyssey Press, 1965.

"Electrical Engineering Units and Constants," *Research/Development,* May, 1965.

Goldstein, Milton, *Dictionary of Acronyms and Abbreviations,* Bobbs Press, 1963.

GPO *Style Manual,* Government Printing Office, 1957.

Guinagh, Kevin, *Dictionary of Foreign Phrases and Abbreviations,* H. W. Wilson and Co., 1951.

MIL-STD-12B *Military Standard Abbreviations for Use on Drawings and on Technical-Type Publications,* U. S. Department of Defense, 1959.

IEEE 260, *Standard Symbols for Units,* Institute of Electrical and Electronics Engineers, 1965.

NASA Publications Manual, SP-7013, National Aeronautics and Space Administration, 1964.

STR-3073 *Electrical Engineering Units and Constants,* National Bureau of Standards, March, 1965.

STWP – Standard No. 1, *Abbreviations for Terms Used in Electronics.* (This is the first of a series to be published by the Standards Council, Society of Technical Writers and Publishers, Washington, D. C., 1967).

"Symbols and Units – UIP Document 6," *Physics Today,* November, 1956.

INDEX